genomics with care

geno with

EXPERIMENTAL FUTURES:
TECHNOLOGICAL LIVES,
SCIENTIFIC ARTS,
ANTHROPOLOGICAL VOICES
A series edited by Michael M. J. Fischer and Joseph Dumit

mics

MINDING THE DOUBLE BINDS OF SCIENCE • MIKE FORTUN

care

Duke University Press Durham and London 2023

Printed and bound by CPI Group (UK) Ltd, Croydon, CR0 4YY
Designed by Aimee C. Harrison
Typeset in Portrait Text and Bitter
by Westchester Publishing Services

Library of Congress Cataloging-in-Publication Data
Names: Fortun, Michael, author.
Title: Genomics with care : minding the double binds of science / Mike Fortun.
Other titles: Experimental futures.
Description: Durham : Duke University Press, 2023. | Series: Experimental futures | Includes bibliographical references and index.
Identifiers: LCCN 2022044510 (print)
LCCN 2022044511 (ebook)
ISBN 9781478020400 (paperback)
ISBN 9781478017233 (hardcover)
ISBN 9781478024521 (ebook)
Subjects: LCSH: Genomics—Research. | Ethnology—Methodology. | Feminist anthropology. | Feminist theory. | Science—Social aspects. | BISAC: SOCIAL SCIENCE / Anthropology / Cultural & Social | SCIENCE / Philosophy & Social Aspects.
Classification: LCC GN345 .F68 2023 (print) | LCC GN345 (ebook) | DDC 572.8/6072—dc23/eng/20230213
LC record available at https://lccn.loc.gov/2022044510
LC ebook record available at https://lccn.loc.gov/2022044511

Cover art: Nathalie Miebach, *Twilight, Tides and Whales*, 2006. Reed, wood, data, 30 × 18 × 20 inches. Image courtesy of the artist.

CONTENTS

ACKNOWLEDGMENTS

That this book exists as a material, mostly coherent thing so many, many years after some of its passages first took shape is evidence of the care bestowed on me for longer than I can remember, from more people than I have remembered to mention here. I'll begin by thanking those first and most memorable caregivers whose fading memory traces only serve to further impress on me their formative, even definitive, force: my dad and mom, Raymond and Alice Fortun, for teaching me to practice the best kind of catholicism, "living according to the whole." I also owe large parts of myself to my brother, David, another public school teacher like our father, and my sisters, Beth and Mindy. I also want to thank my teachers, colleagues, and friends Herbert J. Bernstein and Marcus Raskin, for far too much to recount but, above all, for being my first models of *Menschlichkeit*; the first to have me read Foucault (Herb) and Adorno (Marc) in my undergraduate transition from physics student to historian and philosopher; and the conveners of the conferences on "reconstructive knowledge" in the early 1980s at Hampshire College and the Blue Mountain Center. I think of myself as being hammered into some shape I still maintain at those conferences by the likes of Stanley Goldberg, Doug Ireland, Michael Thelwell, and especially Barbara Ehrenreich and Evelyn Fox Keller, those early feminist analysts of science whose writing talents and styles were an inspiration and models I wanted to work toward. The reconstructive knowledge project also introduced me to Everett Mendelsohn, another formative teacher who first opened and then protected the sliver of space between history and anthropology of sciences for me in graduate school, an opening that Joan Fujimura ushered me through when she came to Harvard in 1988 (along with her husband, Kjell Doksum, the sweetest statistician who ever lived, until his passing in 2021). I am grateful as well to Ken Alder, Mark Barrow, Evelynn Hammonds, and Skúli Siggurdsson, my fellow graduate students there whose care and friendship left an enduring mark.

Although I was only briefly in an actual classroom with Michael Fischer, he along with George Marcus and Sharon Traweek gave me reason through their teaching and writing to want to call myself an anthropologist. It was in his social theory seminar that he held in his first year at MIT that Mike introduced me, through her writing, to Kim. Also present was Rich Doyle, who taught me deconstruction and friendship, the two main drivers of this book. Joan Scott helped me to be comfortable wearing my deconstructive tendencies on my proverbial sleeve, as I do here; I am most grateful to her and to Sylvia Schaefer, who also became a friend in the year we were privileged to share at the Institute for Advanced Study. That year was also when I met Elizabeth Wilson, who has since taught me my most valued lessons in deconstruction, feminist theory, and, above all, friendship; she has been the most valued and generous of readers for my entire career, but never more so than for this book.

Thanks to Lori Hoepner, Jeffrey Thomas, and many other unnamed interlocutors for teaching me about the pleasures, anxieties, and the less directly affective dimensions of research in the life sciences. Thanks to Laura Caghan, Nancy Campbell, Ken Ficerai, Chris Findlay, Kathy High, David Max, and Ned Woodhouse for keeping me well and feeling cared for, and to Vivian Choi, Stefan Helmreich, Jason Jackson, Chris Kelty, Hannah Landecker, Joseph Masco, Heather Paxson, and Kaushik Sunder Rajan for the kind of friendship that's made me less stupid even at a distance. I am grateful to my friends and coworkers on the PECE Design Team, who were also my teachers as graduate students at Rensselaer Polytechnic Institute and at UCI: Lindsay Poirier, Brian Callahan, Ali Kenner, Aalok Khandekhar, Alli Morgan, Angela Okune, Tim Schütz, James Adams, and Prerna Srigyan; thanks as well to Renato Vasconcellos Gomes, Lina Franken, Ariel Hernandez, and Nadine Tanio for being part of PECE and part of my life while writing this book. Other former and current graduate students at RPI and UCI to whom I am grateful are Jon Cluck, Katie Cox, Charlie Curtis, Pedro de la Torre, Thomas De Pree, Sean Ferguson, Gina Hakim, Rodolfo Hernandez, Scott Kellogg, Sean Lawson, Neak Loucks, Richard Salazar Medina, Alberto Morales, Adan Martinez Ordaz, Kaitlyn Rabach, and Guy Schafer. I am grateful also to Simon Cole, Tom Douglas, Paul Dourish, and Gaby Schwab, colleagues at UCI who have shown me nothing but kindness and support. Fran Berman, Peter Fox, Jim Hendler, Deborah McGuinness, and Mark Parsons welcomed me and PECE colleagues into the Research Data Alliance, where I learned to ask about what data needed, and how the exciting boring work of building boring exciting infrastructure was a call to be picked up. And lastly, to Angela our departed cat and Anubis our recently arrived, constant companions who sat with me for the years I sat with this.

Through everything, for everything, my everything: Lena, Kora, and Kim. Kora and Lena never cease to astound me and fill me with the love they give. Being with them as they grew into adults alongside this book gave me the necessary hope and determination to see it through. Kim's gifts to me over nearly thirty years are enormous and innumerable; her gift of Lena and Kora tops the infinite list. This book is only mine because of all of us, and I give it back here to the three of them.

POEM-LIKE *TOLLS* 1 ·

a prelude

Establishment of Dorsal-Ventral Polarity in
the *Drosophila* Embryo: Genetic Studies on the
Role of the *Toll* Gene Product KATHRYN V. ANDERSON,
GERD JÜRGENS, AND CHRISTIANE NÜSSLEIN-VOLHARD

In the course of a number
of mutant screens
in which isogenic lines were established

six totally penetrant
dominant maternal effect
mutations were identified and recovered

(*see Experimental Procedures*).

Females heterozygous
for each of the mutations
produce embryos that develop and differentiate
cuticles of characteristically
mutant pattern.

Four of these
dominant maternal effect
mutations share a common embryonic phenotype,
the *Toll* phenotype. The
cuticle pattern
of $Toll^{D}$ embryos differs strikingly from the
wild-type pattern.

Instead of the characteristic array of
denticle bands ventrally
and fine hairs dorsally,
$Toll^{D}$ embryos have rings or patches of
ventricle denticles
along the entire dorsal-ventral circumference and
lack dorsal hairs altogether.
Other structures normally
derived from dorsal and dorsolateral
anlagen are also missing:
filzkörper, spiracles,
head sensory organs
and the head skeleton are
all absent.

Early
in the development of $Toll^{D}$ embryos,
the pattern
of morphogenetic movements
at gastrulation also
shows a loss of dorsal,
and expansion of ventral,
pattern elements.

Several observations suggest
that there is a direct interaction
between the copies of the *Toll* gene product.
Two of the four dominant
alleles, Tl^{5B} and Tl^{B4C},
behave like amorphic alleles when placed in
trans to a deficiency.

The products of these alleles are thus inactive
on their own, yet
in combination with the wild-type product produce
an abnormal activity.

Two classes
of models could explain
the specific interactions seen between
Toll alleles. One model is that the
Toll protein product is present
as a dimer or multimer
whose activity depends on interactions between subunits.
An alternative
model is that the active *Toll* product
autocatalytically promotes the further
activation of other copies
of the *Toll* product.

The autocatalytic mechanism is attractive . . .

However, the data
currently available
do not allow
us to distinguish
between these two classes of models.

The system that establishes
dorsal-ventral positional information
in the embryo requires the action
of nine maternal effect dorsal-group genes in addition
to *Toll*.

In the absence of
any one of these components, all cells differentiate
according to a dorsal ground state.
The simple model
in which each of these genes controls
one step in a linear biochemical
pathway leading to the production of a ventralizing

morphogen is ruled out
by the double mutations of the recessive alleles of
other dorsal-group genes with *Tl*^{9Q}
since in the presence of *Tl*^{9Q}
ventrolaterally derived structures can be produced
in the absence of
*gastrulation-defective**, *nudel**, *pipe**, *snake*, or *easter**.

The working model we find
most attractive
is diagrammed in Figure 6 . . .
Both the active form
of the *Toll* product and the products
of the other dorsalizing genes
(*gd**, *ndl**, *pip**, *snk**, *ea**)
are required
in a way that we do not yet understand . . .

A Family of Human Receptors Structurally Related to *Drosophila Toll*

FERNANDO L. ROCK, GARY HARDIMAN, JACKIE C. TIMANS, ROBERT A. KASTELEIN, AND J. FERNANDO BAZAN

The seeds of the
morphogenetic gulf
that so dramatically separates
flies from humans
are planted
in familiar embryonic
shapes and patterns
but
give rise to very different
cell complexities.

This divergence
of developmental plans between
insects and vertebrates

is choreographed
by remarkably similar signaling pathways,
underscoring
a greater conservation of protein networks and
biochemical mechanisms from unequal
gene repertoires.

A universally critical
step in embryonic development is the
specification of body axes, either born
from innate asymmetries or triggered
by external cues.

We describe the cloning
and molecular characterization of five
Toll-like molecules in humans
—named TLRs 1–5—
that reveal a receptor family more
closely tied
to *Drosophila Toll* homologs than to vertebrate
IL-1Rs. Spurred
by other efforts, we are assembling,
by structural conservation and molecular parsimony,
a biological system in humans that is
the counterpart of a compelling
regulatory scheme
in *Drosophila*

This signaling pathway centers on *Toll*, a
transmembrane receptor that transduces
the binding of a maternally secreted ventral
factor, Spätzle,
into the cytoplasmic engagement of
Tube, an accessory
molecule, and the activation of
Pelle, a Ser/Thr
kinase that catalyzes the
dissociation of Dorsal
from the inhibitor

Cactus and
allows migration of
Dorsal to ventral
nuclei.

The *Toll*
pathway also controls
the induction of potent antimicrobial
factors in the adult fly; this role
in *Drosophila* immune
defense strengthens mechanistic parallels
to interleukin pathways
that govern a host of immune and
inflammatory responses in
vertebrates.

A *Toll-*
related cytoplasmic domain directs the
binding of a Pelle-like
kinase, IRAK, and the
activation of a latent
NF-kByI-kB complex that
mirrors the embrace
of Dorsal and Cactus.

Components of an Evolutionarily Ancient Regulatory System.

The evolutionary link
between insect and vertebrate immune systems is
stamped in DNA:
genes encoding antimicrobial factors in insects
display upstream motifs
similar to acute-phase response
elements known to bind NF-kB transcription factors in mammals.

Dorsal and two Dorsal-
related factors,
Dif and Relish,
help induce these defense proteins after

bacterial challenge; *Toll* or other TLRs probably
modulate these rapid immune responses
in adult *Drosophila*.
These mechanistic parallels
to the IL-1 inflammatory response in vertebrates
are evidence
of the functional versatility
of the *Toll* signaling pathway
and suggest an ancient synergy
between embryonic patterning
and innate immunity

perhaps the distinct
cellular contexts
of compact embryos and
gangly adults simply result
in familiar signaling pathways and their
diffusible triggers having
different biological outcomes at
different times

Human TLRs and IL-1Rs in Host Defense: Natural Insights from Evolutionary, Epidemiological, and Clinical Genetics

JEAN-LAURENT CASANOVA, LAURENT ABEL, AND LLUIS QUINTANA-MURCI

The immunological saga
of *Toll*-like receptors (TLRs) began with the
seminal discovery
in 1981 that antimicrobial peptides are a key
mechanism of innate host defense
in insects.

This was followed by
the observation in 1991 that the fruit fly
Drosophila melanogaster Toll

and mammalian interleukin-1 receptor
have an intracellular domain
in common. These studies
paved the way for elucidation of
the role of *Toll* in controlling
the synthesis of some
of these peptides in *Drosophila*.

These discoveries soon led
to the identification of a human
TLR, followed by
the discovery of a function for
TLRs with the demonstration that
lipopolysaccharide (LPS) responses were
abolished
in mice with spontaneous TLR4 mutations.

The similarities
between the *Toll* and TLR
signaling pathways
in invertebrates and vertebrates
were initially interpreted
as evidence of a common
ancestry for these defense mechanisms and
subsequently of convergent evolution,
emphasizing their evolutionary
importance.

The 15 years or so following
these findings have witnessed a substantial
rise in interest in
the role of *Toll*
in *Drosophila* immunity,
of TLRs in mouse host defense,
and even of TLRs in diverse other animal species.

Indeed, interest
in TLRs has been such that just about
any immunological phenomenon imaginable—

ranging from host defense and tumor immunity
to allergy and autoimmunity—has been examined from
a TLR perspective.

This phenomenon
has even extended to processes only
remotely connected with immunity,
such as atherosclerosis and
degenerative diseases, and
has also stimulated research into the role of human
TLRs in the pathogenesis of most,
if not all,
human diseases.

Various schools of immunological
thought have conferred different names
on pathogen receptors, including
pathogen associated molecular pattern (PAMP) recognition receptors,
pattern-recognition receptors (PRRs),
innate immune sensors,
and microbial sensors.
Whatever the terminology
used, the underlying
idea is that TLRs detect
a wide range of microorganisms,
discriminating between these microbes and
distinguishing them from self on the basis of
their type, through the detection of
specific, conserved
microbial patterns, molecular patterns, or molecules.

Does this commonly expressed view of
TLRs and IL-1Rs reflect the biological
reality?
Like most immunological
knowledge, it is based mostly on experiments conducted in
the mouse model.
However rigorous, accurate, and thorough
such experiments are, can experimental

findings in mice really provide a
faithful and reliable representation of host defense
and protective immunity in other species,
in their natural setting?

There are differences between species, including several
identified differences between humans and mice, and
immunological generalizations from
a single species may be
perilous.

genomics, double binds, affects

PART I

1 · *fors*

> The obsessive campaign against "jargon" obscures a legitimate anxiety about the language of indeterminacy that mediates politics and theory. Affectivity—fear, ambivalence, terror, shame, disorientation, or dispossession—figures prominently in addressing the subject's identification with, or resistance to, the indeterminacy of change. Affect registers and regulates the subject's ambivalent and anxious responses as it faces what is new, partially known, or without guarantees; at the same time it provides the agent with an imminent sense of sensory and bodily attentiveness to the task of change. To the extent that affectivity is crucial in positioning subjects in relation to contingent and indeterminate circumstances, affect is an acute measure of the time change takes—in particular, the temporalities of transition or transience. · HOMI K. BHABHA, "'THE BEGINNING OF THEIR REAL ENUNCIATION': STUART HALL AND THE WORK OF CULTURE"

A Force for Science

In the winter of 2016–17, the culture of science in the United States underwent what felt like a radical transformation, changing into something far different from the science world in which I had first begun writing about the "care of the data" in genomics, one phrase I've used to name the broad subject of this book.[1] In fact, it felt like an altogether different world with the election of Donald J. Trump to the presidency; I was hardly alone in feeling shocked and stunned. I stress the affective sense of this new world because, in addition to the numerous reasoned arguments that circulated then as to why the imminent Trump presidency would be disastrous for the United States, the amalgamation of those reasons with widespread feelings of dread, disorientation, disgust, and anger moved hundreds of thousands of people to join the Women's March in Washington, DC, and numerous cities around the United States on the day after Trump's inauguration, as a statement against the new political realities. My daughter and I drove to DC to join, and it was extraordinarily powerful.

Even more extraordinary, I think, is that a similar mobilization specifically about science began almost simultaneously. In a politics thread on Reddit, in response to a post linking to a *Vice* article detailing how all references to climate change on the White House website had been deleted immediately after Trump's inauguration, Reddit user Beaverteeth92 made what he later referred to as a "throwaway line": "There needs to be a Scientists' March on Washington." Jonathan Berman, a postdoc at the University of Texas Health Science Center in San Antonio, immediately picked up this line thrown away and created a Facebook page called March for Science; within a few days it had 300,000 members, and quickly grew to 800,000. By Earth Day, April 22, this scattered population had been organized into an array of marches in hundreds of cities around the world, attended by more than a million people.[2]

I donated to the March for Science while it was first organizing, and then when the more establishment American Association for the Advancement of Science solicited me to rejoin after many years of ignoring their imploring emails, I finally re-upped—partly, I admit, because the AAAS said they would send me an electric-blue T-shirt that read in white letters I AM A FORCE FOR SCIENCE (see figure I.1). I wanted to wear it to the local March for Science I planned to attend in Albany, New York, for its obviously affirmative tone as well as the open-endedness it created through its use of monosyllabic generics, apparently obvious and straightforward in meaning but actually asking for closer reading: Force? Not that I *exert* or *convey*, but *am*? What kind of force am I . . . and could this book be? And what would it mean to say that my or this book's yet-to-be-determined force is *for* science?

A partial, provisional answer to be elaborated and complicated: my name for this force is friendship. I try here to be a friend for, and of, the sciences, and particularly the sciences of genomics that I've studied for twenty-odd years as a historian and anthropologist. Mostly, when my scholarly community writes about the sciences they do so not to be "for" the sciences, even if they may not exactly be "against." Only scientists and popular science writers explicitly write *for* science; writing about science from the academy is almost exclusively written to expose something science lacks, misconceives,

FIGURE 1.1 · Science swag. Photograph by Kora Fortun.

or otherwise gets or does wrong, in a mood of dismay or disappointment. All abundantly necessary, and for which there is no lack of material. I've tried here to write more straightforwardly *for* genomics, in part to figure out what that might mean, to see how *writing for* reads. My previous book, *Promising Genomics*, was not decidedly against genomics, but for most readers the takeaway was about unwise investments in genomics (economic, cultural, and scientific) hype, and not obviously *for*.[3]

And *friendly* is a good descriptor for the atmosphere at Albany's March for Science, helped no doubt by the fair weather that is sometimes used as a categorical term for an unreliable subgroup of friends. It was a big but not huge crowd, mixed in ages but almost all White, as are the sciences in the United States, and almost all strangers to each other outside of the small groups of acquaintances and coworkers with whom people arrived. In Albany there were no science celebrities like Bill Nye the Science Guy, who had signed on as a cochair of the March for Science, and no Nobel laureates: it was a crowd of high school science teachers, students, college professors, nature lovers and environmental activists, quality control chemists from local biotech company Regeneron (not yet famous for its anti-COVID-19 antibodies), engineers and physicists from General Electric, and, because of Albany's status as New York's state capital, diverse technicians and scientists from places like the New York State Department of Health's Wadsworth Laboratories, the Department of Environmental Conservation, and many other state agencies and offices of state legislators. These were people for whom science was unquestionably essential to good governance, good environmental standards, good education, and good public health.

Almost everyone at the science marches carried homemade signs, and what is an anthropologist but a reader of signs? Many concerned climate change, others decried the defunding of the EPA, or carried reminders to thank a scientist for their measles vaccine. They ranged from the anodyne ("Science for the common good") to the cleverly anodyne ("What do we want? Evidence-based science. When do we want it? After peer review."). That last one represents what was probably the most popular vein of signs, and the one most indicative of the cultural nerve energizing the mass phenomenon: ridicule and denunciations of "alternative facts" and the "Chinese hoax" of climate change, along with forceful valorizations of a reality to which science provided unique and direct access, via true representation. This feeling was palpable and weird, manifest in the multiple signs reading "I can't believe we have to march for science" or "I can't believe we have to march for facts" (see figure I.2). "Post-truth" may have been 2016's word of the year, but most people still seemed incredulous that we were in this position, where something so supposedly solid, essential, powerful, necessary, foundational, compelling, and incontrovertible as scientific truth had to be protected—through a kind of politicization that also seemed to be its greatest threat.

One historian of science tried to capture the cultural mood by stating that "there's a broader perception of a massive attack on sacred notions of truth that are sacred to the scientific community."[4] I'm not sure conventional anthropological categories of the sacred and profane are actually very relevant here; matters were much more complex, and much more imbricated in the everyday experience of, and work of, the sciences. People I marched with in Albany were more concerned that, say, if scientists on the EPA's Clean Air Scientific Advisory Committee were certain, sure, and/or convinced that, after years of collective work and dedicated analysis, national standards for atmospheric ozone levels should be no more than 70 parts per billion, then that's what they should be, end of story. Holy had little to do with it. And

FIGURE 1.2 · Signs of the Albany, New York, March for Science, April 22, 2017. Photograph by Mike Fortun.

"truth" is a concept poorly suited to understanding what the numerical value "70 ppb" denotes, or what is at stake in the determination—and legislation—of things like ozone standards.

In the sciences of genomics that are the main subject of this book—postgenomics to some, but I just use "genomics" throughout—truths are even more complicated and more difficult to establish, and the concept itself seems even less well-suited to this domain.[5] But "post-truth" satisfies me even less than do claims about scientific truth's sacred status. This book takes readers through some of the processes of how genomicists produce the complex truths of their science, and how those practices changed over the course of about twenty years, as an inward-focused genomics came to turn outward as well, attempting to encompass environmental factors in its analyses. It's a book about the quotidian, daily, mundane practices that genomicists engage in to arrive at those truths before they are elevated to a more exalted status. It's a book about the infrastructures—technical, social, cultural—that are the necessary support for genomics and its truths, the conditions of their possibility that are seen in the margins of our vision, when they are seen at all. It's a book about the importance of data—another key word on March for Science signs—as well as data's insufficiencies, fragilities, and even its impossibilities. Indeed, impossibility will be a key word here because, in brief, it is impossibilities that call for care—not as something exterior and complementary to the sciences, but as an essential quality or mode of their doing.

But before moving on to genomics, and maybe make readers more curious about what impossibility might have to do with sciences and truths worth marching for, let's read one last sign from April 2017—more than one sign, actually. Many signs in Albany, and in numerous other cities, quoted astrophysicist Neil deGrasse Tyson: "The good thing about science is that it's true whether or not you believe in it." You can see a version of this one in the margins of figure I.2. I show this photo to students in my anthropology of science classes, teaching them to read it as an anthropologist, which means teaching them to read as a philosopher, which means teaching them to read as a semiologist. First, as anthropologists, we encounter the cardboard sign on the street, reading it quickly in the pack and flow of a friendly crowd. We get the reference, we have an image of Neil deGrasse Tyson in our heads, the Black scientist we know from watching *Cosmos* or following him on Instagram or coming across him on YouTube, and so we don't need to think too long or too hard about the sign's meaning: it's why we're there, it's what joins us together into a march, this shared sense that science is powerfully and maybe uniquely authoritative, compelling the assent of anyone who isn't insane and/or sociopathic like the new president. Even though I know that science, truth, and belief are each complicated entities, with multiple alternative definitions, and with complicated interrelationships among them, and even though it's my job to teach those complexities, the March for Science wasn't the time or place to argue or explore those complexities and variations. It's a good sign, street-wise.

Although it's more like two (at least) signs, one present there on the street and the other(s) outside the frame, but somewhere in the neighborhood, out of sight but in a funny relationship of support and contradiction. The other metastatement(s) can be framed, outside the first frame, as "Believe me when I say that 'Science is true whether or not you believe in it,' because that statement is true." On the street, everyone was just content to nod in support and move along, avoiding the recursive difficulties. And for those few of us who couldn't help but think about it, for reasons of professional training and inclination, we just swallowed our questions and meta-analyses for reasons of pragmatic efficiency, solidarity, and friendship.

When Neil deGrasse Tyson first stated this aphorism years earlier in 2013 on Twitter, however, it was a different story. Because Twitter is such a good medium for hot takes and quick disagreements, the many affirmative and sometimes fawning responses that accumulated in the thread were punctuated by people finding many ways to take exception.[6] DeGrasse Tyson's Twitter followers registered points made by any number of people that have analyzed the sciences from philosophical, historical, and similar perspectives, and by many thoughtful scientists themselves; they just did it with a lot fewer characters and a touch more style. Some played on the inadequacies of "science" as a unified concept or categorical name imposed on the complicated excess of what humans actually do and how humans actually think; "science" has multiple dimensions and means multiple things, not always consistent, and not always congruent with "true" or "truth":

> Science isn't true, but it is a way to pursue truth. It doesn't reveal truths, but it leads to better approximations.
>
> The idea that science is true is a bit of a stretch, I would say it's the search for truth.

Others had definitely read their Thomas Kuhn or any of a number of historians:

> Or until a paradigm shift occurs and renders all previous knowledge BS.
>
> "Science" has been wrong numerous times throughout history. Just saying.

A few highlighted the empirical-rhetorical reality that science and belief were not mutually exclusive and indeed might rely on each other, occasionally producing significant differences, especially in the more rarefied reaches of physics where deGrasse Tyson himself practiced his craft:

> Some scientists believe in wormholes, others do not.
>
> And yet scientific results are expressed in terms of probability distributions that quantify our degree of belief!

And then there were the tweets that raised more straightforwardly philosophical qualms:

That of course is true of everything true. Belief/non-belief have little to do with true/not true.

Well, I can "make" a "truth" without any science at all. Put a coke can on a table, "coke on table" is "true."

But the best, I think, was the shortest: "That's your opinion. . . . lol." That one captured the unspoken contradiction between the framed statement and its framing metadiscourse—"Scientific truth doesn't require belief, if you believe what I'm telling you I believe"—while adding the "lol" so essential to proper digital expression.

One premise of this book is that these sorts of lol-worthy paradoxes are an inevitable feature of any science, including the sciences of genomics that are the main subject here. In locating and deconstructing them, please remember that I am always lol-ing with them and never lol-ing at them. I am delighted to live in a world that has brilliant, funny, forceful, and righteous Neil deGrasse Tyson in it, and I was glad to March for Science under his sign, even as both are burdened with necessary simplifications, full of omissions, and riddled with contradictions and paradoxes.

I am equally glad to write *for* genomics here, where writing *for*, for me, means detailing these kinds of contradictions, paradoxes, and/or what I will name generally as double binds that structure "science." This book returns over and over to those conceptual-practical places in the complex system of genomics where, as part of its normal and productive operation to establish (believable) truths of and in genomics, something slips, snags, catches, or doubles back on itself as scientists work and think across logical levels—across, for example, the data and metadata levels (see chapters 3 and 4). Or between the systems of proven, reliable technologies and protocols and the newer technical layers that make the system genuinely experimental: capable of producing new knowledge and new phenomena—but also open to complete failure (chapter 5). Or at the limits of the science system as a whole—"genomics"—where new truths beyond those currently known or accepted have to be discerned within or coaxed out of a promised, unstructured future, and the genomics community is at the limits of all its capabilities, technological, conceptual, and rhetorical (chapter 6).

In these multilayered places where a changing, creative, and *limited* experimental science like genomics happens, two interrelated things occur that I try to pay close attention to here. First, in their work and thought, we'll see how scientists find themselves, over and over again, in a position of impossibility. That sounds overly dramatic if not, indeed, impossible: science obviously happens all the time, and scientists continue to be gainfully and happily employed to do it. That science is obviously possible and at the same time somehow impossible is a complicated analysis that will take time to build. The story of Neil deGrasse Tyson's aphorism should give at least some hint of what's involved: on close examination, how we think about, and how scientists think within, "science" (an impossibly simple name for an impossibly large and diverse entity) is marked by rhetorical contradictions, logical paradoxes, and situational double binds that, in principle, resist any attempt at resolution or evasion. And yet sciences and scientists, and our thinking about them, march on. The name in this book for the human practices and characteristics that make both of those things "true"—the formal impossibility of sciences, and the actual doing of them—is care. Far from being radically different or even benignly complementary, science and care are bound so tightly together as to be indistinguishable.

Almost.

Second, these double binds make for some of the more obvious places and moments where rigorous and reasonable scientists experience something affective: sensations on the border of sense, but crucial to it, that become apparent in expressions of surprise, interest, anxiety, or excitement. Sometimes these will take the narrative, and us, into the undervalued and undertheorized dimensions of curiosity and wonder, and what some have called the scientific virtues: amalgams of reason and passion, cognition and emotion, that serve as hallmarks of careful scientists, and careful science. Such affective moments and sensations are an essential part of science—even, maybe, a driver—but are difficult to pick up on or tune into, and just as difficult to name.

On both of these tracks, my dominant approach or method is one of ethnographic *reading*. I've excerpted interviews that I've conducted with scientists, I've analyzed scenes of scientists speaking and interacting, but my primary way of doing cultural analysis is through reading scientific writing. The scientific article is a genre and a technology designed to erase, cover over, or exclude the affective register. It doesn't always work, though, and I hope to show how close ethnographic reading of a scientific article can draw our attention to affective experiences or effects, much like the repunctuating of scientific articles in the prelude (and later in the book) tried to bring forward the poetic infrastructure to these most prosaic of texts. This effort is aided by a proliferation and diversification of the genres and forms scientists' writings take: essay reviews, interviews, and opinion or "perspective" columns are some of the places where an anthropologist can find scientists writing more freely and reflectively in even the most professional of journals. Blogs and social media platforms are another place where scientists are writing more, and more openly and expansively, about their work, thought, and lives. Individual talks and collective workshops are becoming ever more available on YouTube. It's a great time to be an anthropologist of science doing fieldwork in genomics, even as "the field" and one's time in it become harder and harder to differentiate from the rest of an increasingly online life. There is more and more material from which to develop—and this is an overarching goal of this book—a working theory of scientists as a particular kind of human subject, subject to dense and intricate combinations of intellectual, political, social, personal, and emotional forces happening across scales, from the systemic to the most individual and quotidian, in contemporary genomics in the United States.[7]

To begin reading genomics, then, and to begin some further explanation of how the writing here asks you to read . . .

Reading for Catachresis

"Das war ja toll!"

I was arrested by this utterance—let's leave it untranslated for a while—made by Christiane Nüsslein-Volhard sometime in the early 1980s, long before she, Eric Wieschaus, and E. B. Lewis would all be awarded the Nobel Prize for their work in the developmental genetics of the fruit fly, *Drosophila*. The short, simple statement was hardly important to the long, complex, and genuinely important scientific work Nüsslein-Volhard had been doing at the Max Planck Institute in Tübingen.[8] This little speech act was an entirely marginal one, a saying so unremarkable that only the nerdiest of science nerds, like me, would even know of it. The only reason it ever made it into print and so into the cultural archives of the sciences where I could access it years later (just as genomicists now routinely access DNA sequence data archived in their ever-expanding databases) is because this fleeting, off-handed, marginal utterance would be transmuted into the name, *toll*, for one of the genes that is a powerful shaping force in the first twenty-four hours of a *Drosophila* larva's development.

At the time, *toll* the gene was itself not especially noteworthy; it was not one of the genes detailed in Nüsslein-Volhard's and Wieschaus's landmark 1980 *Nature* article—genes with easy-to-read albeit odd names like *gooseberry, patch, even-skipped, hedgehog, hunchback, barrel*, and *runt*. The *Drosophila* genetics community is famed for its practice of conferring whimsical names, a practice honored in the Jabberwocky-inspired poem by Paul Stere, "The *Fushi Tarazu*":[9]

> Twas *brahma*, and the *gypsy pogos*
> Did *toll* and *Krüppel* in the *paired*:
> All *Spaetzle* were the *prosperos*,
> And the *ham numbs engrailed*.
>
> Beware the *Fushi Tarazu*, my son!
> The jaws that bite, the claws that catch!
> Beware the *hunchback* bird, and shun
>
> The *armadillo patched*!

Now that *toll* has resurfaced in this scientist's homage to one of the most famous nonsense poems ever penned, are we ready to venture into the question of what sense to make of Nüsslein-Volhard's utterance? It's also the first of many examples to come of how words, as Lewis Carroll understood so well, don't have to have a referent to make sense—as long as you are willing to play along with them.

Imagine: a young woman scientist, sitting for hours at the microscope, as she had the day before and the day before that, examining minuscule mutant fly larvae one after the other, larvae produced through her and Wieschaus's famed "Heidelberg screen" of carefully pedigreed flies systematically subjected to chemical mutagenesis, looking for . . . something. A sign. Something she hasn't seen before; something no one anywhere has seen before. A difference. A pattern of larval segments and structures unlike previously encountered patterns. It's absorbing and demanding and exciting; it's tedious, repetitive, boring, wearying. Slide after slide after slide, for thousands of larvae. Then—a larva with a back and . . . not a front, but another back. A one-sided larva: all dorsal, no ventral. A lethal mutation, a larva that could never fully develop into an adult fly—the mature fly that, for good reason, had long been the privileged focus of *Drosophila* genetics until Nüsslein-Volhard and Wieschaus came along.

"*Das war ja toll!*," Nüsslein-Volhard remembered saying.[10]

There are slightly different versions of this remembered event and its vocalization; in another, it's condensed to its essence: "*Toll!*"[11] All have the exclamation mark, which I've retained, despite my ambivalence.[12] "That was weird!" or, simply, "Weird!" are probably the dominant translations. "(That was) Cool!" also has currency.

Perhaps it's because he, too, is a developmental biologist sensitive to the presence and power of patterns that Gerald Weissmann, in his own rumination on Nüsslein-Volhard's work and its affinities to Gestalt psychology, also registered the complex pattern of meanings and affects activated by "*toll!*": "Words shouted in the heat of discovery have more than their dictionary meaning. My emigré father used 'toll' when he meant 'crazy,' but also 'curious' or 'amazing'; he used it when he first treated a patient with cortisone. These days German-speakers also use toll instead of 'cool' or 'droll,' 'outrageous' or 'awesome.'"[13]

A "crazy!" translation is reenforced again by Nüsslein-Volhard and her translator when, in another interview for a popular website, she was asked to reflect on a more metascientific level:

> INTERVIEWER: Hat es in Ihrem Leben den Heureka-Moment gegeben? (*Have you ever had a Eureka! moment in your life?*)
>
> NÜSSLEIN-VOLHARD: Immer wieder mal. Das ist ganz toll! (*Again and again. That's what's so crazy!*)[14]

Having arrived at this semiotic juncture, with the "Heureka-Moment" in such close proximity to the "*Toll!*" expression, let me pause to clarify the differences, while stating why this book is so interested in the latter but indifferent toward the first—why, in other words, I care about *toll* and *toll*-like experiences in the sciences, and want readers to as well.

For me, in this book, "*Toll!*" does *not* signal the "*Heureka*!" that is such a predictable presence in so many narratives in and of the sciences, and often considered to be a definitive characteristic of them. "*Toll!*" cannot be translated and should not be heard or read as a confirmatory, self-congratulatory, and satisfying "I have found it!" Although my interest here in this partial ethnography of genomics is in the knotted double-binding of the affective and the cognitive in the experience, practice, and culture of genomics that is also heard in *eureka*, *toll* signs for a different affect-thought bundle. You'll find no Heureka-Moments in the pages to come. I also side skirt any "aha!" moments. "Aha!," an onomatopoeic mark of a sharp exhale, while similar to *toll!* in its near-guttural quality, signs yet a different alloy: the dramatic insight, the rapid coalescence or crystallization of half-baked ideas and dispersed threads of thinking and feeling. *Aha!* and *eureka!* are indeed appropriate signs for the kinds of neural-experiential-cultural events that occur sometimes in the process of doing science (or any other work of thought, for that matter). I'm not dismissing their occurrence or their significance; they just seem overrated to me, too attention-getting—and valued in narratives of science for precisely that reason.

I want *toll!* to activate different signaling pathways. As far as familiar onomatopoeic signs go, *huh!* would be closer to the mark of *toll!* than *aha!* is. *Toll!* signs for a different kind of event in the mind and body of a scientist: less dramatic and more understated; more puzzled and less confirmatory; less conclusive and more initiatory; more ensconced in the murkier thick of activity than in the austere enlightened heights of *aha!* and *eureka!* Aha! and eureka! are clear signals; their drama brings them readily to attention, even forcing themselves onto the analyst, be she scientist, science writer, or ethnographer. *Toll* is harder to tune in from the background noise and static; you need to fiddle with the dials, strain your ears. You need to actively read, and read for, *toll*. I want *Toll!* to mark an unremarkable remarkable event.

But the most important difference of all is this: in the Heureka-Moment, in the *aha!* instant, tension climaxes and releases, cares are lifted and satisfaction sets in. With *toll*, some slight difference or resistance, some flicker in the pattern elicits the smallest spark of interest—*huh, that's weird/cool/interesting*... And it's in that trailing ellipsis that anxious care is not lifted, but *arises*.

And care is another name for science.

Saying that care is another name for science is not the same as saying that care and science are the same. Neither, as this book aims to show, are they all that different. Later I will write their complex relationship as a kind of a double bind, but to keep this introductory part of the book introductory: every time you read "care" here, hear "*toll!*" *Care? Huh, weird*...

I've recounted the story of Christiane Nüsslein-Volhard's non-eureka *Toll!*-Moment for several reasons. It is a story about what happens on the edges of scientific work, in the innumerable, quotidian, passing moments that on some few occasions get noticed and marked—a minimal and ambiguous mark, likely to be overlooked or forgotten or buried in the archive, but nevertheless discernible behind or beyond the bright shine and sharp contours of scientific discovery that attract the most attention. Genomics has gotten a lot of attention, to be sure, and at least some of it is deserved. But I try to move quickly through or around the high-profile events and the genomics celebrities (such as they are) and focus my (and now your) attention on the more quotidian moments from which sciences like genomics, not unlike a fruit fly, develop. I want to call your attention to the infrastructures of genomics—the technical and social *and affective* scaffolding that is too often overlooked but constitutes the capacity for the more eye-catching events and truths.

This *toll* tale is also an introduction to the kind of thing that happened to genes as the infrastructures of genomics were built, at considerable expense, in the last decades of the twentieth century and the first decades of the twenty-first. As "*toll*-like" genes were identified in more and more organisms, from bony fishes to fleshy humans, they could only be thought of as part of a signaling pathway, less an element of information and coded control and more as one node in a distributed biosemiotic structure whose product is a set of possible interpretative outcomes.

And "*toll!*" registers and conveys an *affective* event that, when recognized at all in accounts of the sciences, are regarded as unimportant and out of place in sciences conceived overwhelmingly if not entirely as cognitive or epistemic events. I want the monosyllabic, barely meaningful "*toll!*" to activate complex signaling pathways (*weird!*) in the discourses through which scientists and sciences become meaningful to us, an infrastructure of multiple (*interesting!*) chains of multiple signs that eventually connect your reading (*crazy!*) of "science"—primarily "genomics," in this book—to an equally multiple and complex "care."[15]

Without thinking, we read each of those signs—science, genomics, care—as a proper name, the referent of which might be amorphous and complex and difficult to narrate or write out, but we are confident in their coherence, sure in the sense that if we just thought about any one of them for a while we would arrive at, or at least be able to point to, a secure, stable, and identifiable entity, even if a complex one. Care, genomics, science: signs that, like that sign in Albany's March for Science carrying the DeGrasse Tyson aphorism, make immediate sense. Genomics, science, care: no quotation marks necessary. We feel like we know what they name.

This book asks you to read each of those signs as a catachresis: an improper name for something for which no name is actually appropriate or proper. I hesitate to introduce what my mother would have called a highfalutin term so early in the book, when I am trying to draw in and keep readers for whom such terminology might seem jargony and exclusionary, but it figures too importantly. There is much more about this confounding rhetorical figure below and in chapter 3, but briefly: a catachresis is a word used outside its usual context for something that has no ordinary or proper term. It's like an overly strained or exaggerated metaphor, except that metaphors make for an "aha!" of recognition that reduces tension: *Ohhhh, a genome is like a code, I get it . . .* Catachreses call up a "*toll!*" that foregrounds strangeness and heightens interest: *Science is care . . . wait, what?* Metaphors don't encourage a reader to shake words loose from their obvious or accepted meaning; catachreses do.[16] Catachresis is a rhetorical device for breaking expectations, provoking questions, or producing surprises. An arm of a chair, the foot of a mountain, or (quite pertinent to this book) the horns of a dilemma. *Toll, brahma, Fushi Tarazu*; science, genomics, care. The uncommon trope of catachresis is particularly appropriate in writing about care; I join those many scholars of care who, although for slightly different reasons, have argued that "the most difficult aspect of writing about care is not finding *which* words to use, but dealing with the limits of using *words* at all."[17]

Why begin a book about care in the science of genomics by asking you to read all of those terms differently, as signs hovering awkwardly over a restless, unsettled excess? Why introduce a text about how genomicists are continually elaborating new practices of care in their scientific work, evolving new practices of science in their collective careful attention to the data, analytic methods, and experimental designs that underwrite both fragile and durable genomic truths about asthma and similar complex conditions, by immediately chipping away at the security of your reading? Why compose an anthropology of care, affect, and genomics by drawing attention to their most elusive, unstable, *toll*-like qualities?

Let's return to 1986. In Germany, Nüsslein-Volhard and her colleagues have been identifying developmental genes like *toll* and many others. They were doing *genetics*, a science and practice in which *Drosophila* has long been one of the most important and productive of model organisms.[18] Starting from an organism and equipped with a research infrastructure designed, however multiplicitously, to produce isolated differences, they were identifying and characterizing individual genes through analysis of the physiological or biochemical differences associated with those genes, through controlled reproduction of carefully maintained stocks of well-characterized organisms. This genetic approach—or paradigm, or style—served scientists well for decades and continues to serve them, in multiple organisms from bacteria to mice and fruit flies, and eventually humans, but always one gene at a time.

But genetics was also changing in 1986, and becoming something else. If there were a word for this something else—but as we'll see, in 1986, there wasn't—we would say that genomics was emerging, slowly and unevenly at first, soon picking up speed, driven by what would come to be called the Human Genome Project (HGP). I reread some of the key events in this important history of the HGP in the next chapter, probing for these *toll* moments of weirdness and coolness and the affective responses they elicited from first-generation genomicists. The development of genomics through the HGP, my story goes, far from being an unfolding plan for a preconceived project, was an invention, exploration, and consolidation of a labyrinth-in-progress. And exploring a maze of science while building it, semi-blindly, is a process that both evokes and is powered by a mix of excitement, boredom, interest, worry, surprise, dismay, and other affects. Especially so when developing the science of genomics, so often exalted, is really more the building of an infrastructure, so often regarded as quotidian and dull.

But in 1986 genomics has not yet been named, and there are geneticists only in the world of the life sciences. Let's look once again to the margins of major science events like the HGP and read the less prominent writings of scientists to begin this different way of reading genomics, and of reading naming itself.

A large meeting of scientists was convened in 1986 at the National Institutes of Health in Bethesda, Maryland, to debate the political and scientific wisdom as well as the economic and pragmatic contours of what would later become the Human Genome Project. Should an unprecedentedly large biological project like the HGP even be undertaken? Was it even a biological project (*interesting!*), or "merely" a technological, infrastructural one (*boooorrrring*)? How much federal funding would it take, and would it take away from other (*interesting!*) biomedical research? Should work focus exclusively on DNA sequencing (again, an unspoken: *boooorrrring*), or on a combination of sequencing and genetic mapping (*cool!* and also *crazy!*)? Should humans be the sole objects of research, or should a project analyze other organisms like yeast, fruit flies, worms, and mice, on which a rich foundation of genetic knowledge had already been developed? Discussions were contentious, animated by these affects that, even when noticed or recorded, remained unanalyzed.

At the end of the day, co-organizers Frank Ruddle (a mouse geneticist) and Victor McKusick (a medical geneticist) convened a smaller subset of about fifty people to talk about the launch of a new journal (another fine catachresis: journals are not actually "launched") that would cover a broad and diverse range of scientific work. Not only would it be "a place to include sequencing data and as well to include discovery of new genes, gene mapping, and new genetic technologies," recalled geneticist Thomas Roderick, but also "the comparative aspect of genomes of various species, their evolution, and how they related to each other." Adding even more terms and activities, the journal would encompass "an activity, a new way of thinking about biology," as Roderick put it to an interviewer for the *Journal of the National Cancer Institute*. At the end of the day, Ruddle and McKusick still did not have a name for this new journal to review, collect, and communicate these multiple new lines of work and multiple new styles of thinking that were welling up in biology at the time. So a subgroup of this subgroup went to a bar.

At MacDonalds Raw Bar, Roderick recalled how he and a group of about ten "sat around drinking beer—actually a lot of beer. It was great fun." An enjoyable release for scientists, after a long day of sober and occasionally tense discussion. After at least two and more likely three pitchers of beer, Roderick proposed "genomics" for the journal's title. "I don't know exactly how I came up with the word," he later said. It sounded like genetics, a proper enough name, and Roderick after all was a prominent geneticist at Jackson Laboratory in Bar Harbor, Maine, a famed mouse genetics facility. But it also seems to have been unsatisfying—or at least a scientist's cultivated modesty and tendency toward understatement prevented him from pronouncing it *good*: "We adjourned that evening thinking genomics wasn't a bad name. I thought we had a tentative name for a journal beyond just sequencing and mapping."[19]

To read "genomics" as catachresis is to read it for these multiple, diverse, emergent activities and trajectories of science for which there was (is) no proper singular name. Reading genomics as catachresis is to read it as ever tentative—not a good name, but not bad either. Kind of *toll*, actually. Maybe somebody would someday come up with something better, a name that would encompass more of what's "beyond" the sequencing of genes and the mapping of them to particular parts of a chromosome. But genomics just stuck.

Reading for catachresis unsticks what has become—for not bad reasons—stuck.

Care's not a bad name, either. The importance of care, its essentialness, even, to the fabric and feel of our lives in all their dimensions—labor, justice and equality seeking, illness and health—has become widely noted and theorized by scholars of all kinds, including by many scholars in science studies. I draw on and extend those lines of thought beginning in chapter 3, but also differentiate my catachrestic approach to care from theirs. Put it this way: most of these scholars think care is a good name for this amalgamation of affects and actions that is so definitively constitutive of everything we humans are and everything we do, including the sciences. I think care's not a bad name—and that's good enough. "'Caring' is a banal word," psychoanalyst Stephen Mitchell has noted, "for a complex affective involvement."[20] In the pages to come, thinking of care as catachresis turns our attention toward the complex affective "object relations" that scientists have not only with people but also with bloodless objects like genes, machines, and data—lots and lots of data—but still leave you thinking and feeling unsettled, uneasy or (to use another word that's not a bad stand-in for care) anxious, moved to read care again and again.

Thomas Roderick considered "genomics" a not bad name for an exceedingly heterogeneous and still multiplying and shifting field of scientific action and infrastructure: it's a good enough word, people get it, and it allows everyone to get on with their work. Care, for me, catachrestically, is not a bad name for similar sorts of excess, a pervasive and shifting mode or mood or affect or style of thinking-doing that remains, at least in part, "beyond": it's a good enough word, people get it, and it allows me to get on with the writing and everyone else on with the reading. Just don't forget that in my reading it doesn't actually have a referent proper to it, or in the language of paradox that structures care (see chapter 3): embrace it, and don't get too attached to it.

Reading for catachreses in the words and worlds of genomics is different than reading for metaphors. Numerous scientists, philosophers, historians, feminist theorists, and others have read for how the sciences are shaped by and/or rely on metaphors: a genome is a text, for example, or in some more florid versions, "the Book of Life."[21] Statements in scientific discourse about sperm "penetrating" eggs, or immune systems "targeting and destroying" microbial "invaders," are taken as evidence of the improper intrusion, however inescapable, of political, cultural, or otherwise nonscientific concepts, beliefs, or values concerning gender differences, militarism and violence, and so on into the domain of nature proper. Such critical accounts of scientific thought and discourse have helped us understand how the sciences and their doing involves more than strict adherence to an austere, positivistic reason through the purging of all connotations from a set of precise denotations. Reading in this vein, however, comes to a conclusive end once metaphor's two registers of the tenor ("a genome") and vehicle ("a text") are linked. There's nowhere else for reading to go once it has arrived at this kind of one-to-one mapping. And one gets the sense that, at least as far as metaphors in sciences go, there's something lacking, you're missing something, or you've settled for second best. Metaphor may be fully satisfying in the literary realm—"All the world's a stage" works beautifully and feels complete—but metaphorical analyses of the sciences carry a sense of having missed a direct mark, a sense that a transparently true representation has escaped our grasping minds and we've substituted (perhaps temporarily) a second-best replacement. An improper poetic human sensibility has been swapped in for the prosaic no-nonsense literalness that is proper to both science and nature.

Reading for catachresis doesn't shift one domain's sense into another's; a catachresis, having no proper relationship to either, cascades through a set of pathways and our reading along with it. To read for catachresis is to read for complex movements through disseminating structures, to read for shifting patterns and possibilities, to read for impossibilities and double binds those places in a text or in a culture where contradictions emerge, and logic stumbles or slides around, as with the Neil deGrasse Tyson sign at the March for Science.

Caring for Genomics

As long as you did not read it too closely, the Neil deGrasse Tyson quote about the self-sufficient, self-evident truth of science seen on numerous signs at numerous 2017 Marches for Science functioned fine. The dominant cultural reference there, no doubt, was the truth of climate change as determined and stated by climate scientists. The research budgets of the National Oceanic and Atmospheric Administration (NOAA) or the National Center for Atmospheric Research (NCAR) were obvious causes for concern and calls for protection in early 2017, but there was also clear awareness of and advocacy for federal support of the environmental sciences undergirding the regulatory agenda of the Environmental Protection Agency, and similar sciences whose importance seemed clear. I doubt genomics was on anybody's mind but mine. I saw no signs and heard no frets concerning federal funding for the National Human Genome Research Institute, and I am absolutely sure that no one had lain awake the night before worrying about how the truths of genomics were being cast as a Chinese hoax. Genomics did not seem to be high on the list of things people thought it urgent to march *for*, if it was on those lists at all.

Why, then, is it so high on my list? Why ask you, reader, to care about genomics and its truths? Why am I trying to figure out what it means to write *for* genomics, toward an ethnographic account that says *yes!* to the *toll!* truth-making practices genomicists have developed over the past twenty years to better understand complex diseases like, say, asthma? The "vast machine" necessary to produce truths about rapidly changing planetary climate conditions (to use Paul Edwards's name for the expensive and immense global infrastructure of skilled experts, computer models refined over decades, and diverse instruments of data collection and analysis that are all necessary for knowing the dimensions and dynamics of the even vaster geo-atmo-hydro-bio-anthropo machine that sustains us)—it's understandable why that should compel our attention. The relevance is clear and urgent, and it's doubtful anyone would argue against greater public literacy about how to "read" climate science, or about how to interpret its arguments and claims, or about being able to evaluate and judge its complexities, strengths, and weaknesses. But genomics? Writing *for* such a generously funded, hyped, and hegemonic research endeavor might seem superfluous at best and misguided at worst, considering the burgeoning scholarly literature critiquing its simplifications, its self-exaltations, its continuations and extensions of racialization, its commodifications. In sum, genomics has been overvalued, and more critical appraisals are required.

All of which is true, and it's a literature to which I have contributed. My previous book, *Promising Genomics*—an ethnography of speculative practices in the worlds of genomics, finance, and biotech corporations in a few of the volatileXstable lavaXlands of genomics (chiasmi, as I called them there, at least some of which were double binds), including Iceland, the US Securities and Exchange Commission, internet chat spaces, and orca rescue missions—was also about multiple ways of assigning value to genomics.[22] That book has been read not only as a critique of Iceland's scandalous deCODE Genetics, but read in such a way that deCODE became a metonym for genomics itself. As a result, the "promising" of the title was usually reduced, even if for understandable reasons, to "saying more than one is authorized to say." Promising was nothing but overstating, whether honestly or dishonestly, knowingly or unknowingly.

But even as I was researching and writing that book, my fieldwork in genomics took me to more places than Reykjavik or the corporate filings databases of the SEC; fieldwork in genomics is as territorially multiple, heterogeneous, and excessive as the field of genomics itself. Just as I had to learn to hear and read offhand references to Keiko the killer whale as an understated but significant part of the deCODE events, so too did I learn to pick up staticky signals about the importance of something like "care" in genomics that I would only later try to dial in more distinctly, as I learned to think and *care* about care and affect from friends and scholars also dedicated to these subjects. And there seemed to me no place in the narrative of *Promising Genomics* to explore, develop, and convey these other ways to read genomics.

I was learning, too, to think and care about care and affect from genomicists and other scientists themselves. That obvious anthropological methodological point should not be lost. There was no place within the plot and genre constraints of *Promising Genomics* to come to terms with my many other encounters with genomicists, and the data-driven excitement that was palpable when they spoke with me in transdisciplinary workshops we attended together, when I visited them in their labs, when I listened to how they critically evaluated research proposals at funding panel reviews, and as I continued to read their writings in professional journals. I tell some of those stories here, many revolving around asthma as an emblem of the kind of "complex conditions" to which many thoughtful and careful scientists are directing and I would even say *devoting* their attention.[23] As I was coming to better understand more about how genomicists thought and felt about scads of data, thought and felt about the many vicissitudes of experimentation and the production of new genomic truths, I was also coming to better understand care, and its importance not only to but *in and of* genomics, and in and of the sciences more broadly.

Genomics with Care is written in the spirit of experimental ethnography to convey how the sciences, and the doing of science, is a matter of the tight, intimate, but uneasy relationship between thinking, caring, and affect, as evident in the colloquial understanding that "thoughtfulness" and "carefulness" are, for all practical purposes, the same thing. Or as I put it more accurately—also enigmatically and annoyingly to some, I know—in *Promising Genomics*: sameXdifference. And part of this experiment is to convey the sameXdifference of science and care in writing not only for genomics, but writing also for genomicists themselves. When much "experimental" writing in anthropology takes the form of a poeticization readable to a valued but increasingly limited professional audience, the experiment here involves writing *for* by writing outward rather than inward, privileging the prosaic over the poetic, the learning reader over the learned writer.[24]

We—we careful scientists, we thoughtful humanists—already seem to know that such differences are both minimal and consequential. Put differently, and in the rhetoric of chiasmus: experimentalists already know, when being truly thoughtful about what knowing is, that they care to know when they know to care. In their *longue durée* history of the changing concepts, practices, technologies, and visualizations that together have constituted the slowly changing forms of objectivity in the natural sciences, Lorraine Daston and Peter Galison refer to these kinds of culturally patterned personal qualities in scientists as "epistemic virtues," a term that also yokes together the conventionally distinct domains of science and ethics. For scientists to know in such a way as to achieve objectivity required a certain kind of subjectivity, a scientific subject capable of producing authoritative truth. My analyses and aims are close to theirs, but although I think that there is something of the "epistemic virtue" to "care," there is also something more to it, something still in between knowing and sensing, traversing bioscience, bioethics, and biopolitics—and biology.

To open the question of that "something more," chapter 2 tells a sketchy early history of genomics as it came to be through the US Human Genome Project in the 1980s and 1990s through a kind of biological frame, asking, what can we get by reading key events in the development of genomics for affect, rather than for the sociological, organizational, political, personal, or technological dynamics that have been the organizing frame for most accounts? Stephen Hilgartner's sociological history of the Human Genome Project, for example, is oriented around the changing norms and protocols for controlling, publishing, and sharing genomic data, and the conflicts and negotiated agreements that resulted.[25] My reading of this period is also oriented around data, but asks not how it was regulated but rather, what affects did genomic data provoke in the embodied subjects of (largely male) genomic scientists? The short answer: genomic data produces *toll!*-like affects, subtle but dense and variable amalgams of surprise, excitement, boredom, and anxiety. Reading for signs of *toll!*-like affect—in scientific articles and editorials, in the science and popular media, in oral history interviews with prominent genomicists—helps expose the labyrinthine structure and process of genomic inquiry, and of scientific research more broadly: an experimental process, in both laboratory and social worlds, that constantly responds to the unpredictable twists of material and conceptual constraints as they emerge out of, and through, previous stages of work.

I explore the binding of genomic research by these limits that inevitably develop in what sociologist Andy Pickering calls "the mangle of practice" (another sign for the twisting, turning development of scientific work and thought in the labyrinth of its own ongoing construction) in more detail in chapter 3, where some of these binds are shown to be double binds: contradictory or paradoxical messaging at different levels of the experimental genomic system or situation, both of which must be and cannot be heeded.[26] This experience of impossibility—or (im)possibility, possibly—is a fundamental truth of the genomic condition, and marks places where "care" is enacted as a necessary, ineluctable aspect of doing genomics, or indeed *any* science. Care has now been theorized extensively in feminist and science studies; chapter 3 extends and diverges slightly from that work to explain why care is not only complementary to the practice of science but also constitutive of it and, from some angles, identical to it. Care is a name for what genomicists do when they encounter, as they inevitably will, the limits of formalisms, algorithms, and a complex, twisted material world.

Care is also a name for the amalgam of affect and thinking that genomicists experience and project in these encounters. The "minding system" was what psychologist Silvan Tomkins named these conglomerations of affective and cognitive states and processes of the human brain. Chapter 3 focuses in particular on two affective registers Tomkins analyzed, surprise-startle and interest-excitement, as especially important to the *toll!*-like responses that may be—I am hypothesizing—the basis or underbelly of an epistemophilic impulse or drive shared by all humans but taking perhaps a more acute or evident form in scientists. I take up this hypothesis again in the postscript, but it really awaits future work aimed more explicitly at generating more ethnographic data on these matters of affect that few anthropologists of science, including myself, have so far cared to develop and explore. Chapter 3 uses Tomkins's conceptualization of the human psyche's "minding system" to introduce and advance the encompassing narrative of the book: by "minding genomics" over several decades, through practices of careful thoughtfulness and thoughtful carefulness, genomicists developed ways to more carefully produce genomic truths out of bio-technical-social complexities that were becoming ever more so. Here are analytic stories, then, of *how genomics got better* as genomicists minded—cared for, made smarter—the material and conceptual labyrinths emergent from their own doings.

Each of chapters 4 through 6 is oriented around a different name for name-exceeding "care": curation, scrupulousness, and solicitude name different patterns of care traceable at different levels of genomics as a pragmatic, technical, conceptual, and cultural system. These should be read as "artificial" separations, made to ease analysis, writing, and reading, and to keep them all moving forward. A thoughtful, careful genomicist, the overall text should show, is always doing two and usually all three of these at once. Here again the rhetorical trope of catachresis helps out: there is no "natural" definition or reality of "care" that would then have to be "artificially" broken down (or up). Curation, scrupulousness, and solicitude are neither more nor less forced, improper names for the essential practices of genomics than is care. Their different etymologies and connotations, pragmatic and metapragmatic uses, however, can bring out the different qualities, styles, targets, and temporalities associated with those different patterns, making them markers of the particularities of a phenomenological experience of care that is thoroughly, confoundingly, and refreshingly relational.

Chapter 4 is oriented around the patterns of relations that make up curation, specifically the curation of data. Data has always been troped as solid and foundational to the sciences, and has become an object of intense attention, desire, and hope in the contemporary cultural moment in science and beyond. We'll read why data, as analyzed by a growing number of archivists, librarians, information scientists, and anthropologists, is better written as (meta)data: a complex, differentiated, or relational entity rather than a simple, solid, sovereign one. For data to fulfill its promise as foundational to genomics and all sciences, it always needs to be refounded, again and again: it requires relationships to other data, conventionally differentiated as metadata, and needs the continual care of curation to be cleaned, assessed, ordered, rendered findable, made interoperable, and, perhaps most importantly, maintained and archived. Like good dental hygiene, the need for curation is constant and repetitive; data calls for careful practices to be done over, and over, and over again. Data issues impossible, contradictory double-binding demands that, if given language in a psychoanalytic setting, would read like, *Don't touch me, I'm firmly self-sufficient; you'll always need to take care of me.* Or, *I don't need you; please help me.* Chapter 4 explores these double binds in the hot pursuit of more, more, *more* genomic data in the 1990s and 2000s, and describes some of the vital but underappreciated curation work of the individuals and organizations that care for genomic data in multiple ways. Curation, then, is the pattern of care manifest in the relations between a scientist and her data.

Chapter 5 adjusts the dials to draw out patterns of *scrupulousness*, that epistemic virtue of the scientist whose usually unacknowledged affective dimension comes out in the common expression, "painstaking attention to detail." Doing genomics with care, minding it, here means taking great pains to attend to every possible thing that might have gone wrong, or might still go wrong, in one's experimental system: a necessary but bad assumption, spotty samples or (meta)data, a slight deviation from protocol, a finicky piece of equipment, an experimental design in need of update or revision. A scruple, as we'll see, was the name thoughtful Romans gave to the tiny pebble you put in your shoe to create a minuscule but constant source of discomfort, a material reminder of the small doubts, troubles, and worries that should always intrude on one's awareness with every passing step, eliciting and focusing concern.

And therein lies the double bind of scrupulousness: a call of conscience speaking in the softest whisper, a source of uneasiness and hesitation that is—maybe—practically insignificant and you should just suck it up and carry on, or it may magnify or cascade and become debilitating. The dominant double bind in play here can be expressed as an uneasy modification of the popular self-help book series trope: *Do/Don't Sweat the Small Stuff.* To be a careful genomicist or any other kind of scientist, you *must* sweat the small stuff—and it's all small stuff, things you must be anxious, worry, care about. And the simultaneous counterdemand, usually unspoken but always sensed, "You'll never get anywhere, worrying about tiny things like that. And most other people find you insufferable." Scrupulousness as a name for "care" bears its own contradiction, simultaneously virtue and, through simple magnification into "overly scrupulous," vice.

Doing science well has always demanded high degrees of scrupulousness, but in "high-throughput genomics" with its ever-growing (meta)data analyzed in an ever-growing number of combinations and relationalities, even the smallest concerns can multiply, compound, or cascade in unexpected ways that can be difficult to localize and address. Chapter 5 discusses how, in the mid-2000s, in genomics and in many other sciences caught in the avalanching excesses of Big Data of their own making, what counted as truths or at least "significant results" were too often failing what was supposed to be the crucial test or criteria of scientificity: reproducibility. The ability to cheaply produce and analyze DNA sequence data from increasingly large study groups made genome-wide association studies (GWAS) enormously popular in the late 2000s, as well as apparently powerful in their ability to detect small genomic differences within large populations of people. Those differences promised to better account for complex conditions like asthma, schizophrenia, diabetes, and the like. Almost as quickly, however, GWAS seemed to raise as many doubts and questions about its powers, utility, and, most significantly, replicability. Genomicists responded by creating new forms, and heightened levels, of scrupulousness, critically examining every component of the (post)genomic experimental system, from data algorithms to journal publication mechanisms to even broader cultural interventions. A related development in this new social landscape of heightened care was the growth of consortia—new networks of individuals researchers, organizations, and eventually "networks of networks" to share data, methods, standards, and analytic strategies to better attend to the greatly increased number of "moving parts" in play when complex conditions such as asthma became the focus of genomic research.

Genomic research became still more complex, however, as it attempted to attend to the even greater complexity of geneXenvironment interactions as shapers of the clinically evident complexities of conditions like asthma. It is not my partiality for the impossible chiasmus of the double-binding "X" that makes me write "geneXenvironment interactions" this way in chapter 6. Along with the simpler acronymic alternative of GEI, GxE became one of the typographic conventions some genomicists used to signify the important shift in the late 2000s to this greater level of complexity. Genes, conceptually as well as pragmatically, had long been entities that did not have an environment, or for which "the environment" just did not need to be, and indeed *couldn't* be, part of the research equation for understanding genes and gene function. There really wasn't a place for the environment in the experimental systems of genomics before the turn of the millennium. But then genomicists began trying to make that place, where genomes and environments could be analyzed together. The "X" marks perfectly both the new interactions between these once-separated domains and, not just the difficulty but, I'll argue, the *impossibility* of their joint analysis. We'll also see people doing impossibility . . . and that's kind of exciting.

As a pattern of care, the solicitude of chapter 6 that is part of doing the impossible is like an inverted image of scrupulousness: rather than attending to each of a multitude of components, solicitude embodies a care given to the multitude itself, as an ensemble or assemblage. Solicitude is the pattern care takes in the relations between genomicists, collectively, and their overall collective enterprise—not between genomicists and their experimental systems, but between genomicists and genomics itself. "To solicit" means to shake a structure in its entirety, to wrap one's arms around a whole—all of a research program, all of asthma, all of genomics, all of an environment, and all of their interactions and all of the efforts to know them—and bend it, stretch it, scrunch it, and rattle it to feel where and how it moves, to sense which parts are loosely in play and which parts hold more rigidly, which parts hold and which parts give, and how even distant parts of the overall structure might affect each other. The recurrent, and recurrently deferred, desire and hope for a holistic understanding of "wholes" in the life sciences—whole organisms, whole ecologies, the entire evolution of "endless forms most beautiful"—is something evident in each and every "-ome" word coined, and each and every "-omics" effort to reach or define it: proteomics, transcriptomics, metabolomics, exposomics. (It's this impossible desire and hope for a whole that may also animate anthropology.)

Solicitude is care taken from and given to the impossible conjunction of both the *promise* of such wholes and the *threat* of their nonarrival, dissolution, or disclosure as only one piece of a still larger whole: genomes and genomics disclosed the need for proteomes and thus proteomics, which disclosed the need for metabolomes and thus metabolomics, which disclosed the need for exposomes and thus exposomics, which . . . is where chapter 6 takes place. It's a place where genomicists, networked in curatorial consortia, scrupulously examine every aspect of their collective enterprises, in order to better solicit the entire political economic technological cultural conceptual apparatus of a "Science" aimed at understanding the whole health (the terms are tightly intertwined if not entirely redundant) of diverse bodies with diverse developmental histories in complex bio-socio-technological environments that are both shared and wildly differentiated. Personalized medicine, in other words.

To judge only the extent to which genomicists succeed or fail in this effort, or at what point in time success or failure might be decided in the future, would be to miss the double bind of threat and promise inherent in the effort—to overlook how solicitude can and does happen even without such a decisive judgment. And we would miss another kind of time that the sciences run on or within, an exhilarating and anxious time of being always "on the verge." This is the time of care.

Writing for Genomicists

A depiction of scientists as carefully soliciting the impossible wholes of -omics and asthma, while scrupulously attending to their logical and experimental apparatuses powering and powered by curated data, risks the kind of romanticism that remains attached to the legacy systems of "care" that I discuss in chapter 3. I acknowledge and take that risk, betting that a deliberately friendly depiction of genomics and genomicists doesn't have to end up being naively lovey-dovey. Euro-American culture hardly lacks accounts of science and scientists that span the limited range from the merely reverential to the fawningly adorational. Once the apparent exceptions are accounted for—variations on the mad scientist and the corporate-captured sciences—the options seem narrowed to somewhere between "I love science!" and "ifuckinglovescience.com," a social media entity of this era that started in 2012 as a Facebook page with a thousand followers before morphing into its own website full of clickbait stories and a sidebar of ads, followed by 22 million just three years later.[27]

In chapter 7, I try to hedge my bets by sketching one last pattern of care, friendship, in which an impossible double bind becomes a powerful, prosocial bond. Through friendship I take up some of interrelated questions driving this book: What kind of relationships do scientists have with the objects of their science—a fruit fly, a *toll*-like receptor, a fragment of DNA sequence information or a database full of it? What kinds of affective attachments do they have to those objects, to the equipment in their labs, to the hypotheses formed with them, to the truths advanced through them, beyond the noncommittal detachment that is the culturally and epistemically sanctioned relationship? To their near and distant colleagues, and to their science writ large? Why, when analyzing the numerous consortia and networks and other collective groups that are now an indelible feature of the many worlds of genomics, do we not think of these collectives in their affective dimensions, patterned through friendship?

When we are assigning value to infrastructural endeavors like genomics, asking questions like "Was the Human Genome Project worth the $3 billion investment?," what difference do those affective investments make to our assessments? What would it mean to value knowledge infrastructures according to how well they elicit and encourage some hypothesized epistemophilic impulses or object-relation tendencies, according to how likely those infrastructures are to make someone shout or even just murmur "*toll!*"? "What sciences," as Banu Subramaniam and Angela Wiley ask in a similar vein, "would we put on our proverbial boots to march for?"[28]

Regarding some previously published parts of chapter 2, the bioinformatician Russ Altman characterized them as a "marvelous meditation." I hadn't thought of my style of writing in that way, but it made sense to me when he named it such. I take great satisfaction that this assessment came from a genomicist, and hope this book, too, works for a wider readership of scientists in the same way. It was as kind a compliment as it was a perceptive analysis of style and its effects: although I advance "arguments," make "claims," and take up "positions" here concerning genomics, science, and care, I recognize that, for better and for worse, I am less comfortable in and committed to those customary scholarly styles (especially dominant in much of science and technology studies) than I am in the genre of something like the meditation. Anthropologist Stephen Tyler casts contemporary ethnography as a "meditative vehicle," something that readers come to "neither as to a map of knowledge nor as a guide to action," but because it "provokes a rupture with the common sense world and evokes an aesthetic integration whose therapeutic effect is worked out in the restoration of the commonsense world."[29] Or, in a different vein, Sarah McNamer characterizes the Christian meditations of the Middle Ages: "gentle invitations to the reader to enter into imagined and embellished scenes," taking readers through a "step-by-step realistic narrative" designed, through deictic rhetoric and direct appeals to readers (you!), to produce "affective piety."[30] Affective piety has many risks, as I've already suggested and will say more about, but they seem to me worth taking if only to better explore and understand them.

By trying to write *for* genomics, I am also trying to write *for* genomicists, and for scientists more broadly, in the name of friendship or something like it. Anthropologists often say they are keen to write for their "interlocutors," those with whom they speak and to whom they listen and about whom they write, although the extent to which they actually do, let alone successfully, is open for discussion. I have written deliberately to be read by scientists, especially the undergraduate science students that I teach, while remaining true to concepts and practices of the humanities that may be unfamiliar to them, speaking patiently and honestly about the difficulties and differences as they arise. As friends do.

I tried to deliberately write to be readable by scientists who aren't much in the habit of reading anthropology, wanting them to recognize something of themselves in these pages, and perhaps be surprised by that recognition. Maybe I'm dreaming. I want all readers, too, to put down the book more interested in genomics and its data practices, in the infrastructures and practices of care necessary for crafting genomic truths, and to finish reading having become somewhat better friends to the many scientists who mind these and all the other things that depend on them, and who do so thoughtfully and carefully. In the process, I hope to show how one might still be "critical" and that friendship, rather than fleeing politics, has a politics of its own, new relations and actions urgently needed in our thoroughly technoscientific, anthropocenic world.

So, friends, *he apostrophized*, stay with me . . .

2 · labyrinth life

AFFECT EXCESS INFRASTRUCTURE

Anxiety, then, is in the first place something that is felt. We call it an affective state, although we are also ignorant of what an affect is. · SIGMUND FREUD, *INHIBITION, SYMPTOM AND ANXIETY*

Epileptics do not startle. . . . Their experienced world is different in this one fundamental way. If epileptics had in addition lacked fear and rage, their world would have become even more different than the usual humanly experienced world. They experience a pistol shot as sudden but not startling. A world experienced without any affect at all, due to a complete genetic defect in the whole spectrum of innate affects, would be a pallid, meaningless world. We would know that things happened, but we could not care whether they did or not. · SILVAN TOMKINS, *AFFECT IMAGERY CONSCIOUSNESS*

This chapter begins by rereading some episodes in the scientific and institutional development of genomics in the 1980s, when the Human Genome Project was first being discussed and then initiated, through its (ritually announced) completion in the first years of the twenty-first century. Rereading the history of the Human Genome Project is something I've been doing for a long time—in fact, since before there even was a Human Genome Project—and most of those rereadings involved an attempt to swap in what I thought was a more appropriate, less catachrestic name: a lowercase "genomics project," for example, or "Projecting Speed Genomics."[1] Part of the intent was to shift the emphasis from the end goal, completing "the genome" (and it was always about more than the human), to the genomics infrastructure that enabled that and so many other ends. Shifting readers' attention to the building of genomics infrastructure also attempted to break up the sense that an organized, coherent "Project" was what was most interesting and deserving of attention. It reenforced an ideal of science as fundamentally a carefully planned and rationally strategized activity, when the truth is much more complicated.

The complexities and contingencies of doing science at any scale, from the lab or the field to the state-sponsored enterprise, are grounds well-trod by scientists themselves, and by historians and sociologists of science. One of the earliest works in the latter category was Bruno Latour and Steve Woolgar's *Laboratory Life*, depicting the activities in Roger Guillemin's lab at the Salk Institute. Based on their experiences, they analogized the process of experimentation to a game of Go, in which early, "almost entirely contingent" placements of black and white discs gradually become more ordered and thus constricted, and it "becomes less and less easy to play anywhere," some moves become "almost necessary," and eventually many areas and even the entire board are "definitively settled."[2]

In the rest of this chapter, we'll see similar patterns develop on a larger scale than the laboratory, where the ordered pattern of the Human Genome Project is cast as the eventual outcome of distributed moves in a disordered maze being built from the inside out. But there's an additional dimension in my account of the building of genomics infrastructures. *Laboratory Life* would come to exert a powerful influence on large portions of the science studies field, with its stated disinterest in and indeed disdain for anything "cognitive," which for Latour and Woolgar slid seamlessly into "the mind."[3] It was an understandable and in many ways productive move in a disciplinary metagame of Go, capturing and stabilizing a large scholarly territory but rendering it blind to "the mind" of the scientist, for which there was no room.

No less contingency-forward but far more attentive to matters of the mind, including not only the cognitive but also the affective forces composing it, is geneticist Francois Jacob's autobiographical account of the scientist at work and in play:

> The march of science does not consist in a series of inevitable conquests, or advance along the royal road of human reason, or result necessarily and inevitably from conclusive observations dictated by experiment and argumentation. I found in science a mode of playfulness and imagination, of obsessions and fixed ideas.... Those in the front ranks displayed exotic blends of passion and indifference, of rigor and whimsy, of naivete and the will to power, in a triumph of individuality.[4]

What Jacob calls "exotic blends" of opposites, I will go on to analyze as double binds, but the point of our patternings is the same: to better portray and understand the complexities of both science and the scientist subject engaged in and by it. My rereading of genomics' developmental pathways for some of the microforces of affect begins with early articulations of the "boring" quality of DNA sequencing as a form of scientific work, and even the boringness of DNA sequence data itself.

And Yet

One of the most frequently read and referenced of statements regarding the "boringness" of DNA sequence and sequencing is Walter Gilbert's 1991 *Nature* opinion piece, "Toward a Paradigm Shift in Biology." Gilbert is a Nobel laureate for inventing (with Allan Maxam) a DNA sequencing method that was briefly popular, founder of Biogen (one of the earliest biotech companies, and still successful), and an early vocal proponent in the mid-1980s of a sequencing-dominated vision of a "Human Genome Project"; his short article would be frequently cited (often derisively) for its imagined future when people would carry around CDs (!) holding their sequenced genomes and be able to say, "This is me!" But before it ever got to that hyperbole, the opening paragraph conjured a mood, albeit a somewhat mixed one: "There is a malaise in biology. The growing excitement about the genome project is marred by a worry that something is wrong—a tension in the minds of many biologists reflected in the frequent declaration that sequencing is boring. And yet everyone is sequencing. What can be happening? Our paradigm is changing."[5]

Marking the times then, Gilbert now messages those of us grappling with similar recurrent questions about change in the sciences, and in scientists, over time: when analyzing affects and their shifts, we're always presented with complex amalgams. Malaise can be mixed with excitement, excitement will be shot through with worry, and any of these component affects will be growing or diminishing, becoming more or less intense, but always in some kind of psychic stew. A careful, good story of scientific change cannot take the simple form, "genomics was boring, then it became postgenomics, which is exciting."

But *something* was changing already in 1991; scientists could feel it even if they couldn't quite put a finger on it, provoking Gilbert to ask the historian's question: What can be happening?[6] Sequencing is boring, "and yet" everyone is sequencing. There is excitement, "and yet" there is worry, boredom, malaise. DNA sequencing is *toll!*/droll/weird-leveraged-with-sarcasm and yet it is *toll!*/cool/amazing. (Watch for Gilbert's "*and yet*" throughout this book, italicized to emphasize the simultaneity of difference, and sometimes contradiction, that Gilbert marked here.)

Maybe Gilbert's statement simply begs for differentiation and specificity: *some* biologists in 1991 were bored with genomics, *some* were excited by it, and these different subject positions and their affects correlate with their social position: Nobel laureates, full professors at elite institutions, and (ex-) CEO's of successful biotech corporations may have enjoyed more freedom and privilege to find more excitement in genomic speculation than the assistant professor, postdoc, or lab worker worrying about the tedious and repetitive labor of grant writing, marker characterization and development, and quotas of DNA sequence to be fulfilled. Your *toll!*-age may vary.

Some such social analysis is certainly warranted, as will be seen below. But I wouldn't want to explain away too quickly the force and logics of complex affective states, the possibility of their being widely shared, across difference, and their importance for understanding scientific change over time.

Having used affect and collective mood to set the historical stage and open his essay, Gilbert never again invoked it. Affect faded into the background and became invisible, giving way to an equally complex amalgam of arguments about the reigning biological paradigm and the one Gilbert thought he glimpsed emerging from it. This mixture of arguments was framed by a statement that was neither inside nor outside Gilbert's text, an exclamation that appeared only in the article's subtitle; it seemed to emanate from an editorial rather than authorial position: "The steady conversion of new techniques into purchasable kits and the accumulation of nucleotide sequence data in the electronic data banks leads one practitioner to cry, 'Molecular biology is dead—Long live molecular biology!'"[7]

Who at *Nature* wrote this line? Almost certainly not Gilbert, who never "cries" anything along the lines of this Franco-Anglo reassurance of uninterrupted sovereignty, surely echoed over and over by the peasantry with at least a droll hint of Pythonesque sarcasm and a *toll!* roll of the eyes: *yeah yeah, long live the king*... Gilbert's own trope for paradigm change (the one that clearly occurs inside his authored text) was more definitive in its invocation of a "break," and more Sino- than Euro- in historical connotation: "The view that the genome project is breaking the rice bowl of the individual biologist," he suggested, was one biologists needed to get over.[8] The great leap forward into the genomic future would require breaking the attachments scientists had to individual R01 grants from the National Institutes of Health (NIH) that were supporting their independent laboratories, and learning to enjoy working in large collectives, centrally organized, planned, and funded. Not surprisingly, there was some counterrevolutionary grumbling (see below).

Despite his attunement to the revolutionary potential of genomic kits and data, however, Gilbert did not envision increased surprise, excitement, or any *toll!*-like affects as outcomes characterizing the new era. It was all business. "The tenfold increase in the amount of information in the databases will divide the world into haves and have-nots," Gilbert predicted, "unless each of us connects to that information and learns how to sift through it *for the parts we need*" (emphasis added). Genomics is, in this view, predicated on fulfilling what we already know we need and want, just faster and more efficiently. There is little sense that the *needs* themselves, let alone genomic desires, could, should, or would be transformed.

"Sequencing is boring" was a frequent declaration in the debates in the mid- and late-1980s leading to the institutionalization of the Human Genome Project (HGP). James Watson's personal aversion to anything boring was broadcast most loudly in his 2007 memoir *Avoid Boring People*, but this has been a persistent trait of his; his dismissiveness or hostility toward Rosalind Franklin was certainly a product of complex gender differences, but surely some part of that involved something along the lines of *she was uncreative, couldn't see the big picture, didn't know how to interpret the data she was obsessed with, was, in short, boring.*[9]

So Watson's judgments concerning the boringness of DNA sequence and DNA sequencing can be considered indicators of broader cultural patterns. Early in his brief but influential directorship of the National Human Genome Research Institute, an organizational role that translated symbolically as "leader of the HGP," Watson spoke in 1990 (around the time of Gilbert's article) and gave the problem of nonexcitement a slightly different twist:

> The people who wanted to do it [the HGP] were all old and almost retired, and everyone young was against it, because they figured if we did it, it would take money away from their research. So all the people you nor-

> mally would expect, because they're going to do something, were against it, and all the people, you know, who really almost stopped [doing] science, were in favor of it. Now that includes me: I was really in favor of it, as was Paul Berg. And you could say that the objective was a wonderful objective. What's more important than this piece of instructions? But everyone else felt essentially frightened. It was going to be big science, it was going to be very boring—just determine all these letters—so anyone who would do it is someone you wouldn't really want to invite to dinner anyways.[10]

Both DNA sequencing—"just determine"—and DNA sequence itself—"all these letters"—were troped as tedious and mind-numbing. And it wasn't simply sequencing and sequence that were regarded (wrongly, in Watson's view) as boring; the development of high-resolution genetic maps had provoked a similar affective response, at least among some NIH administrators and colleagues on review panels: "The trouble about getting these genetic maps, was that doing it was very boring, and in fact David Botstein had put in a grant application and had been turned down by NIH: it was too dull to be good science. But in fact it was a sort of tool that you really needed."[11]

Not only are affects always amalgams—it seems that Botstein, at least, found some excitement in the boring work of developing better genetic maps—but these affect-amalgams are always assembled to epistemic objects and their larger cultural webs. This is part of the reason why affects tend to disappear from view: debates about *tools* and *big science*, for example, are well-recognized concerns of historians, sociologists, and philosophers of science and technology—and of scientists and engineers, too. So what does it really matter if Watson, Botstein, or any other scientist is bored, when more social and collective things are at stake, things we already know how to document, understand, and value?

With a promise to return to such questions, let's continue to follow Watson's 1990 talk, which prompted an interesting exchange with Matthew Meselson, the Harvard molecular biologist renowned for numerous scientific achievements (like determining, in what are often described as "beautiful" experiments with Franklin Stahl, the semiconservative replication pattern of DNA):

> MATTHEW MESELSON: Jim, in *Drosophila*, I don't know of a single gene that has been gone after intelligently, that hasn't been cloned with a little effort, even though we don't have the complete sequence of *Drosophila*. So I gather that with humans, it's different, because we can't do genetic crosses and certain other manipulations as well with humans—but they might come along. So I would like to hear you explain why [this project is so] necessary for humans.

Watson's initial response—"I think it's necessary in *Drosophila*, just because to get them all to cost—if we can do it at one-tenth the cost that it's being done in your lab, eventually it will be cost-effective. It won't be cost-effective if you do it at five to ten dollars a base pair, but if you do it at fifty cents a base pair, you'll get it out"—was hardly satisfactory to Meselson: "That's a different reason," about economic efficiency, Meselson argued back, not doing science "intelligently"; he restated his objection:

MESELSON: Not a single important gene that anyone has gone after intelligently has failed to be cloned and sequenced.

WATSON: Yeah, but there is a lot until we do it that you don't know the existence of, and the question is, if you actually see the total thing, will you be surprised and get interesting scientific insights? And my guess is you will, but that's my—

MESELSON: That's a different reason than the one you gave. That's a good reason.[12]

For Meselson, unlike Watson and Gilbert, productivity and efficiency were not particularly good reasons for dedicating $3 billion in public monies to something like the HGP—but to "be surprised" by "the total thing" and get new "interesting scientific insights"? *That's a good reason.*

Again, "good reason" may mean only a reason shared widely (enough) in the scientific culture of which Meselson and Watson are (elite) members: *Unexpected surprises? Sounds good to me! And since le roi est mort, vivre le roi! and let the rest of 'em eat reagents.* But what if it really is a good reason—by which I mean, what if there were a shared understanding that a society that contains, and cultivates, scientists wanting to be surprised and interested, rather than simply efficient producers of knowledge, is an actual social good? That's a question I'd like readers to hold.

It's also worth remembering that even non-elites enjoy their affects along with their reagents, so I use my *toll!*-like probe to pull up one more remembered event concerning these pre-postgenomic years. Going into my basement and accessing my dead-tree database from my history of science graduate student days, I found the interview I conducted with Robert Moyzis, then a leading scientist in the US Department of Energy's genomics programs. Everyone reading this probably knows the name James Watson; some may even be familiar with Matthew Meselson, if not for his work on the mechanism of DNA replication then perhaps for his laudable work on the control of chemical and biological weapons.[13] I am betting you never heard of Moyzis, whose research then focused on telomeres, the ends of chromosomes whose primary function was the apparently rather boring (but essential!) job of just keeping chromosomes intact. I would bet even more lavishly that you would not have heard of him in 1991, as he was just beginning his rise up the DOE scientific-administrative ladder. We had been talking about the history of early meetings sponsored by DOE that are often credited as "precursors" to the HGP, meetings that he had helped organize in the mid-1980s as a midlevel scientist at Los Alamos National Laboratory:

> MF: Can I just interrupt here for a second? It sounds from what you're telling me that the sort of standard account we get about mutation rates, and the technologies for detecting very small mutation rates, was not as much of a concern or a goal as building resources.
>
> ROBERT MOYZIS: . . . I think in some ways that was an after the fact justification, in the sense that—I mean, keep in mind that DOE has always been historically interested in those problems. . . . So that's not incorrect to say that that's what DOE's interests are, because that's always been DOE's interest and probably will remain DOE's interest. . . .
>
> [But] I think the history was more, "Hey, I'm real excited about this, this is a good idea; some aspects of this maybe are bigger science by biology's standards, therefore we're going to need organization, structure; average academic lab doesn't have that, DOE does." And then lastly, "Why the hell is DOE involved in this?" "Oh, well, you know, we want to study mutation, et cetera et cetera." And I don't really think that that business about "gee, we want to understand mutation et cetera et cetera," really started happening until the criticism started. . . . In fact in my mind, the kind of fruitcakes who kind of dreamt up this project, myself included, that's not where they were coming from. . . . I mean, the genome, or DNA—it was kind of like this challenge to see if you could put the jigsaw puzzle together, as sort of an intellectual challenge. And many of the people at that Santa Fe meeting, I think, had that kind of mentality.
>
> . . . I think it's really after [the 1986 Cold Spring Harbor meeting] that a lot of the apologizing and sort of rewriting of history to say, what's the scientific justification that the Department of Energy is involved in this?, really began. So you've got a stretch there of perhaps almost months where, from my viewpoint, that issue wasn't even discussed. It was still, we can do this, it's worth doing, and it might actually be some

fun doing, for a lot of crazy reasons. And I think it was certainly always discussed that there would be all these biomedical payoffs, but a lot of the initial players I don't think were even looking at it from that perspective.[14]

For Moyzis, the "intellectual justifications" for doing a Human Genome Project, while "not incorrect," were nevertheless secondary to "fun" and related "crazy [cool/awesome/exciting/*toll!*] reasons." Promises of "biomedical payoffs" and other such rational justifications were certainly crucial to packaging, branding, and selling the HGP to its funders in the US Congress, but it was the "intellectual challenge" of assembling a massive "puzzle" that caused it to be "dreamt up." (It's worth recalling that, in the conventional Kuhnian paradigm of scientific change, "puzzle-solving" is the mundane, perhaps boring work of "normal" nonrevolutionary scientists, not a challenge for fruitcake revolutionary ones.) Following Moyzis, if we can say that cognitive reasons for planning an organized genomics project were "not incorrect," we might also say that those reasons are *not* not joined to affective forces.

I've used these few episodes and recollections to characterize one aspect of resistance to a centrally organized effort to develop the sciences and technologies of genomics in the 1980s as predicated on its being "boring"—that is, not eliciting the affects I'll discuss more below, those psychologist Silvan Tomkins named surprise-startle and interest-enjoyment. Resistance in the scientific community to the Human Genome Project was fairly widespread, with a far more complex quality than simply "it's boring," and with its own history, in which *some* resisters who were critical of *some* aspects of *some* of the changing project definitions that were put forward in the mid- to late-1980s came to be supporters of the HGP as its institutional and scientific definition emerged from various expert committees and bodies.[15] These affective threads were woven together with institutional turf politics and their attendant mix of scientific and ideological arguments about the shape and form of the HGP. In those expert panel discussions, which resulted in a less sequence-obsessed and more mapping-inclusive project as well as the inclusion of various "model organism" genomes (fruit fly, yeast, mouse, etc.) in addition to the human, the Department of Energy and its scientists were often troped as rather mindless, good only for tool-building or engineering-type infrastructural problems (sorting cells, building and shipping chromosome libraries, banking but not analyzing data, etc.), while the forces of the National Institutes of Health were ones of creativity, able to pose and answer actual biological research questions. In fact, though, there were many DOE scientists who, like Moyzis, sensed something crazy, intellectually challenging, and fun within the technological quotidian.

But the dominant troping in US biomedical research culture continued to be that DOE=boring, NIH=interesting. This partly explains why NIH would come to control twice as much money as DOE and be regarded as the "lead" agency throughout the more than ten years of the HGP. The affects associated with different sciences and scientists registered as cultural patterns, patterned in turn with the sociopolitical events of funding and organizational politics: NIH scientists were about discovery, invention, excitement, leadership, and (in a quieter voice) fun, while DOE scientists . . . well, they really might as well be engineers, right? What with the tools and the rote routines and the plodding workaday whatnot . . .

As mentioned earlier, one of the most persistent strains of criticism at this formative time revolved around a perceived threat that an expensive, centralized genomics program would pose to individual R01 research grants, the central pillar of the NIH extramural funding program. On this issue, too, a cultural valuation of boring and dull was attached to large scientific collectives, which stifled individual creativity and interest. Scientists who had expressed such reservations about a genome project were asked to testify at a US Senate hearing, where US Senator Pete Domenici (Democratic senator from New Mexico, home of Los Alamos National Laboratory, and one of the key congressional advocates of the Human Genome Project along with Senator Al Gore) confessed with an evident degree of exasperation that he was "thoroughly amazed . . . at how the biomedical community could oppose this project":

> I cannot believe that you are going to insist on business as usual in this field. It is beyond my comprehension, I repeat, beyond my comprehension. . . .
>
> You cannot sit here and tell me that in all of the research that is going on with the marvelous individual investigators . . . you cannot tell me there is not more than $200 million, that if we even asked you to go look, you would say probably went for naught. . . . People had a lot of fun. Scientists had a lot of exciting mental activities. But it is inconceivable that out of $7 billion in grants in this very heralded R01 peer review approach . . . that there is not at least $200 million or $300 million that even one as dedicated to the field as you are could not go out there and look at and say, maybe we do not have to do this. . . . [There] are too many scientists in other fields that are using hardware and new technology and new techniques that support this as a tool, that I cannot believe that you really oppose it.[16]

Here we have some evidence to support the hypothesis that "fun," or the more clumsily phrased "exciting mental activities," is an important driver of scientific activity. Even US senators—whose understanding of science, among other subjects, is notably limited—understand this. But Domenici turns this on its head to make an argument that playtime—in which "marvelous individual investigators" got paid with public dollars to try to surprise themselves—playtime was over, and even if that time was simultaneously its opposite ("business as usual") it was nevertheless time to get on with the more serious, un-fun, and boring work of tool-making, manufacturing the "hardware and new technology and new techniques" of the Human Genome Project.

But some people get excited about tools. "Number one," Leroy Hood emphasized to members of the US Senate in 1990, "the human genome initiative is about developing new technologies.... It is going to create a fantastic infrastructure for biology and medicine."[17] The mix of the fantastic and the infrastructural was weighted, understandably enough, toward the more quotidian. "The genome project is a biological infrastructure initiative," HGP chronicler Robert Cook-Deegan wrote in 1994, and "investigators using genetic approaches to explore the biological wilderness need to start building some roads and bridges."[18] In the early 1990s both Cook-Deegan and I were in conversation with genomics infrastructure planners like Hood and Eric Lander (at MIT's Whitehead Institute in those years), who are probably the most responsible for this revisioning of the HGP. Lander stressed the infrastructural as a kind of counter-rhetoric to some of the more fantastical tropes then (and still) going around. "I guess I found it very hard, very troubling that people were cramming [genomics] down people's throats as 'the Holy Grail,'" he told me when I interviewed him in 1992:

> I eventually stood up and said, Come on, it's Route One we're talking about. This isn't so fancy. We are talking about building a highway, for god's sake.... Yes, it's very important, but it's infrastructure, and I thought "Route One" conveyed better than anything else, a) how you did the project—that you didn't do it in small pieces, but b) that it wasn't such a grandiose thing either. And so it conveyed both of the things, namely... people were proposing building this, by everybody go out and build a mile of highway on either side. And that's crazy, because you're going to have too many gaps and things like that. But the other aspect of it is that it somehow demystified the mystique of the human genome, and put it in what I still think is an appropriate context: it's an infrastructure

project. It's a very important infrastructure project. You try to build infrastructure [pause] efficiently. And that ought to be our basic metaphor.[19]

We should appreciate a scientist's explicit attention to the importance of metaphor, but more importantly I recount this exchange in support of my contention that genomics would be better minded if the idealized and capitalized Human Genome Project was instead more widely understood as a matter of infrastructure. And to heighten the impropriety of even the best metaphors like "infrastructure," I'll add that I wish Leroy Hood's "fantastic infrastructure" was our culture's popular catachresis for the surprising excesses of quotidian genomics.

Centers Are No Fun

These work-play, serious-fun, science-tool, fantastic-infrastructure double binds ("exotic blends" for the Jacob fans) continued to structure the HGP in its earliest stages, when the questions of organizational form were first confronted. Prior to that, however, Shirley Tilghman (a prominent mouse geneticist) became one of the few women to be enrolled into some of the early and more contentious discussions about whether and how to do such a project, let alone how to organize it. Asked later if she had been drawn in "as an advocate or as a neutral thought leader," she replied bluntly: "I was drawn in as a person who has two X chromosomes. That's the truth."[20] Tilghman joined the National Research Council Committee formed to evaluate the different plans for the proposed project and make a recommendation to Congress about how the HGP should be defined, and if it were worth doing. At that point, additional biocultural criteria for her continued inclusion in these high-level planning sessions became evident.

> When I received the membership list of the committee, it became crystal clear why I was on the committee, which was full of people who had never sequenced DNA. It turned out I was the only person on the committee who had actually sequenced DNA with my own two hands, and that's because I was the youngest member of the committee by probably 10–15 years. And I think I was on it because they had put this very distinguished group together and said "oh my god we need at least one woman." And I got the short stick. And that's how I got into genomics.[21]

After the NRC committee endorsed a plan that included sequencing the human genome—along with the genomes of other organisms, along with developing better physical and genetic maps for these organisms, and along with the main but rarely stated goal of building genomic infrastructures—Tilghman was asked to join another committee to work out actual organizational strategies for doing it and managing the differences of opinion and interests they raised: as Tilghman put it, "to start putting meat on the bone." This committee, with several more women now involved, met at Cold Spring Harbor in 1989 and found itself having to negotiate (again) disagreements about whether monies should go to a few large sequencing and mapping centers, or into the more established, distributed, and biologically driven model of research based on individual RO1 NIH grants. Linked to those different cognitive positions on organizational and funding policies were a set of questions concerning scientific style and, ultimately, affect. "Biology had never been done the way physics is done," Tilghman recalled:

> We are a cottage industry; the creativity of our work comes out of small laboratories doing investigator-initiated work, and we're [planning] to spend a lot of money on boring, boring science. . . . And then the final criticism, I think, was that we were going to generate all this DNA sequence that we wouldn't be able to understand what they were telling us.
>
> One of the revelatory moments for me was going over to talk to the physics department at Princeton, who had invited me over to explain all this to them. And I offered that as one of the criticisms of the committee, and I could see that they were all frowning at me, like, "huh?" And I asked, why are you reacting this way? And they said, "Oh my god that's all we do! We just generate data and that's what gets us to start thinking about what they might be good for." You know: sky surveys. So that's when I had that sort of "aha!" moment, that it's possible that the data itself will be stimulatory. Which it has been.

Tilghman's impromptu, para-ethnographic "aha!" moment (supporting, I would like to note, the differences I drew in chapter 1 between aha! and *toll!*) allowed her to step outside the thought styles of her scientific community, and the affects assembled with them, to be stimulated by the stimulatory potentials of genomic data to come. Tilghman spoke to these affective data effects again toward the end of the Human Genome Project, as we'll see in chapter 4. But at this time, when the work was just getting organized, Tilghman spoke in no uncertain terms to *Science* magazine about how "the actual work" of the large genome centers that were now part of the official plan, at least sidelining the "cottage industry" approach, would be "technician-oriented, hard-slogging, and not much fun."[22]

Toll-Like Receptors and Pattern Recognition

Interpretive analysts are habituated to read for these kinds of multiple signaling pathways, in which catachrestic signs like "fun" or "*Toll!*" or "huh?" can spark a pattern of different meanings. Funnily enough, the gene *toll* itself turned out to display a similar pattern: *toll* turned from being known as a developmental gene in *Drosophila* larvae to a gene crucial to immune response in the adult fly. And then the pattern of *toll*'s effects differentiated even further, as "*toll*-like receptors" turned out to be a previously unanticipated "innate" part of human immune systems as well, evolutionarily ancient but scientifically brand spanking new. That surprising story turned on being able to search the databases of boring exciting DNA sequences accumulating even in these early years of the Human Genome Project.

We read in chapter 1 how Christiane Nüsslein-Volhard and Eric Wieschaus first showed *toll* mutants to be involved in the control of dorsal-ventral polarity in the fruit fly embryo. In the early to mid-1980s, *toll* became known as one of a number of genes in *Drosophila* that helped create the patterning effects of early development—making fronts and backs to larvae, tops and middles and bottoms, differentiated and spaced segments. *Toll* the gene became an "epistemic thing," the name biologist, historian, and philosopher Hans-Jörg Rheinberger gives to new objects made knowable—and manipulable—through the work of scientists inventing and using experimental systems, like the "Heidelberg screen" Nüsslein-Volhard and Wieschaus had built. Such "objects of knowledge," in Karin Knorr Cetina's equivalent term, are "characteristically open, question generating and complex," "process and projections rather than definitive things": "Objects of knowledge appear to have the capacity to unfold indefinitely. They are more like open drawers filled with folders extending indefinitely into the depths of a dark closet."[23]

Toll, first identified and characterized as a developmental gene, and named as such, offered new folders to explore in the mid-1980s, as Nüsslein-Volhard and a cascading network of scientists began pulling on its cabinet drawer: Where is this lethally mutated gene on a fly chromosome? What does it "produce," "express," "code for"—catachreses all—when it has not been changed by "the Heidelberg screen" into its lethal mutant form, damaging the fly's bimorphism, making it one-sided, halting the patterns of development and life? How does it become activated or, in the common parlance of genetics that mixes the electrical with the sexual, "turned on"? What happens next? And what role might *toll* play in an adult fly? Epistemic objects like *toll* are also affective objects, eliciting the anticipatory excitement of the unprecedented and the unknown: you want to pull on that drawer and see what's back there in the dark.

Like most *Drosophila* geneticists in the pre-genomic early 1980s, Nüsslein-Volhard and Wieschaus were more interested in questions of the biochemical and organismal effects of *toll* than they were in the actual DNA sequence of this still-puzzling gene. Sequencing DNA at that time was still expensive, tedious, and, more importantly, not immediately relevant to their kind of breeding-and-mapping genetics research. It wasn't until 1988 that Kathryn Anderson (a close colleague and former postdoc of Nüsslein-Volhard's) cloned *toll* and showed that this gene coded for a transmembrane protein, a large molecule partly within a cell, partly outside (see figure 2.1).

"Transmembrane protein" is one name given to these structures, the most plainly descriptive and probably the most widely used. (Anyone who paid even minimal attention to more recent news about the unique properties of the COVID-19 coronavirus will recall how its protein spiked structure enabled it to attach to a different transmembrane protein, the ACE2 receptor, and enter human cells.) Biosemioticians, a relatively small peripheral group of primarily theoretical biologists who conceptualize organisms as engaged in a constant process of interpreting, generating, and processing signs, think of them as "semiotic bridges." I'll expand on the differences marked by these naming schemes later; for now, "bridge" can better suggest the vital function served by such structures like this one that the gene *toll* gives rise to: they bridge an inside and an outside, allowing something—information? messages? signals? meaning?—to cross a membrane dividing (and therefore also connecting) two domains. These transmembrane semiotic protein bridges contact an extracellular environment on one side of a layer of (essentially) lipid molecules, and an intracellular genome on the other; they are where the "gene-environment interactions" discussed later in this book take place.

The 1980s were still early years in the development of an understanding of what is usually called the mechanisms of cell signaling: how cells . . . communicate with each other and with an outside environment through a series of molecules, including those like the *toll* protein that, traversing the cell membrane, occupy a key position in these labyrinthine signaling networks.[24] But precisely what signals were being sent and/or received by *toll* and its associated proteins, by what mechanisms or pathways, and with what end result, remained largely unknown. It would be another five years before the next set of cool, weird *toll!* results were registered by working scientists, recorded in lab notebooks, and reported in scientific journals. I continue with that parallel story of signaling and communication, a story of *toll*'s proliferating forms,

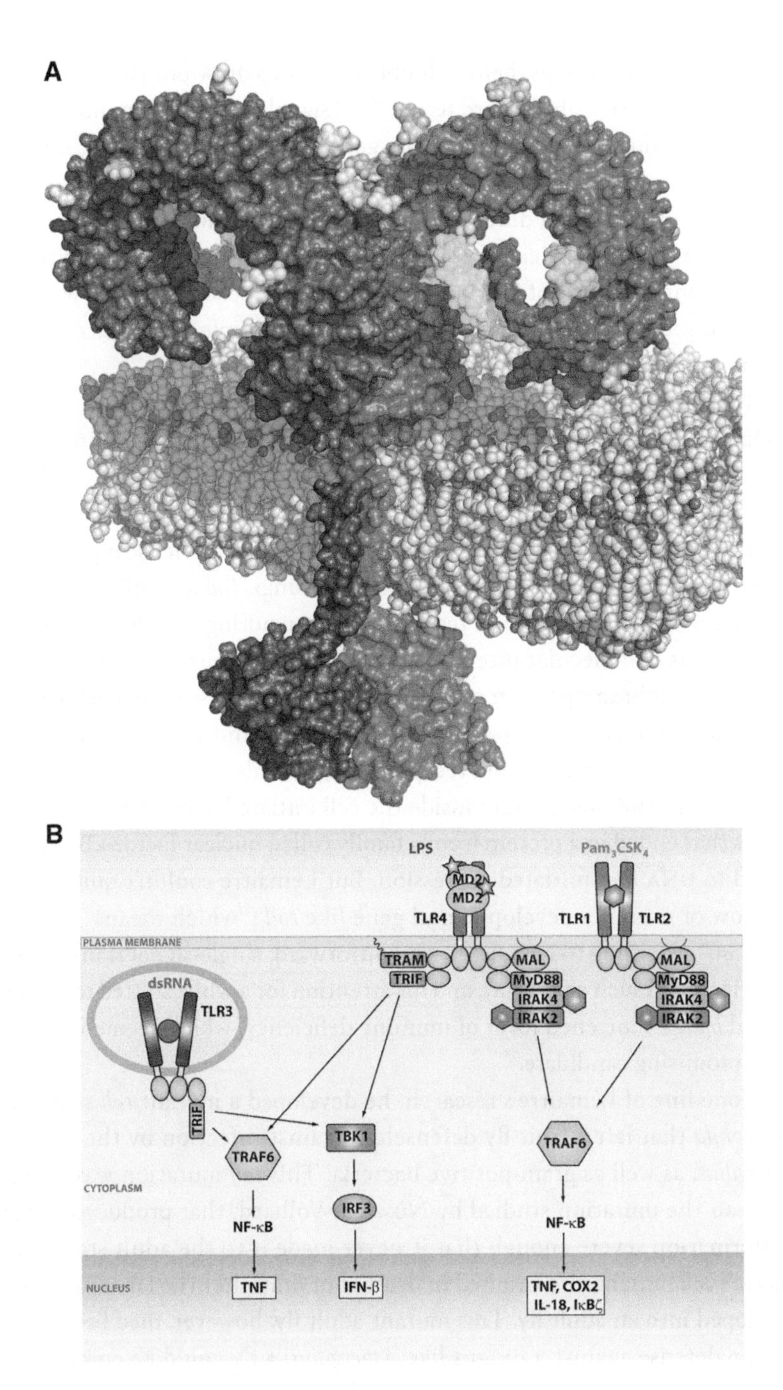

FIGURE 2.1 · *Toll*-like receptor and its signaling pathways.

functions, and meanings, before doubling back to draw out the *toll*ness of these seemingly straightforward terms, "cell signaling" or "communication."

Between 1993 and 1996, *Drosophila* geneticist Bruno Lemaitre, immunologist Jules Hoffman, and other scientists established the important role of the gene *toll* in a different signaling pathway, not a developmental one but an immunological one—and, even more surprisingly, not only in the signaling pathways of flies but in those of mammals as well. *Toll* was disseminating, differentiating, and becoming more multiple, both conceptually and materially.

My streamlined account here draws on the more complex story told by Lemaitre, at the time working in Hoffman's lab in Strasbourg, France, but in contact (and competition) with researchers in Sweden, Germany, and the United States. These researchers were all exploring the network of genes and proteins that made flies produce their own antibiotics, a defense mechanism triggered when a fly was infected by bacteria or fungi. *Toll* was still "a developmental gene," but some of these researchers began noting "parallels" between the patterns of molecular interactions happening through and around the *toll* transmembrane protein in adult *Drosophila*, and similar patterns of molecular interactions happening through and around another transmembrane protein, the interleukin-1 receptor, in mammals; in both cases, the end of the transmembrane protein inside the cell initiated a molecular chain of events that ended in a protein from a family called nuclear factor-κB, which bound to DNA and initiated expression. But Lemaitre couldn't quite work out how or even if a developmental gene like *toll* ("which means 'cool' in German," according to Lemaitre's straightforward, single-signal translation) participated in such a pathway, and his attention for a while shifted to a gene named *imd*, a shortened form of immune deficiency, which seemed to be a more promising candidate.[25]

In one line of Lemaitre's research, he developed a mutant *toll* strain of *Drosophila* that left a fruit fly defenseless against infection by the fungus *Aspergillus*, as well as gram-positive bacteria. This *toll* mutation was different than the mutation studied by Nüsslein-Volhard, that produced larval malformation severe enough that it never made it to the adult stage; Lemaitre's *toll* mutant also resulted in death, but only after the larva had fully developed into an adult fly. This mutant adult fly, however, died because it had no defense against a fungus like *Aspergillus*; a fly could be completely colonized by a fungal invasion—a demise dramatic enough, and a discovery important enough, to land the electron micrograph depicting it on the cover of *Cell* in September 1996 (figure 2.2).[26] Through the careful work of scientists

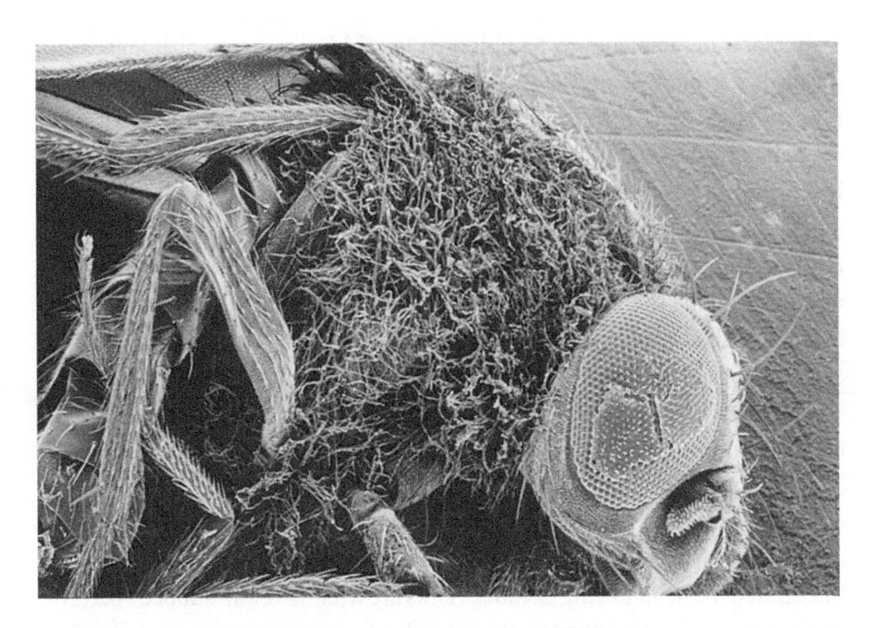

FIGURE 2.2 · *Toll*-mutant *Drosophila* colonized and killed by fungi.

attracted to its question-generating, opening-file-drawer character, *toll* the "developmental gene" in *Drosophila* had unexpectedly become *toll* the "innate immunity gene" in *Drosophila*.[27] The same gene was a different gene.

But wait—it gets *toll*er. The *toll*-drawer opened further, continued to generate still more questions, promised still more surprises, elicited still more unanticipated knowledge, driving scientists and scientific change. *Toll*—or something *toll*-like—would next make a surprising appearance not only in the fly's immune system, but in that of humans as well.

Charles Janeway led a renowned immunology lab at Yale University that had no truck with fruit flies, only humans. Work in the Janeway lab would lead to characterizing the first human gene coding for what would become known as a "*toll*-like receptor."[28] Janeway's group began this work by searching the DNA sequence databases that were just starting to show the rapid growth that the HGP was promising, probing for genes with potential immunological significance. They didn't really know what they were looking for. They were expecting at the time to find something called a "C-type lectin domain" encoded in the clone they were browsing for. Ruslan Medzhitov later recalled their group's "initial disappointment" that their database search did not produce that result. What they found instead was a human DNA sequence that (weirdly) matched a fly gene—*toll*.[29] It didn't make sense at first.

But later in 1996 Janeway had his own "*toll!*"-like moment—although technically it was more of an *aha!*-like moment. At a grant meeting of the Human Frontiers in Science program that had funded both their work, Jules Hoffman happened to show Janeway the electron micrograph he and Bruno Lemaitre had made of "a Toll-receptor deficient fly overgrown with aspergillus hyphae" (the one that would soon appear on the cover of *Cell*, figure 2.2). Janeway realized the "stunning discovery" that Lemaitre and Hoffman had made: not only had *toll* the fly developmental gene turned out to be also *toll* the fly immunological gene, but a very similar *toll*-like gene seemed to be present, across a huge evolutionary distance, in people.[30]

Janeway's lab opened a new phase in the understanding of the mechanisms of innate immune response in humans (rather than the adaptive immune system already familiar to us), and popularized the new acronym for *toll*-like receptor, TLR, that "as an abbreviation" was "fast becoming as famous as RNA or DNA," evidenced by over seventeen thousand papers that would be listed in PubMed in the following decade.[31] TLRs are what Janeway dubbed "pattern recognition receptors" (PRRs), these kinds of transmembrane proteins that bridged an outside to an cellular inside and its complex signaling pathways that include numerous other molecules as well as those more widely famed RNA and DNA.[32] Such signaling pathways are hallmarks of genomics now, where agency and action are distributed throughout cells and organisms rather than centralized in the gene (which, like *toll*, can have more than one function), and reassuring linearity (*booorrrring*) gives way to confounding complexity (*exciting!*).[33]

The "environment," previously excluded or otherwise controlled for in genetic and genomic sciences, was starting to become in the mid-1990s a meaningful part of that distributed action, albeit a part that was difficult to research and understand. One area where those difficulties became especially evident, and also especially interesting to researchers, was research on asthma, where initial expectations for identifying a few "asthma genes" were quickly abandoned, yet the disconcertingly dramatic increases of asthma prevalence globally nevertheless seemed to be patterned in ways connected to the patterns of TLR differences in populations. (This part of the TLR narrative, in which genomic epidemiology becomes a search for what Gregory Bateson called "patterns that connect," picks up again in chapter 6.)

More important for this chapter's narrative is the action initiated by the boring interesting DNA sequence information plucked from the US government-run database, four years before even a "first draft" of a complete human genome sequence would become available, that is, before "the

Human Genome Project" is conventionally said to have been completed (in 2000). Janeway's lab didn't just pick out a part from a "parts list" being compiled by the genomicists of the HGP; these researchers were bridging the domains of genetics and immunology, and opening up an exciting and entirely new branch of the latter, initiating an unexpected line of research into the previously unknown territory of the innate immune system. It's nowhere near Holy Grail-dom, but it wasn't a bad demonstration of the kinds of surprising effects one could get from building boring, quotidian genomic infrastructure.

Genomic Mandala

If the idea of loads of DNA sequence information in the mid- and late 1980s elicited a *toll!*-like affective mixture in which "boring" mixed with but still outweighed "interesting," it didn't seem to take long for that affect-amalgam to flip its compositional forces. By the mid-1990s, only a few years into the distributed, dedicated, federal-tax-revenue-supported, multi-organismal sequencing and mapping infrastructural efforts shorthanded as the Human Genome Project, its being hit with a torrent of sequence information—and it indeed seems to have been dramatically, "stunningly" physical—would elicit surprise and excitement, with barely a tinge of boredom.

An informative marker for this shift was the 1995 publication of the full sequence and map of *Haemophilus influenzae*, signifying for some "the real launch of the genomic era."[34] The paper represented work done by J. Craig Venter (already eliciting both love and hate in the genomics community[35]) and thirty-nine coauthors, including Nobelist Hamilton Smith and Venter's then-spouse Claire Fraser.[36] The majority of them worked at TIGR, the Institute for Genomic Research, a nonprofit organization that Venter had founded a few years earlier. Science writer Carl Zimmer noted that the publication of the 1.8 million base pair sequence landed "with a giant *thwomp*," disrupting what he wryly called "the dark ages of the twentieth century, when a scientist might spend a decade trying to decipher the sequence of a single gene." Even though there weren't "a lot of big surprises" about the microbe itself, Zimmer recalled that "what was remarkable was the simple fact that scientists could now sequence so much DNA in so little time." Moreover, that simple remarkable fact was transmitted in an instant, through a "kaleidoscopic wheel" mapping all 1,740 genes (see figure 2.3): "It had a hypnotizing effect, like a genomic mandala," reflected Zimmer, and "looking at it, you knew biology would never be the same."[37]

Biology, at least the genomics part of it, had indeed begun to change, powered by a number of factors. There was Venter's and TIGR's new "shotgun" sequencing approach, and this was its first test and triumph. The *H. influenzae* genome was simply blasted apart in its entirety (rather than chunking it up first into an ordered set of smaller clones that could then be individually sequenced and reassembled), and the pieces were then sequenced and reassembled by brute computational power and software finesse. With this successful demonstration of their technique, Venter turned his attention toward the vastly larger human genome, and set out on the path that in a few years would put him and the new corporation he started, Celera, in open and bitter competition with the publicly funded HGP.[38]

And this raises the other change in biology worth noting: because the US NIH thought Venter's approach would not work, his grant application

FIGURE 2.3 · Genomic mandala: the complete genome of *Haemophilus influenzae*.

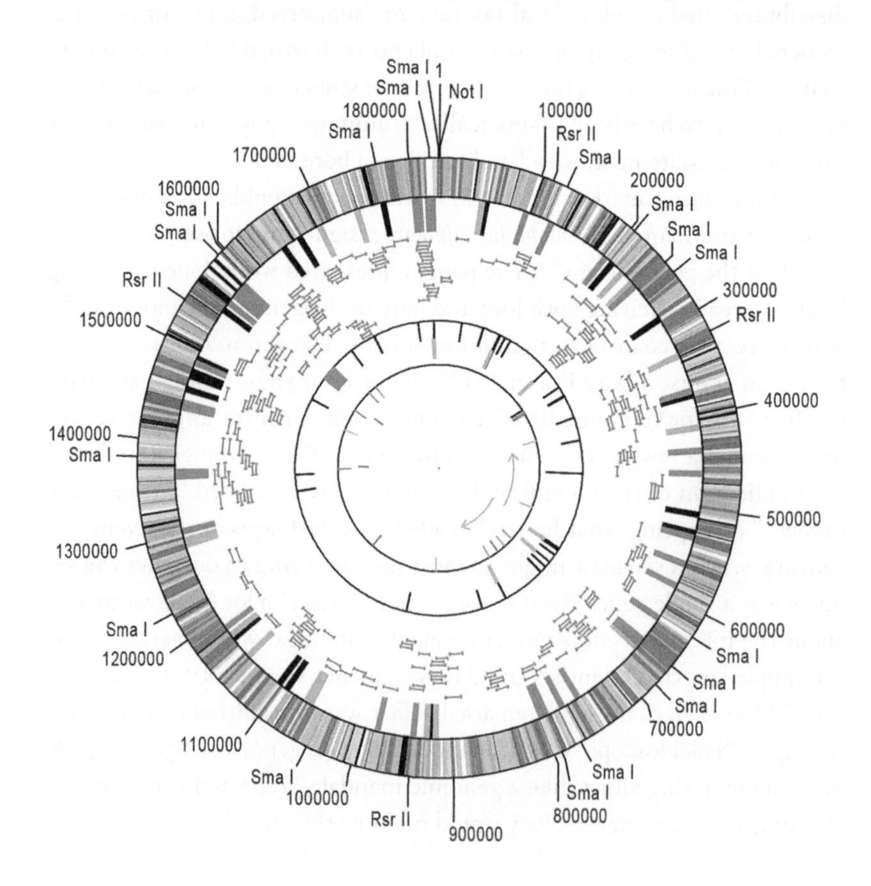

there for this project had gone unfunded. The production of excitement, boredom, and their combination in the sciences requires capital, often in large quantities. Such was the case with genomics, and in the case of *H. influenzae*, the nonprofit TIGR got the necessary capital from its for-profit partner organization, Human Genome Sciences Inc., one of the first and, for a good while, most successful if not actually profitable genomics corporations. (It was eventually swallowed up in 2012 by GlaxoSmithKline for $3 billion in cash and debt, about the same price usually tagged onto the entire HGP.[39]) Venter had helped establish both TIGR and HGSI in 1992, a joint-and-separate operation that qualifies, at least in my reading, as a chiasmus. And in a further twist of the publicXprivate chains, TIGR's success with its shotgun sequencing approach made the NIH quickly change its tune; it announced a new $20 million program to test the same approach on the human genome. Venter immediately began writing another grant application.

Like many scientific papers, the *H. influenzae* one makes for a boring read, page after page of methods accompanied by long lists and mapped chains of every one of those 1,740 genes, and barely a trace of affect anywhere, as per the genre convention that has prevailed in the scientific literature for more than three hundred years now. (Even I can't find any poetry in its prose.) Anything hinting at the affective was relegated to the accompanying news article in *Science.* "Lo and behold, the two ends joined," lead author Robert Fleischmann was quoted in recollection; "I was as stunned as anyone."[40] It is possible to be surprised even by what your reasoned intellect tells you to expect, to be stunned that your best laid plans and the machineries for realizing them actually worked.

Other expressions of surprise and excitement came from a more removed distance: "It's going to be just fascinating to see the whole genome"; "I'm really thrilled. This is epic-making." The *Science* writer used these and other quotes to convey the real sense of excitement pervading the new genomic atmosphere, as microbial genetics joined human genetics "at the top of the cool science hierarchy," and a host of surprising findings and tantalizing questions were immediately noted and raised from this single paper.[41]

I use this episode, hugely significant and symbolic to the genomics community, as itself a kind of mandala for the numerous events that had begun to occur within genomics writ large that I don't have the space or stamina to detail, and for which most readers are unlikely to have the patience. Fleischmann's "stunned" condition echoes Charles Janeway's affective state produced by his probing of the nowhere near complete GenBank of human DNA sequence data, and his surprise at finding a new human gene involved in an entirely unexpected system of immunological responses and signals, because its sequence matched one from a fruit fly. There are approximately 1,739 similar events, let's say, that I am gesturing toward here with a story first of a single gene, *toll*, and now of a single organism's entire genome. In each of those stories, once-boring DNA sequence data, which only some biologists had considered to be worth generating speedily and massively, were being transformed into a source of continual surprise and excitement.

And yet—let's not forget to invoke Wally Gilbert's marker phrase for a time out of joint, the simultaneity of paradox, and the amalgamation of any affect with others. Surprising difference attenuates to sameness; boredom reasserts its persistence, in sequence. And since genomics is so marked by dynamics of acceleration, all this affective drama began happening more quickly too. So even as genomicists "witnessed aspects of microbial diversity beyond what had been previously appreciated" in the weeks and years following the kaleidoscopic hit of the *H. influenzae* genome, and even as new human *toll*-like receptors began multiplying like flies (we are up to thirteen now), and even as 1,738 other like events unfolded from about the mid-1990s to 2000 and beyond, in scientific fields from virology to forestry, the mounting number of these surprises generated by the new sequencing capacities and new comparative analyses did not take long to simply become "widely accepted" features of the increasingly postgenomic landscape and lifeworld. An affective condition apparently as infectious as *H. influenzae* itself, Carl Zimmer (inspired by Michael Eisen) dubbed this variant the "Yet-Another-Genome Syndrome."[42]

Highlighting genomic surprises would become something of a dull routine by the time the HGP had been ritually marked as completed in 2000. "It appears now that hardly a week passes without some new insight into the 'genome' taking us by surprise," drolly noted biologist Richard Sternberg.[43] The complex signaling pathways had switched, somehow, and "*toll!*" became . . . well, "*toll!* [sarcasm]." Despite or maybe because of their frequency, and the growing frequency with which they were thus noted in passing, surprises still remained largely extraneous "color" to the main story about the "completion" of the HGP and the continuing, accelerating production of genomic sequence and genomic knowledge. But the funny thing was that this completion (at least what was pronounced in 2001 as complete) only made clear how insufficient genomic data was and how absolutely necessary the subsequent "annotation" and "interpretation" of that genomic sequence data would be. After considering affects like surprise at greater length in the next section, I return to the funny things that the Human Genome Project was producing, despite and because of itself, driven to a finality that would never actually arrive.

Affect, Imagery, Consciousness: Surprise, Amazement, *Toll!*

Money and power/knowledge: that's how I and many others writing about genomics in the fin de *deuxieme* siècle would have summed up what was driving the HGP. Sociologists, anthropologists, historians, and analysts of all stripes felt generally uneasy, and this came out in a burgeoning scholarly literature that expressed that uneasiness in thought that was, in a word, "critical." It seemed necessary and well warranted, I have to say. The intellectual mood in the humanities and social science communities oriented around genomics was dominated by anxiousness and worry, identifiers of care (as we'll see in the next chapter) to be sure, but further sedimenting into a sense of disdain and disappointment. At the very least, few if any of us were anywhere near as excited about genomics as scientists themselves seemed to be.

But the more I mulled over events and stories like the few narrated above, especially stories regarding "raw" DNA sequence data, with its funny double status of crucial and negligible, exciting and boring, the more I wanted to entertain another hypothesis: genomicists produce, accumulate, and contemplate DNA sequence not because they are driven (just) by and toward money, power, and/or reductive knowledge, but also because they are driven by and toward surprise and excitement. If DNA sequence is fundamentally about difference—A is different than T, C, or G; a different percentage of farm kids with a particular *toll*-genotype are diagnosed with asthma than are farm kids with a different *toll*-genotype; a *different* different percentage of farm kids with that particular *toll*-genotype are diagnosed with asthma than are *city* kids with *the same toll*-genotype—then those kinds of differences based in and around DNA sequences, accumulating and compounding with all manner of biocultural forces for science must somehow stimulate something—amazement? excitement? *toll!*-ment?—in genomicists beyond some coldly sensed awareness that their power/knowledge has increased.

My hypothesis here is that there are extensive, if subtle and elusive, connections between what is almost always troped as an overwhelming "avalanche," "flood," or "rush" of data found in contemporary genomics (and other technoscientific fields, often grouped under the rubric of Big Data), and the creative but mostly tacit epistemic virtues that I present here in terms of "patterns of care": the artful and anxious *curation* of large data sets, the *scrupulous* design of experimental protocols to leverage that data into analysis, and the *solicitous* grappling with excessive data, multiple analytic interpretations, and their multiple misfires, all multiply entangled. All of this in turn is bound to a palpable sense of excitement and eager anticipation that suffuses the inhabitants of the data-intense genomic science space. That sense of excitement, anticipation, amazement, or heightened *toll!*-ness is a forceful driver of scientists and sciences. Even if only one of many such drivers, the affective deserves to be better accounted for if we are to better understand why genomics (or any other science) is so popular, fast, ubiquitous, and powerful.

By "extensive connections," I mean *biocultural* connections: dense entanglements that would include physiological and neural elements, and thus also evolutionary ones, which we have as yet crude ways to assay and map. We know, thanks to genomics, that today's genomicists—and today's anthropologists, for that matter, but we'll get to them later—share an evolutionary history with zebrafish, *Danio*, a model organism popular for exploring the biology of basic neural development and functioning. The genes that give rise to our glycine receptors and transporters are slightly mutated versions of similar genes in *Danio*, conserved over millions of years of evolution—and having lively glycinergic synapses turns out to have *maybe* conferred some small selection advantage. Or so the genomic evolutionary story seems to go. Zebrafish exhibit a very pronounced "startle response" to a sensed change in their environment—the looming shape of a predator swimming into view, say, or fingers suddenly tapped against the side of a glass laboratory tank—and this response is being used to model "startle disease" in humans, in which seizures can be provoked by a sudden touch or noise. It's a response that can be stunted and dulled by exposure to lead or other environmental toxins.[44]

So I don't think it's irresponsible to hypothesize that genomicists, like zebrafish, and like anthropologists, can be startled by some change or difference, even a small one, and that such an event happens beyond their conscious awareness. Change or difference registers on an organism, it *affects* (even if, like Freud, we are not entirely sure what that means) not only its consciousness, but its neurophysiology somewhere on consciousness' border, or just beyond it. So here I begin to sketch some hypotheses toward a theory of a scientist-subject of genomics, who is or has something in addition to a consciousness—an unconscious, maybe, although you might prefer a "body" or "glycinergic synapses." This *something* is multiple and distributed, across registers from cultural to natural and back again, resistant to naming in any fully satisfactory way, but might at least be said to include the fundamental affects that psychologist Silvan Tomkins named "surprise-startle" and "interest-excitement."

Reviewing the burgeoning literature in fields from philosophy to film studies that constitutes a swelling "affective turn" is beyond the scope of this book. That extensive scholarship encompasses numerous positions, theories, and flights of speculation, as well as the disputes that inevitably arise in any such turn of scholarly events. Making no claim to comprehensiveness, I instead bricoleur my limited analysis from the work of trusted friends and colleagues who have found Tomkins's work on affect in the 1950s and 1960s well worth rereading and retheorizing.[45]

Tomkins described eight (sometimes nine) basic affects: interest-excitement and enjoyment-joy (the two positive affects), surprise-startle (the only neutral affect), and the negative affects distress-anguish, anger-rage, fear-terror, shame-humiliation, and dissmell-disgust. The doubled terms are meant to convey the range of intensity in which the affect could be felt, or make itself present (remembering that we don't yet know what it might meant to "feel" an affect, what grammar or sense might be appropriate to this sense-ability). Affects in Tomkins's writings are distinct from the more complex emotions that are co-assembled, along with scripted cognitions, from those affects. Emotions, in short, are more akin to what some contemporary anthropologists would categorize as "biocultural," while for Tomkins, affects—to the extent that it makes any sense at all to speak of affects in isolation—are decidedly neurophysiological.

Not *simply* neurophysiological, however. Affects are always bundled—with other affects, with drives like hunger, and with cognition as well. In a characterization that will take on fuller import in chapter 3, Tomkins spoke of the combination of cognition and affect as a "co-assembly" that he called "the minding system." If anything could be said to be "innate" in humans, it is this co-assemblage: "I propose an ancient term, mind in modern dress, the minding system. Minding stresses at once both its cognitive process mentality and its caring characteristics. The human being then is a minding system composed of cognitive and affective subsystems. The human being innately 'minds' or cares about what he knows."[46] So although the discussion below is cast mostly in terms of affect, keep in mind this "minding system" of care that is our eventual heading.

To say that affective responses are "fundamental" to our lives is something of an understatement; for Tomkins, we are shot through with affect, composed of and by them; even a moment of our existence without them is unimaginable. "A world experienced without any affect at all, due to a complete genetic defect in the whole spectrum of innate affects, would be a pallid, meaningless world," Tomkins wrote in the massive multivolume *Affect Imagery Consciousness*, the title of which signals the unpunctuated differences in our experiential existence. "We would know that things happened, but we could not care whether they did or not." It's worth noting that the conventional, sanctioned subject of science is supposed to have, or at least aspire to, exactly that kind of detachment, in the name of objectivity.

Tomkins later reiterated this connection between a capacity to be affected and a capacity to care: "With respect to the density of stimulation and neural firing, then, the human being is equipped for affective arousal for every major contingency. The general advantage of affective arousal to such a broad spectrum of levels and changes of levels of neural firing is to make the individual care about quite different states of affairs in quite different ways."[47]

Since it's still not clear what an affect is, let's leave the meaning of "care" relatively opaque, too. For now, it's good enough to follow your sense-making habits and presume affect and care are distinguishable, definable things, good to have or do, as we continue with a closer examination of two particular affective registers in Tomkins's schema: surprise-startle and interest-enjoyment.

"Surprise-startle" may be the only "neutral" affect in Tomkins's schema, but it is the affect that comes first; surprise-startle "is ancillary to every other affect," Tomkins writes, "since it orients the individual to turn his attention away from one thing to another."[48] Surprise-startle is a "circuit-breaker," a shatterer of habits that prompts a "re-orientation." Surprise is "a general interrupter to ongoing activity"—such as straining your eyes over the twenty-seventh fruit fly embryo of the day under your microscope, or algorithmically aligning the DNA sequence of one more chromosome microfragment with that of another in the publicly funded and publicly accessible database you've been laboring on for the previous eight years, or any of the other routine and often "boring" tasks and procedures that constitute so much of scientific activity. The smallest difference triggers a minimal surprise to spark the barely ambivalently meaningful *"Toll!"*

Later we'll see some "surprises" of the Human Genome Project, as recollected and expressed by a number of its leading figures. These surprises register what Evelyn Fox Keller calls "the funny thing that happened on the way to the Holy Grail": an interruption of the expectation that "it might be a big code, but it's still just a code."[49] The surprise was that the code was always already scrambled, or at least noisier or more open to multiple determinations than one dreamed. It was funny—less *ha ha* funny, more *toll!* funny—to realize that every center and every command and every imagined mastery was decentered and disrupted and subject to deferred interpretive orders. More generally, though, surprise-startle might be said to be absolutely central to all "Western" "modern" sciences, that is, experiment-dependent sciences since 1600 CE. Surprise-startle is the affect generated by any effective experimental system that successfully produces what it was designed to produce: something unexpected, something "unprecedented."

For Hans-Jörg Rheinberger, an "experimental system" counts as a research system only if it can produce something "unprecedented"—something recognizable enough to current knowledge and technologies that it can be produced and then continue to be reliably reproduced, *and yet* something different enough, something one wasn't expecting and doesn't quite know what to make of, or even what question one might need to ask next of it. Rheinberger contrasts *research systems* to *testing devices* on precisely these grounds: testing devices, although useful and even essential, will only produce or confirm what is already known, and raise no troubling questions. A research system needs that kind of stability as part of it, but it needs something more: it must produce "epistemic objects" that exceed the knowledges and even the questions of the present. The research system must produce a remainder, in other words, an excess produced by "eliciting differences" from matter without destroying the "reproductive coherence" of the system. Differential reproduction is the intimate combination of this stable predictability, and the generation of a surprising, "unprecedented" remainder that cannot yet be accounted for. An experimental system, says Rheinberger quoting molecular geneticist Francois Jacob, is a "machine for making the future." It is a generator of surprises.

Rheinberger limits the surprise of the unprecedented to the cognitive realm: "epistemic objects" are the product of experimental systems valued by his analysis, not "affective objects" valued as well for their generation of startle-surprise. But without the latter effect, Tomkins helps us to recognize, why would we even care enough about an epistemic object to seek to generate it in the first place? The experimental system and "the minding system" need each other and feed off of each other. This brings us to the "positive" affect that the "neutral" one of surprise-startle becomes compounded with, and amplifies: interest-excitement.

Like surprise, interest-excitement according to Tomkins is elicited by something that couldn't be simpler—difference, as registered by "neural firing":

> Consider interest. Any sudden movement (which was neither sudden enough to startle nor sudden enough to frighten) which was steep enough in its acceleration to produce a correlated acceleration of neural firing could innately activate interest or excitement. Interest and excitement are the same affect, differing only in intensity. Consider now how the same

neural profile could be produced by learning and "meaning" without the necessity of a homunculus or "appraisal" process. Suppose upon reading a book the novelty of an idea activates information processing at an accelerated rate. This would initially amplify and thus maintain "thinking" by innately activating excitement. If this now exciting implication keeps inferential processes alive at the same accelerated rate, the individual will then again be rewarded with a burst of excitement at each new expanding set of conceived possible implications of the original idea. So long as the combination of successive inferences and recruited affects sustains yet another inferential leap, the individual's interest will remain alive. When he runs out of new possibilities, he will also lose interest.[50]

Along these lines, to be a genomicist is to be interested by even the smallest difference—a quantum of *toll*! The giga-wads of difference produced routinely in a genomics research experiment is not just neutral information uploaded, downloaded, and regarded dispassionately; it excites and incites a desire for still more difference, sooner, faster. There are, of course, many more sources of interest-excitement in genomics than raw sequence difference, and the social, cultural, and cognitive contexts in which those differences are read (i.e., interpreted, differently) are enormously productive of scientific desire as well. We'd be foolish to disregard or discount those additional layerings, but I am here exaggerating the force that the simplest of differences can impress on our affective capacities to remind us to account for affect when we are accounting for the many drivers of any "big data" science like genomics. That brightening of the eyes, that contraction of the facial muscles pulling the sides of the mouth upward into a growing grin that I have seen so many times on the face of a genomicist at the merest mention of a new data set, or increased sequencing rate, is not just an effect of sober cognitive assessment. The greater the difference—in intensity, or in quantity—the more neural activity, the more interest-excitement. And the more interest-excitement, the more one cares:

> The interrelationships between the affect of interest and the functions of thought and memory are so extensive that absence of the affective support of interest would jeopardize intellectual development no less than destruction of brain tissue. To think, as to engage in any other human activity, one must care, one must be excited, must be continually rewarded. There is no human competence which can be achieved in the absence of a sustaining interest, and the development of cognitive competence is peculiarly vulnerable to anomie.[51]

In chapter 3, I turn more to care, but I'll add here one more brief quote from Tomkins for a better sense of how he differentiated, yet linked, affects, emotions, and drives, and on how thinking about affect and care can also help us think about thinking, and genomics, and sciences:

> This affect [interest-excitement] supports both what is necessary for life and what is possible, by virtues of linkages to subsystems, which themselves range from concerns with the transport of energy in and out of the body, to concerns about the characteristics of formal systems such as logic and mathematics. The human being cares about many things and he does so because the general affect of interest is structurally linked to a variety of other apparatuses which activate this affect in ways which are appropriate to the specific needs of each subsystem. While excitement is sufficiently massive a motive to amplify and make a difference to such an already intense stimulation as accompanies sexual intercourse, it is nonetheless capable of sufficiently graded, flexible innervation and combination to provide a motive matched to the most subtle cognitive capacities.[52]

In the remainder of this chapter and in later chapters I present more episodes in the development of some of the "subtle cognitive capacities" of genomics with which the affects of surprise and interest co-assemble, and which in turn reinforce those cognitive capacities. In other words, I'll discuss how genomicists learn to better mind their own "minding system"—how they come *to be more careful, to care more* for data and other crucial epistemic objects, to be simultaneously surprised by, interested in, excited by, and more thoughtful about the objects and objectives of their scientific work. "I use minding rather than knowing," Tomkins wrote, "to preserve the function of knowing as caring, or minding."[53] Knowing caring minding, then, is the central concept of *Affect Imagery Consciousness*.

But as important as the affects just presented are to the "minding systems" of humans, it's as important to listen once more to Tomkins as he also cautions against taking any such names too seriously or rigidly. "Matter, life, and mind are nouns," he wrote, "suggestive of differences in substances, that we employ because of our incurable visual-mindedness, which favors extended spatial entities that endure in time."[54] To counteract such suggestions of different substances, Tomkins privileged "process terms—mattering, living, minding—in order to conceptualize fixed structures as the exceptional special case of rates of change that are relatively slow, rather than essentially static structures that change as a special case." It's vital, in other words, not to think that terms like "interest-excitement" refer to some mental module or static structure, but are instead meant to elicit awareness of a process, transaction, or relation—or, as he put it elsewhere, a "conversation":

> Knowing is nested within minding as that is nested in living and mattering. In minding, a society of conversationalists continually enlarges its conversing via multiple interdependent monologues, dialogues, and pluralogues. Conversations include languages as special cases, but a hand that shapes itself around a ball and is itself shaped by that ball is no less a conversational encounter. A scientist who puts an experimental question to nature is awaiting an answer from nature in a conversational encounter in which both talk and listen, though the success of the conversation depends more on the asking of the question than on the initiative of the more taciturn Mother Nature. Although she will give a straight answer to the right question, she will speak somewhat ambiguously to ambiguous questions. Further, her answers may be interpreted so that the conversation enables more and more productive conversation in the future, or they may be so misinterpreted that the questioner is led into a cul de sac, from which he must initiate a radical detour in the conversation if he is to find his way again to more illuminating conversation in theory construction that can be cumulative.[55]

Interesting as Tomkins's broader metaphysical and ontological thinking might be, I only want the quote above to emphasize what we could call his processual, relational, or anti-essentialist view of all phenomena, from the "natural" object of an experiment's outcome to the "knowing" that designed that experiment in the midst of ongoing pluralogues in a community of scientists to the "minding" that nested that "knowing" within other relations and processes arrested and named as "care" and "affect." Catachreses everywhere.

Note, too, the allegory of experimentation-as-maze that crops up here, subject to detours and cul-de-sacs. In the next section I discuss the labyrinthine qualities of experimentation in general and the Human Genome Project in particular, but let me conclude this section by saying something about a detour of my own, in which my thinking and writing was going down a path toward the related terms *maze* and *amazement*, ending in a cul-de-sac that my conversation with Tomkins led me out of.

It was tempting, for example, to position my analysis within a genealogy of the concept of "amazement" as it has played out in philosophy and in science since Aristotle. Katherine Park has narrated significant chunks of that history, in which a mental state of "amazement" or "wonder" has been theorized as a foundation of, spur toward, or sometimes impediment to scientific thought and practice, varying with historical period and local conceptual culture. The risk in such otherwise insightful analyses is in essentializing such states as stable, identifiable, and inherent human capacities. I've made *toll!* my refrain, rather than *amazing!*, *wonderful!*, or *exciting!*, for its ambivalent minimalism—an infrastructural utterance, if you will, coming from somewhere underneath or in the interstices of grand emotional structures like wonder. I want *toll!*'s strangeness and the multiple signaling pathways that it activates to keep your reading from settling too easily and deeply into what is a nearly irresistible pattern of identification and satisfaction.

Whatever affects are, and however isolatable they might be, we scientists have a "tendency to experience multiple, contradictory affects at the same time, assembled with one another as well as with cognitions and drive states," Adam Frank and Elizabeth Wilson remind us. Such "complex emotional assemblages are embedded in scripts that serve to negotiate our motivational realities," they caution, and in our pursuits of a "revised notion of emotional expression," we are well advised to avoid "falling back into an idealized, self-authenticating interiority."[56] Amazement, wonder, and plain old interest-excitement as well are not some biological, mental, or even conceptual substance or capacity that some people might have more or less of than others; they are "not bad" names that draw our attention to a complex set of events and relations. Reading Tomkins and those like Wilson and Frank who engage him in conversation helped me negotiate my way toward a more *toll!*-like understanding of amazement, and toward a more disseminated but, I think, more accurate appreciation of "minding" as a process or conversation lighting up the multiple signaling pathways of affect, care, cognition—and much, much more.

Surprises of Minding the Human Genome Project

Let's start to end this chapter by noting a few of the other objects or events, in addition to the accelerating accrual of sequence information—or sequence difference-that-makes-a-difference to flies and humans and transnational research programs—that elicited surprise and interest among genomicists, and in more and more areas of scientific research. These affects are always entangled, as we'll see clearly here, with complex cognitions, cultural scripts, social and institutional investments, and other features of the emergent "minding system" of genomics.

To its credit, when Cold Spring Harbor Laboratories introduced an oral history collection, "Genome Research," to its webpages, it didn't just add obvious topical sections like "Mechanics of the HGP," "Challenges of the

HGP," and "Gene Patenting." Interviewers also thought to ask its interviewees, "What surprised you the most?" (Although a very few women were interviewed on the other topics for the website, on this one it was all male respondents.) In this series of fireside chats (metaphorically and literally—the fairly conventional interviews were mostly shot over a few days in front of a fireplace at the Cold Spring Harbor Laboratories), the affective register was often right on the surface—genomics was turning out to be surprising, interesting, exciting, *toll!*, anything but boring. None of these developments were on Walter Gilbert's 1986 list of "the parts we need" that he imagined would be provided by accelerating the development of genomics.

Perhaps the biggest and most publicized surprise was the much, much lower value than expected for the number of human genes in a human genome. Scientists had known that any such estimate was a gamble, which was why a betting pool called GeneSweep was organized in 2000 as the "first draft" sequence was being prepared. You could get in for $1 in 2000 and $5 in 2001, but it would cost you $20 in 2002; the rules said a winner had to be named in 2003. Conventional wisdom clustered around 100,000 genes, but some estimates went as high as 300,000; only a handful were below 30,000.[57] The strong skew toward the higher numbers was an effect of the dominant scientific sense of the time: since a gene was supposed to do one thing, more genes must be required to produce more complexity. It only stood to reason.

Sequencing-innovator Bruce Roe, who led the group of genomicists at the University of Oklahoma that first sequenced an entire human chromosome (chromosome 22), answered this way:

> That's an easy question. I put my dollar down on 120,000 genes. And for somebody to tell me that we only have twice as many genes as a worm, twice as many genes as a fly, you know, that's kind of disconcerting.... So that's one of my big surprises. The other big surprise was that there are genes overlapping genes. And genes inside of genes. Here we have this huge genome that only one and a half percent encodes for, and these genes overlap each other, you know. Why would you ever do that? Well, you know, I didn't design it. We're just looking at what the designer did.[58]

That last sentence there is rather surprising, isn't it? Maybe "the designer" is merely a hastily uttered shorthand for "several million years of natural selection." Or maybe then National Human Genome Research Institute director (later director of the entire NIH) Francis Collins isn't the only scientist who believes that genomes are written in "the language of God."[59] In any case, the GeneSweep winner was eventually declared to be Lee Rowen, who bet on the lowest estimate of anyone, 25,947 genes; in addition to splitting half the pot (her take was $570) with the only other two bettors under 30,000 genes, she also received a signed copy of James Watson's *The Double Helix*. (I imagine her, upon being handed this infamous account of sexism and underhanded competitiveness in mid-twentieth-century Euro-American science authored by the sexist and racist Watson, muttering "*Toll!*" in a deeply sarcastic tone.)

In his interview clip Robert Waterston mentioned another now well-known surprise of genomics, "that fifty percent of the worm genes are shared with people, and fifty percent of human genes are shared with worm or something like that. It's just astounding."[60] One example of how DNA sequences reoriented understandings of evolutionary history and theory is my earlier story of *toll*-like receptors, as researchers followed genes, gene functions, and gene sequences from fly to human and beyond, reorienting much of immunology in the process. Books with titles like *Your Inner Fish* multiplied on the shelves of Barnes and Noble, popularizing similar genomic kinship narratives, fueled by this epistemic/affective object of DNA sequence information.[61]

But I found the reflections of chemist-turned-genomicist Maynard Olson to be among the most honest and interesting, for the kind of metaperspective on the sciences of genomics they offered:

> There are a lot of things that surprised me about it. You know, history always looks so clear in retrospect, but not so clear in prospect done.
>
> I thought that it went much more rapidly and much more smoothly than I could have imagined. And it did so for a whole bunch of reasons. I mean, the problem was enormous. . . . Not conceptually difficult but practically an immensely difficult problem. . . . We were just mismatched. We didn't have good enough techniques. We didn't have anywhere near enough strong investigators. The whole computational infrastructure didn't exist. Most of our ideas about how to proceed were wrong. There was no overall organization that we had any kind of experience with. There was just the idea of building one. The problems just seemed rather overwhelming but, of course, exhilarating. I was never pessimistic although I was always restraining people that said, you know, this is going to be easy. . . . I think that if you had asked me how long it was going to take I would have been off by a decade or so. And even then I would have felt I was being optimistic. Because again I just couldn't see the path. And I couldn't see the path because the path wasn't there. A lot of other people thought that they could see the path but if you go back and read in detail what they said, that didn't turn out to be the path. Those were a lot of dead ends.[62]

In the Labyrinth

"I couldn't see the path because the path wasn't there." Olson's take on the entire history of the Human Genome Project can be read as a strong confirmatory signal for Rheinberger's complex figuring of experimental systems, not only as a "machine for making the future," but as an emergent, enlightening, and bewildering labyrinth:

> An experimental system can be compared to a labyrinth whose walls, in the course of being erected, simultaneously blind and guide the experimenter. The construction principle of a labyrinth consists in that the existing walls limit the space and the direction of the walls to be added. It cannot be planned. It forces one to move by means of checking out, of groping, of *tatônnement*. . . . The articulation, dislocation, and reorientation of an experimental system appears to be governed by a movement that has been described as a play of possibilities *(jeu des possibles)*. With Derrida, we might also speak of a "game" of difference. It is precisely the characteristic of "fall(ing) prey to its own work" that brings the scientific enterprise to what Derrida calls "the enterprise of deconstruction."[63]

I am scaling up Rheinberger's analysis of experimental systems here, which he more carefully restricts to those material assemblages within laboratory walls—for example, the fly-test-tube-mutagens-microscope-genetic map assemblage of Nüsslein-Volhard and Wieschaus—that is both stable enough to provide completely reproducible effects, yet open enough to produce unexpected results and questions. Or, as Bruno Lemaitre, the other *Drosophila* geneticist whose work on *toll* was highlighted above, phrases it in a way that resonates with what is analyzed in chapter 5 as the double bind of scrupulousness: running an experiment or a laboratory is a matter of "taking care of details" and at the same time "trying to leave a space open to hazard."[64]

We are justified in considering entire laboratories as experimental systems in this sense, I think, and even entire systems of laboratories assembled into a years-long project that has to grope its way forward into an overwhelming unknown whose shape is constantly shifting along with the defined parameters of the project that are in the process of having been invented. The only Ariadne's thread here is the one that will have been laid down after the labyrinth has come to pass—and it won't be of much use, since there's little sense in going back. Unless you're a scientist in front of a camera, or an anthropologist and historian archiving and watching with care.

Why should anyone care at all about the affective dimensions of the sciences, and of scientists in their invention of and immersion in experimental systems? My first response as an anthropologist is simply to say that I'd like to understand and appreciate these dynamics of "labyrinth life" better—they interest and even excite me. The lab is not only a "center of calculation," to use another Latourian term; the lab is a labyrinth and a maze of affect. The surprise and excitement generated in the doing of science is an affective effect of its labyrinthine qualities, where (as with any encounter with an excess) one is alternately overwhelmed and exhilarated.

To be clear: I am not groping my way toward an "affective explanation" of genomics or the Human Genome Project. But I am trying to conjure up the walls of another labyrinth entirely, this text through which we can follow these genomics subjects as they navigate their own labyrinths, in groping pursuit of numerous emergent ideas, objects, and events—and follow the ways in which affects come into play in this "game of difference."

A second response concerns the linkages between affect and care that have been touched on a number of times already here. Care is another "subject effect" that the labyrinthine nature of postgenomics produces within its builders, and one that also requires us to navigate our own analytic labyrinth as it emerges, as a "play of possibilities" whose appropriate terms, grammars, and syntax will come only with its future. The words "affect," "interest," and "care" have occurred together in several places above, but their qualities and interconnections require more . . . careful articulation.

The terms have occurred together, and they blur together, but they are not interchangeable. Care is not an affect, like interest or surprise, even if it can't occur without them, even if their interrelations are extensive, intricate, and deep. To reiterate part of an earlier quote from Tomkins, "To think, as to engage in any other human activity, one must care, one must be excited, must be continually rewarded."[65] However we end up understanding how the elements of the genomicist's minding system work together—and we're going to need a lot more ethnographic data for that project—the amplifying effect of affect will be important:

> The primary function of affect is urgency via analogic and profile amplification to make one care by feeling. It is not to be confused either with mere attention as such nor with mere response as such but with increased amplification of urgency—no matter how abstract the interpretation of its stimulus and no matter how abstract or specific the response

which follows. . . . Without affect amplification nothing else matters, and with its amplification anything can matter.[66]

My interest in the affects of surprise-startle and interest-excitement in the sciences concerns this amplifying effect, their ability to incite a sense of urgency that can be a driver not only of an individual's research trajectory, but of social projects like the HGP and "beyond," into today's genomics. If "matters of concern" (to use yet one more Latourian construct here) are indeed a crucial concern of those of us in the science studies community, we could use a richer understanding of the forces affect exerts on how those matters come to matter so much, so urgently. For that we might need something that begins to approach what Bernard Stiegler calls a "general organology," an analytic that would help us understand the "psychotechnological and sociotechnological apparatuses" by which we come to pay attention to objects—that is, to care:

> Attention, which is the mental faculty of concentrating on an object, that is, of giving oneself an object, is also the social faculty of taking care of this object—as of another, or as the representative of another, as the object of the other: attention is also the name of civility as it is founded on *philia*, that is, on socialized libidinal energy. . . . The latter constitutes a system of care, given that to pay attention is also to take care. . . . Such a system of care is also a libidinal economy, wherein a psychic apparatus and a social apparatus hook up.[67]

So, if the HGP was a labyrinth of surprise and interest (and so much more that I have glossed over), we need to take better account not only of its labyrinthine qualities, but of the felt urgencies and anxieties of its simultaneous building/emergence and navigation. A groping *tatônnement* is one thing, difficult in itself and maybe even impossible (as I will further hypothesize later on, turning the labyrinth into a double bind in the process)—but an anxiously excited groping *tatônnement*? How does that not devolve into flailing clumsily and violently, or staggering thoughtlessly and rudely? This is where the difference of care from affect really makes a difference. For it's not only affect that is being amplified in the doing of science; that cybernetic looping of the minding system amplifies the dynamics called care as well.

So what might it mean to care, not after or alongside, but as part of the doing of surprising, exciting, labyrinthine science? Who is the scientific subject who, having been surprised, having found herself or himself in a new territory of vast data and only slightly less vast possibility, interesting and exciting possibilities that feed back into themselves—who is it who is able to care for her or his science? What else would a "general organology"—an ethnoneurobiopsychosemiosociotechnical analysis—have to include to account for the increasingly urgent care demanded by an increasingly urgent genomics?

3 · double binds of science

The restoration of balance between the intrapsychic and the intersubjective in the psychoanalytic process should not be construed as an adaptation that reduces fantasy to reality; rather, it is practice in the sustaining of contradiction. . . . Relatedness is characterized not by continuous harmony but by continuous disruption and repair. · JESSICA BENJAMIN, *LIKE SUBJECTS, LOVE OBJECTS*

Dreams of Data

"I sometimes see them in my dreams," mused David Roy Smith. "The colorful peaks and troughs, the sharp, crisp waves spread across my computer screen, the rolling nitrogenous mountains, each with its own nucleotide sitting solidly on the summit."

That is fine prose by any standard, probably some of the finest to appear in *Frontiers in Genetics (FiG)*, and almost certainly some the finest ever written about DNA sequence data. Smith—bioinformatician, genomicist of protists and their organelles in particular (yes, some organelles have genomes of their own)—was waxing nostalgic about electropherograms, "those beautiful but

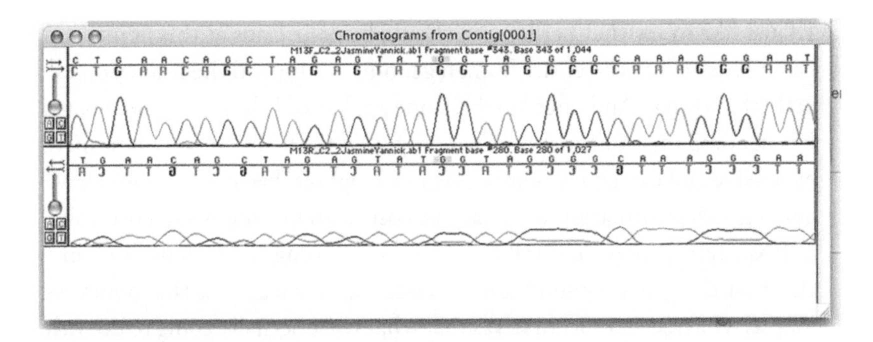

FIGURE 3.1 · Chromatogram (a.k.a. rolling nitrogenous mountains) produced through Sanger method of DNA sequencing.

oh so 'old-gen' bioinformatics data generated from automated Sanger sequencing machines, such as the Applied Biosystems 370—the geriatric of genome sequencers." Most readers have probably never seen "rolling nitrogenous mountains"; they looked like figure 3.1.[1]

The ABI370, the first automated DNA sequencer, came out in 1987, a successful commercial product of National Science Foundation–supported research. A human genome project was still just an argument among bioscientists and science agency administrators, and even the "not bad" name, genomics (chapter 1), was barely a year old. One set of contentions concerned how much to invest—in terms of funding, scientific person-power, and emotional-cognitive attention—in sequencing all three billion DNA base pairs in a human genome, and how much to invest in mapping chromosomes and chromosomal segments with genes and other, less informative but still very useful, markers. Chapter 2 showed something of the complex pattern of affects that informed these differences: sequencing was boring, tedious, mechanical, dull, routine, and robotic—and also afforded a dramatic scale and narrative arc, a whole Holy Grail genome at the end of that three-billion-step quest. It was the sequencing that made the whole thing epic, and thus more understandable to members of Congress holding the keys to the public money vaults, and more amenable to the media dramas of "racing" public and private efforts toward a climactic White House media event.

At least some bioscientists understood these dynamics of the cultural political economy very well. At the earliest "public" debate about these issues, a 1986 meeting at Cold Spring Harbor, in the middle of the often contentious discussion among the hundred or so molecular biologists and geneticists, one of them stood up to argue in terms that were so politically savvy as to be quite prescient:

I'd just like to say that perhaps we should be politicians at this point, and call it sequencing the entire human genome, and spend the money on exactly that thing. [Audience laughs, some applause.] I think—no, no, no, it's like sending men to the moon. Sending men to the moon was extremely expensive and extremely pointless [some laughter] because we could have got as much information with half the cost, with machines—scientifically, at least. But if we go and tell Congress we're going to do something like this that the general public can understand, they'll give us the money, as long as we agree among ourselves that what we're actually going to do with it is maybe sequence the one percent of the genome that's interesting and try to develop [audience laughs], try to develop technologies that in fact will allow us to sequence the human genome in 1999, just in time to meet the deadline. [Audience laughs.][2]

So sequence-questing was good political optics but, at least in these early years of genomics, expensive and slow. And funnily enough, at least a little pointless. Project architects decided to make significant investments, therefore, to encourage and support the development of sequencing technology as essential genomics infrastructure, making it faster and cheaper.

That all worked pretty much as promised, it must be said. By the late 1990s, row upon row of the successor machine, the ABI3700, churned out DNA sequences in these kinds of "sharp, crisp waves" in cavernous, industrial-style spaces in major Euro-American cities: in Reykjavík, where deCODE Genetics was catching a cresting wave of venture capital and Iceland-mania; in Cambridge, Massachusetts, and Cambridge, United Kingdom, and St. Louis, Missouri, where teams of genomicists in government-funded labs pushed toward a ritual but not actual completion of the Human Genome Project in 2000; and in Rockville, Maryland, where the privately funded Celera Corporation raced the US and UK government scientists—while also selling those very same machines (Celera and Applied Biosystems were bound together as a corporate entity) back to those public agencies whose funding had been instrumental in their development.

By 2013, when Smith was writing his essay for *FiG*, the HGP was long finished; Celera had been cannibalized by another biocorporate entity; deCODE had surged and declared bankruptcy and then risen again, zombie-like, to be acquired in a putative $451 million deal with the biotech giant Amgen; and "next-gen" sequencers like Illumina's HiSeq, NextSeq, and MiSeq had thoroughly colonized the laboratory ecologies once monopolized by the now-antiquated, "old-gen" ABI370.[3] In Smith's nostalgic reminiscence of those earlier days, we can continue reading, at the most quotidian infrastructural level of genomic data, for signs of surprise-startle, interest-excitement, and other *toll!*-like affects in genomics that will orient us toward its care and minding:

> As a grad student, I spent countless hours pruning, editing, assembling, and occasionally oohing and awing [*sic*] over Sanger sequences. These 800-nucleotide genetic snippets intrigued, inspired, and motivated me. They contained just enough data to pique my interests—a novel exon, strange repeat, or foreign gene—and always left me craving a bit more: one additional sequencing read to extend that PCR product, find that stop codon, or join those lonely contigs. Usually, it would take weeks or months to get that extra read, and when it arrived I would savor the experience, exploring and analyzing it like a new book from a favorite author.
>
> After I devoured the data, I would say to myself, "If only I could get my hands on a great number of sequencing reads from my organism of interest then all of my genomic woes would be over."[4]

In chapter 1 I began to develop the hypothesis that bioscientists like Smith could and should be understood as biocultural, neuropsychobiological, caring-knowing subjects who form complex affective-epistemic attachments to the objects of their sciences, even when those objects are as supposedly indifferent and neutral as data—here a series of colored peaks capped, not by "nucleotides," but by simple signs, the letters A, T, C, and G that signify nucleotides to a trained reader like Smith. The chapters that follow extend and thicken this hypothesis into a kind of object-relations analysis of scientists as embodied subjects, continuing a project in psychoanalytic theory begun long ago by feminist scientist, historian, and philosopher Evelyn Fox Keller.[5]

Such a curious thing, data, that it can elicit such affective responses.[6] Thanks to a growing literature in science studies (and to scientists like Smith), we understand better how data in the sciences are misapprehended with words like raw, brute, cold, and hard, when in reality (i.e., in practice) they are quite complex, rich, and fragile.[7] If data, following Gregory Bateson's formula about information, is a difference that makes a difference, then one of the differences that it appears to make—at least at the present moment, and at least for those who have been encultured and learned, like Smith, to respond in such ways—is a difference on the affective register of its human producer, interpreter, and "devourer." Data intrigues and inspires, piques and motivates, provokes oohs and aaahs (and sometimes awes), and more than the occasional minimal, inscrutable *toll!* Smith's language here accords nicely with Tomkins's affects of surprise-startle and interest-excitement discussed in chapter 2; it suggests modest modulations rather than the full-blown drama of emotions, something so subtle and resistant to naming that its sign is breathy and guttural: *ooh, aah*. His language also suggests how these small affective changes can become part of a dynamic pattern of object relations: rather than raw, data verges on haute cuisine, with subtle flavorings to be savored and relished, sparking not only anticipation but craving, an increasing appetite for more—and more and more.[8] Scientists' ongoing relations with data congeal into powerful circuits of affect and care, my hypothesis goes, complex feedback loops that run through and on biotechnological platforms—a term meant to include *us* as well as an ABI370 or an Illumina MiSeq. It's this circuit (data-surprise-interest-excitement-data-surprise-etc.) that, in my view, deserves more ethnographic and analytic attention as a driver of scientific work or truth production more broadly—what Freud and Melanie Klein after him called the epistemophilic impulse or instinct, to be placed and theorized alongside (for Klein, at least) the death and life drives.[9]

To return to our more immediate subject, however: Smith's fond recollection of a time, not long past, in which these circuits of object-relations hummed at an energizing rather than enervating rate, was just the narrative setup for the predictable fall: "Naively, I believed that the more sequencing data I had, the more productive I would be. Be careful what you wish for from the genome gods. The onslaught of next-generation sequencing (NGS) technologies and the access to previously unfathomable amounts of genomic data have made me dizzy, disillusioned, and anything but efficient."[10]

Epistemically, then, the development of genomics and related -omic technologies in the 1990s leads to more complex and more refined understandings of organisms. But affectively? Expecting ("naively," maybe) the "parts list" of organisms-as-codes that could be deciphered and employed "productively," genomicists encountered instead organisms-as-texts—texts as in literary texts rather than instruction manuals, molecular texts requiring multiple readings to activate their multiple interpretative possibilities, tissue-y texts calling for close reading practices that are "anything but efficient." That's a surprise, and exciting.

Affectively, then, the surprises of what genomics was becoming in the 2000s strengthened those epistemophilic object relations—powered, I remind us again, by many hundreds of millions of research dollars—to produce more productive technologies that produce more data that produce more surprises . . . and so on. But it all became a bit much, in my hypothetical narrative, and a different set of affects began to register, for Smith and others—more anxious and disorienting ones. Still, the epistemophilic drive kept kicking in, and ratcheting up. "My mind is gradually overheating from an accumulation of NGS [next-generation sequencing] reads," continued Smith,

> What's worse is that I'm still sending more samples for sequencing. It's become my default setting: when in doubt, sequence. If a colleague drops by my office and says, "Smitty, you interested in milkweeds?," my first response is, "You betcha. Let's send some for sequencing." Student asks: "Professor Smith, do you have any ideas for my honors thesis?" "Hmmm," I say, "how about we sequence another green alga." Grant money left over, what do I do? You guessed it: two for one RNA-seq at the campus sequencing facility. And if the data come back contaminated or the quality is poor? Easy, I sequence more! It's gotten to the point where I should begin my conference presentations with, "Hello, my name is David and I'm a NGS addict."[11]

Smith's editorial served as a kind of diversionary amusement for most readers of *FiG*, no doubt, even if some of us are inclined to read it for the psychoanalytic possibilities it holds. So although he doesn't offer any suggestions about what a bioinformatician might do to interrupt these addictive cybernetic loops, he has given us some clues to work with far back at the beginning of his nostalgic trip. The "oohing and awing" he did as a graduate student, as you can verify above, was only an "occasional" event, and came only after he had "spent countless hours pruning, editing, assembling."

The surprise and excitement in "old-gen" genomics, then, was bound to another set of practices that (sticking to the affective register) might be called "non-startling" or "non-interesting," or just simply boring. No such list of counterforces, not even one of just three terms, accompanies his playful but plaintive narration of the "next-gen," genomic research scene of addiction.[12] But such counterforces do exist in genomics, and they are a necessary supplement to its equally necessary, but also unruly, affective circuits. These are the practices of care I have been touching on, and which the second half of the book presents at greater length. Before going there, however, we'll need to dedicate some attention to "care" itself.

Strong Theories of Care and Why They Need Weakening

When I first began writing about "care" in relationship to genomics in 2005, I didn't really know what I was thinking. An undergraduate philosophy seminar at Hampshire College in 1980 devoted entirely to reading Martin Heidegger's *Being and Time* imprinted in me his idea that *Sorge* (care) was absolutely fundamental to human being(s). Much later, as I was beginning to think about data and data practices in genomics, I was turning to Michel Foucault's ideas about the "care of the self" (see chapter 4). Wanting more, I read widely on "care" in the many years since, and learned from feminist theorists in many areas, especially in science studies, who in the intervening decades since 1980 had developed strong theories of care—maybe a little too strong.[13]

The scholarship on care that began to develop in the 1970s and 1980s may owe the most to the political philosopher Joan Tronto, in conversation with Carol Gilligan's work in psychology on the "different voice" women used in ethical reasoning. Of the many truths about care these scholars put forward, a particularly important one is that care is work, forms of labor that are almost always undervalued and over-gendered, most notably in the health-care and domestic-care domains. The laboriousness of care practices, in and out of the genomic laboratory, with and without the maternal codings that have often informed them, will be a constant current here. And although she was not working within Heidegger's phenomenological line of thought, Tronto certainly understood care to be an essential human capacity—indeed, essential in the sense of a product of evolution: "On the most general level, we suggest that caring be viewed as a species activity that includes everything that we do to maintain, continue, and repair our 'world' so that we can live in it as well as possible. That world includes our bodies, our selves, and our environment, all of which we seek to interweave in a complex, life-sustaining web."[14]

With the exception of the "species activity" line of thought, this one sentence presages most of the points science studies scholars would go on to make in great detail: care is a doing, a web of practices aimed at the repair and maintenance that, although often neglected and devalued, are essential to a good life and a good world.

But the dominant sense of care, even when generalized and extended to an environment or "world," almost always concerns a person, or some group of people, often marginalized in some way. They require our attention, concern, or care because they fall outside of some normal: impoverished, ill, frail, shunned, racialized, or simply too young, too old, or just too much *not* the self-sufficient rational white male regarded as the (unspoken or uncoded) norm. That's who needs or asks for our care, these other human beings who don't *quite* fit the stringent standards conventionally construed to constitute the "human."

So, for example, Daniel Engster's later (2005) "rethinking" of "care theory" delimited definitions of care in terms of three "virtues": attentiveness, responsiveness, and respect. Care, for Engster, "in sum," included "everything we do directly to help others to meet their basic needs, develop or sustain their basic capabilities, and alleviate or avoid pain or suffering, *in an attentive, responsive and respectful manner*."[15] But not exactly "everything." Engster wanted his definition of care to be "narrow enough to exclude practices . . . not usually associated with caring, such as house building and plumbing." If you were building a house for houseless people, that particular act of housebuilding (and presumably plumbing) would qualify as care, "because it is encompassed by a larger caring aim. It is not the activity per se that classifies it as caring, but the aim and virtues."[16] Care is to be understood, by these terms, as a matter of direct aim or intent—"everything we do directly to help others"—together with the affective virtues tied up with those aims, and there needs to be a human in the picture somewhere, not just pipes.

Much of "care theory" theorizes care in this way, focused on the moral and ethical registers, and the social and political concerns to which they are tied. These are valuable conceptualizations that have thickened and enriched our understandings of what it means to think and act in worlds in dire need of repair. But care can be and deserves to be understood more expansively, and scholars in science studies encourage us to do so.

In this literature, "care" is first extended toward other living creatures. We care for any Other, not only those of our own species but the Others of our "companion species." Pain in any species begs for care, not "to share the dogs' suffering, or that of participants in today's experiments," writes Donna Haraway, "not to mimic what the canines go through in a kind of heroic masochistic fantasy but to do the *work* of paying attention and making sure that the suffering is minimal, necessary, and consequential."[17] Carrie Friese writes eloquently about how this kind of benevolent care affects the biology of laboratory rats and is an essential practice in translational medicine; her analysis of the "uncanniness" of care, its doubled nature as both familiar and strange, embraced and disavowed, parallels my analysis here.[18] Care retains the sense of work under these expanded sensibilities, an attentive practice that costs its practitioners something—time, physical and emotional effort, money—and is most often directed at a more generalized suffering that extends beyond the species boundary of the human.

Still, this expanded purview of care stays on the moral and ethical registers, and the social relations extending from those; the epistemological is not yet on care theory's table. Haraway extends the usual senses of care still further, beyond the ethical and toward thinking, knowing, and reasoning. "If we know well, searching with fingery eyes, we care. That is how responsibility grows."[19] Such "felt reason," she points out, "is not sufficient reason, but it is what we mortals have. The grace of felt reason is that it is always open to reconsideration with care."[20] Similarly, Maria Puig de la Bellacasa strives "to articulate a nonidealized vision of care that is meaningful for matters of thinking and knowing," a vision that acknowledges that "care is somehow unavoidable: although not all relations can be defined as caring, none could subsist without care."[21]

In this rich vein of work, the material engagements named as "care" are directed toward other humans and other creatures, but they are also evident in our relations to plumbing and other infrastructures, like genomics infrastructures. These infrastructures sustain our species and our companion species—all of planetary life, in other words—and infrastructures need constant maintenance and repair.[22] Especially in such infrastructural cases, as we will see, the embodied practices of care, their "felt reasons," are "infused with experience and expertise and depend on subtle skills that may be adapted and improved along the way when they are attended to and when there is room for experimentation."[23] Here, too, I work from the perspective that all "relations of thinking and knowing require care," as Silvan Tomkins would have agreed, and that "caring is more than an affective-ethical state: it involves material engagement in labours to sustain interdependent worlds."[24]

Genomics with Care extends such work to, on the one hand, trace care's relational force still more widely and deeply, to find it already being practiced in the sciences at the most molecular level, where people care about and, more importantly and interestingly, care *for* everything: a data set, a segment of DNA, a cell line, a piece of lab equipment, a culture and system of social institutions supportive of thoughtful scientific inquiry—the "fantastic infrastructures" of genomics. On the other hand, this book also works against such strong, all-encompassing theories of care, and delimits care's scope to situations structured, as I detail later in this chapter, by double binds, paradoxes, and impossibility. To play this double game of expanding and delimiting, strengthening and weakening, let's double back to reading for catachresis.

Care, Catachresis, Contradictions

It's easy to get carried away with care theory. It's easy to get carried away with anything, but "care" is such an old and familiar word that, once its importance is recognized, as care theories have helped us to do, it magnifies even further in scale and scope. Reminding you to read "care" as catachresis, an improper name for something that knows only impropriety, is one way to resist this effect. And as long as we are using five-dollar words, care should also be understood as a paleonym. A paleonym is part of a kind of legacy system, an outmoded program (in computer sciences) or apparatus (in engineering communities) embedded within a larger system that one is not able, or at least not willing or ready, to replace: it would be too costly, or it is too deeply imbricated in too many unknown parts of the larger system, or it works well enough so that it's not worth the trouble to upgrade, or it's so rooted in habit or routine that you're not even aware of it, let alone of the need to rethink or replace it.[25] So a paleonym is not just a very old word, as the name would suggest; it is a very old word that, like a segment of DNA that can be translated into a *toll*-like receptor, *has been culturally (evolutionarily) conserved*, with its extensive connections and pathways gradually being conscripted into service in new systems of semiosis, new patterns of meaning-making. Just as the *toll* gene started off in insect immunity pathways hundreds of millions of years ago, later becoming involved in larval development, and then differentiating and multiplying into numerous other species and systems, so too has the history of care's development in Indo-European language made it increasingly entangled in our conceptual and cultural systems, often in powerful ways that are nevertheless difficult to access and read, from Roman myths to medieval Christianity to German phenomenology to feminist psychology and philosophy to the interdisciplinary fields of science studies.

In other words, "care" is a paleonym that orders our conceptual world in both recognized and unrecognized ways, in conscious and unconscious meaning-patterns that are too big, too multiple, too associated with habits of thought, too transcendentalized and essentialized—in a word, too *strong* to make "care" completely trustworthy to think with. And we absolutely need to think care better. Care itself presents us with a double bind: we have to embrace the term, and we have to resist its assurances. In that sense "care" is much like "reason" or (more narrowly) "science," those other paleonymic legacy-system terms that we are wise to hang on to (as if we had the choice) and that need our concerted unsettling.

The theories of care developed in recent scholarship that I outlined above are compelling because they are strong theories. Strong theories "unify a wide range of disparate objects"; they tend toward broad, confident statements and are comfortable in generalities and idealizations.[26] Weak theories, in Silvan Tomkins's words, are "driven by and developed to account for . . . very specific experiences which are neither intense enough nor recurrent enough to prompt the generation of more than a crude general description of the phenomena themselves."[27]

Driven by and toward, and in the style of, strong theory, I can write: Care is what happens before cognition, before reason, before any kind of thought. Carefulness and thoughtfulness are for all practical purposes simultaneous and synonymous. You don't, and can't, think about something you don't or can't care about, aren't already caring and thinking about. As Ludwik Fleck, writing from the Polish margins of the early twentieth-century Euro-American empires of science, noted from the perspective of a practicing scientist turned socio-historico-conceptual analyst of science, the "concept of absolutely emotionless thinking is meaningless."[28] We can sense people, creatures, and things, we can make sense of people, creatures, and things, only because we already at the same time care about people, creatures, and things. Care is what binds us to people, creatures, and things, a binding tie we can't think or do without. Care names an essential modality of our relational being-in-the-world, a fundamental feature of humanity, an embodied capacity of our natural-cultural inheritance as ex-apes.[29]

At this magnificent scale, care operates in our biosemiotic legacy systems to produce two powerful effects, still dominant in strong care theory: (1) care is an integral, essential thing or phenomena, interior to us even if it is said to be "relational"; and (2) integral care remains entirely Other to those other essential, foundational words, science and reason. It's these strong legacy forces and their effects that most merit weakening.

The concept of care as currently used (one more time: laudably and necessarily) in science studies and other scholarly literatures often remains as a stable, proper, and *recognized* singular entity *within*: we are full of care, we are careful. As part of that, care tends to be understood and presented as a primarily (if not exclusively) warm, kindly, gentle, well-meaning, *positive* human comportment. Not surprisingly, gendering is part of this pattern. If care in these dominant understandings and usages were to be given an emoticon, it might appear as a comforting feminized face with the slightest but still noticeably reassuring smile under tenderly opened eyes, all framed in an open, inviting, and kind countenance verging on the beatific. It's not only good to care, it *feels* good to care, comforting in its comfort, and it's written all over your face.

Etymology is always a good way to begin reading otherwise, to try to interrupt these tendencies. One signifying thread of care stems from the twelfth-century Germanic *kara*, lament. It is this oldest set of care's many associations that, even when they are recognized in strong care theory, tend to end up marginalized, underemphasized, or forgotten altogether; I do it, too. But before it named anything else (*kara*) care named sorrow, grief, mourning, lamentation.

Although I never really cared for his writing, it's hard to go wrong with Chaucer (c. 1386): "Lat hym care and wepe and wryng and waille" (*OED*). To care is to be uneasy, unsettled, uncomposed, and uncomfortable. "Anxious" would not be a bad word to substitute for "care," in the same sense that "genomics," as we learned in chapter 1, is not a bad word for the other excess this book is trying to think along with care. The earliest meanings that care—"in no way related to Latin *cura*," the *OED* emphatically reminds us—signed for, were "a burdened state of mind arising from fear, doubt, or concern about anything; solicitude, anxiety, mental perturbation."[30] This face of care might appear as pursed lips under slightly flared nostrils and narrowed eyes topped by a furrowed brow. Since "affects do not always come in tidy packages," as Silvan Tomkins reminds us, we need to somehow superimpose these emoticon faces in our imagination when we are reading and thinking about "care."[31]

The other effect of strong care theory it's important to take down a notch is its Othering from that other integral entity, science, along with a romanticization. Goethe's *Faust* illustrates the kind of romanticism of care that is so deeply enmeshed in our Greco-Roman-infused cultural systems of sense-making, particularly concerning the sciences, that is impossible to completely avoid and can only be read against, from within. Condensing that sprawling narrative down to its thematic core, Faust's compact with Mephistopheles is not simply about his drive for science, reason, and power, which is what we remember (if we read or remember it at all). Faust's bargain is also driven by his desire, as bioethicist Warren Reich writes, "to be care-free, that is, free of the disturbing anxieties of care that the pursuit of his goals would entail in working with ordinary human resources." When "Care" (*Sorge*) reappears at the epic's end, personified (and feminized) as a gray crone, the "eternally anxious companion," she denounces Faust, which "has the effect of bringing about Faust's turn from burdensome care to the uplifting solicitude of positive care."[32]

These mythemes—care as fundamental to humans, as doubled but usually resolving to uplifting positivity, as a feminine maternal force or essential spirit completely other to the masculine force or spirit of reason and science—became philosophemes over time, and came to structure the cultural space in which most of us have thought (cared) about care. We (Indo-Europeans) can't simply put it all behind us, but can only think inside that space, so need to do so . . . carefully. It's something of a double bind, about which more later.

In the end, strong care theory feels confident it has named something with an identity. I hold that confidence a bit more lightly, reminding myself and readers that "care" is a catachresis, a sign that foregrounds or exaggerates its semiotic disconnect from the "object" that syntax alone does so much to reestablish, and not only through Faustian tales. And I confound strong care theory's foundational differentiation of care from science, its relegating of these two integral entities to different domains, which are then shown to be in relation.

Gayatri Chakravorty Spivak, extending from the work of Jacques Derrida, thinks and writes often about and through the trope of catachresis. One instance is particularly apt here and worth considering in detail, as she attempts to explain it through recounting a conversation she had with a cell biologist:

> Here is why I have to use the word catachresis. I was recently having a discussion with Dr. Aniruddha Das, a cell biologist. He is working on how cells recognize, how parasites recognize, what to attack in the body. I asked him why he used the word recognize, such a mindy word, a word that has to do with intellect and consciousness. Why use that word to describe something that goes on in the body, not really at all in the arena of what we recognize as mind? Wouldn't the word affinity do for these parasites "knowing" what to attack? He explained to me that no, indeed, the word affinity would not do, and why it is that precisely the word recognize had to be used. (I cannot reproduce the explanation but that does not matter for us at this moment.) He added that the words recognition, recognize, lose their normal sense when used this way; there is no other word that can be used. Most people find this difficult to understand. And I started laughing. I said, yes, most people do find it difficult to understand, what you have just described is a catachrestic use of the word recognition. In other words, no other word will do, and yet it does not really give you the literal meaning in the history of the language, upon which a correct rather than catachrestic metaphoric use would be based.[33]

Discussing the catachresis of recognition, the cell biologist and the literary scholar recognize something of each other's mindy-ness, each other's world, each other. In my imagination, I hear Spivak's laugh like Nüsslein-Volhard's *toll!*: *Aren't signs weird crazy cool? And isn't it even weirder cooler that the two of us . . . recognize that, across our pronounced differences, each in our own mindy way?*[34] Metaphors are prone to positionality, irony, and conflict; catachresis edges us toward limits, laughter, and friendship.

Keep this in mind in the pages to come as "care" appears over and over, pulled in from the scientific articles and attendant discourses of genomics that are central to my research. By asking you to read "care" as catachresis I am asking you to resist those stable, familiar meanings that will inevitably occur to you—have already occurred to you. There are no true examples of care, a radically unstable concept. And the concept is vitally necessary—because it is radically unstable. So each time "care" appears on a page I hope you will pause, interrupt your habitual understanding, and read differentially—that is, carefully.[35]

And yet: I will be naming acts, events, concepts and technologies as manifestations of care, as part of its patterned weave. Genomic databases are enactments of curatorial care. An article in *Nature Genetics* can be a technology of scrupulous care. A panel meeting at the National Institutes of Health becomes a collective act of solicitous care. There is no proper term for what is going on in or with each of those events, but they insist on a name so I give it: care.

And: science. To say that everything from data to culture is care, or is "in need of care," is in some sense just another way of saying these things are "in need of science." That's not to say that care and science are the same thing, or that they represent interchangeable discourses. It's to say that we still need to develop multiple ways of naming, talking, and writing about the confounding relationalities of the worldly lives of the sciences and of scientists. It's to say that discourses of science and care are both essential to thinking and engaging well with those meshy, mindy, material relationalities, and that both discourses will in some sense fail. It's to say that pushing the differences within and between the discourses of science and those of care promises a better grasp of their interworkings, their codependencies, and their mutual misapprehensions. Mapping, narrating, or diagramming those dense and shifting entanglements is another game of cat's cradle that Donna Haraway has done so much—more than almost anyone, really—to encourage us to become expert in playing. My primary strategy of play in the rest of this chapter and in the book beyond is to begin multiplying the names of care—curation, scrupulousness, solicitude, friendship—as a way to simultaneously trick out the otherwise expansive grayness of "care," make it more down-to-earth, while bringing the intra-actions and infra-structures that bind it to "science" into play.

One last effect of strong care theory to be loosened: by making "care" out to be so fundamental to our ways of thinking and doing anything, or by making it a feature or consequence of an equally broad "relationality," our involvement with people, with other living creatures, and with every aspect of a world on which our collective lives depend and for which we must collectively care, we risk overlooking some of the more peculiar qualities of that world, some particular patterns of "relationality," that make especially—maybe uniquely—urgent demands on and for care.

I'll put forward a hypothesis that I hope will both strengthen *and yet* weaken the growing care literature: it's because our cat's-cradle lifeworld is chock-full of double binds that we care.[36] The sciences are an excellent place to see such double binds in action, and thus to see how particular patterns of care exemplify an always necessary, usually effective, and occasionally brilliantly creative means of acting within double binds. Subsequent chapters describe how each of care's specific patterns that are manifest in the sciences (curation, scrupulousness, solicitude, and friendship) lead to and/or are consequences of some double-binding structure or situation. Here, still writing generally about magisterial "care," I'll first describe some of the more formal features and effects—the logical and psychological *patterns*—of double binds.

Double Binds in/of the Sciences

Double binds and double-bind theory are most strongly associated with their renowned formulator, anthropologist Gregory Bateson, and the collaborative that coalesced around him in Palo Alto in the 1950s and 1960s. The double bind is often used loosely (but productively) to characterize any set of contradictory, or even just competing, demands. When thinking about the sciences and scientists, however, I have found it analytically good to stick more carefully to Bateson's more exacting conceptualization: double binds arise from "paradoxes of abstraction" that occur when our thinking and doing move, as they must, across different logical levels or, if you prefer, from one set of things to a more encompassing set.

Mathematician and cybernetician Norbert Weiner introduced Bertrand Russell and Alfred North Whitehead's work on logical kinds and set theory in their *Principia Mathematica* (1910–13) to Bateson, who took those general propositions about the impossible relationships or statements that *must* occur between logical levels and used them to think about a wider range of communicative situations involving animals, especially *people* animals. In the psychoanalytic and therapeutic settings that most preoccupied Bateson's group, the relationships and communications under analysis were intense familial ones that could produce phenomena as extreme as schizophrenia. But Bateson used these double-binding patterns to think across a whole range of complex systems, and not just that one called the family.

Russell and Whitehead's theory of logical types is as interesting as it is difficult; multiple articulations and permutations of it abound, not always in agreement, but all have to contend somehow with some version of self-reference or recursiveness. "The town's barber shaves every man who doesn't shave himself" is one of the most succinct statements or situations of the kind that led to the theory. Does the barber shave himself? Saying yes invalidates the statement; saying no invalidates the statement. (Saying the barber is a woman is a different kind of solution to a different problem.) Russell and Whitehead's resolution of the paradox was to amend set theory to recognize two different sets of statements, one containing statements concerning people who shave and another containing statements about people who are shaved—and to recognize the first set of statements as operating on a different, more abstract, or meta-logical level *about* the second set. (Note the ambiguous relationality of the "about" here; it will recur below when we come to the subject of metadata, "data about data.")

Any system that features these multiple levels of abstraction, and not only mathematics or logic, is prone to this kind of paradox, contradiction, or, as Bateson and colleagues wrote in "Toward a Theory of Schizophrenia" in 1956, double bind:

> Our central thesis may be summed up as a statement of the necessity of the paradoxes of abstraction. It is not merely bad natural history to suggest that people might or should obey the Theory of Logical Types in their communications; their failure to do this is not due to mere carelessness or ignorance. Rather, we believe that the paradoxes of abstraction must make their appearance in all communication more complex than that of mood-signals, and that without these paradoxes the evolution of communication would be at an end. Life would then be an endless interchange of stylized messages, a game with rigid rules, unrelieved by change or humor.[37]

Processes of abstraction are certainly a prominent feature of the sciences, as is their need for multiple forms of sense-making and communication that are more powerful and complex than a glower or a smile. So I think it is well-warranted to use double-bind theory to analyze the relationships between scientists and their objects (including data), the paradoxes and communication disorders that arise between them, and the possibilities for change (not to mention humor) that they harbor. And change does happen, however difficult or even *impossible* such change might be. Indeed, I'll argue—hypothesize, propose, promise—later, impossibility is essential to change in the sciences.

But for now, as we did with "care" above, let's start with double binds as they are evident in familiar human relations, before going on to extend the analysis to other forms of relationality between humans and other living and/or nonliving things. Like care, double binds are thoroughly relational, always involving at least two participants. One of the participants is named the "victim" of the double-binding situation. (Later, I'll identify this victim as the scientist.) The other participant in the situation issues a demand, order, or "injunction." (Later, it will be data making the demands, although "reality" or nature would fit the same structure; in part 2 we will see additional demands on scientists that arise from a crossing of additional logical levels—from data to experiments, from experiments to research programs, and from research programs to "science" at its most abstract and general.) This primary demand or injunction is often negative—for example, "Love me and I won't punish you"—but doesn't have to be. Simultaneously, however, the demanding participant communicates another order, at a more abstract or meta level than the first, and contradicting it—for example, asking for love and promising not to punish while adopting a defensive body posture and contemptuous facial expression that conveys the nonverbal message, "I hate you anyway and will make sure you know it." Or, more verbally but no less logically confounding, "You can't really love me if I have to ask you to do it." This secondary command may also be negative but, again, doesn't have to be—or they might be mixed messages, impossible to decide if they are negative or not.[38]

There are two other conditions necessary to double binds for the Bateson group, but let's pause here and ask: How might this apply to scientists and data (like "rolling nitrogenous mountains" of DNA sequence information)? But even before we can untangle that relationship, we need to understand that "data" itself is not a unified entity; data harbors its own relationality, its own intra-actions, and in effect *differs* from itself in ways that immediately establish these double-binding logical or communicative paradoxes. Think first about the ubiquitous and vocal calls for more scientific data of all kinds, in our time of Big Data. If you know even a little about the plethora of data-demanding, data-generating, and data-sharing projects that are now in vogue and in progress, in and out of the sciences, you know that these projects also demand the creation, curation, and sharing of metadata or, as that term is frequently formulated, "data about data." Both formulations should make it clear that there are different logical levels within "data"; data contains its own abstraction; data is also metadata or data about data, that is, data, which is also metadata, which is . . . *und so weiter*. Or, as the title of a recent book of critical essays on data phrases it, raw data is an oxymoron.[39] Without appropriate metadata—where the data came from, what instruments produced it, what analyses have been performed on it, what uncertainties inhere in it, and more—data is useless, or at the very least far less useful and reliable in the making of truths than it could and should be. Data is always shot through with other data; data is always (meta)data, or meta/data, or metaXdata.

You can begin to see the paradoxes and contradictory demands that make the scientist-data relationship an intense one, to say the least. If data could speak like a person in a relationship, it would be issuing to a scientist a tangle of primary orders that would sound like, *I am "given," purely and solely given. You mustn't cook me, I must remain given, raw. I've given you everything, I'm all you've ever needed, the only thing you need. You don't need anything else, you just need more of me. I don't need you, and you're lucky that I give myself to you.* But in truth, since data is always already (meta)data, alongside those orders it issues another set of commands, more abstract, more implicit, maybe disavowed or repressed, but certainly in absolute conflict with the primary set: *I'm nothing without you, always in need of your attention and love. To be complete I need you, I can't stop needing you. And I'm never enough, you always want more of me, and more than me.*

I've kept these somewhat anthropomorphic projections abstract, but flesh them out later in the book. The anthropomorphism is clearly an exaggeration, meant to heighten the general dynamics of the data situation in the sciences today. These dynamics are both logical and psycho-logical, summoning up affects that we have to learn to read better—not to demonize scientists as "mad," and not even to "humanize" them as "people with feelings, too!" (although this would not be such a bad secondary effect). My goal is to bring out the (infra)structures in play in (meta)data and all across sciences like genomics today, to advance a view of the sciences and scientists as essentially and powerfully *careful*: a dense, hot amalgam of affect and cognition (for starters) that needs to be reckoned with, attended to analytically and culturally, respected and even admired as a response—indeed the only possible response—to impossible, double-binding situations.

To further understand why I think that's an urgent conceptual, cultural, and even political challenge today, let's continue with the analysis of double-binding relationships as Bateson's group first articulated them. There are two additional structures—strictures—that compose the relational space in which the paradoxical messages sketched above take place. A third injunction or order, spoken or unspoken, prevents the "victim" from simply exiting the situation: "You can't leave me!" For genomicists today, this command is utterly clear and compelling: there is no possibility of ever wanting, and of ever having, *less* data. Genomicists will henceforth, from here on out, work in an era not of Big Data, but of Bigger and Bigger (Meta)Data. Avalanching Data, to foreshadow chapter 4. That's just the reality of the double-binding situation genomicists find themselves in—which brings us to the fourth and final characteristic of the double bind: it iterates, and in iterating, escalates. Big Data is simply scalar, a cross-section of the vector of Avalanching Data.

To continue with analyzing the double-bind pattern in today's sciences, it's worth reiterating Paul Gibney's summary of the double bind at work in a family:

> a · the individual is involved in an intense relationship in which he or she feels they must get the communication right;
> b · the other party is expressing two orders of messages, and one denies the other;
> c · the "victim is unable to comment on the contradiction," i.e., he or she is unable to make "metacommunicative statements" that might help to resolve the mess. These situations, endlessly replayed, result in an individual unable to read accurately the context of messages, and unable to communicate effectively or coherently. In short, he or she will live in a world of disordered messages, where active and appropriate deciphering will be experienced as dangerous, and possibly only known, as a nameless, felt, perpetual angst.[40]

I don't think I would get much argument from most scientists today if I described their relationship to data as "intense," on any number of levels. And the strong desire to get the "communication right," to getting data to speak clearly to you, to speak truth to you: I think they would recognize this, too. I know I've just begun to give good reason for how data sends conflicting messages and how, coupled with the intensity of an inescapable situation that iterates, this produces a double bind that is experienced both affectively and cognitively, but I promise further empirical—that is, ethnographic—substantiation.

In saying that the situation iterates, with no possibility of resolution, and that we scientists are thus consigned to a time and place of Avalanching (meta)Data, I hope I've also made it clear that this should not be read in terms of "technological determinism." Although technology plays an important role, the situation is much more complex and distributed than that. Indeed, I think the work of historians of science makes it pretty clear that scientists have always lived in an era of Big (meta)Data—astronomers have been living that pattern for centuries, and they are just one obvious example—but it's the double-binding nature of this situation that makes it repeat with increasing intensity and drives Avalanching Data.

For the Bateson group, because these communicative paradoxes of abstraction were double-binding and impossible to resolve, they were schismogenic—producing difference and change—and sometimes schizogenic, producing pathological splitting. Their hope (and it largely remained a hope) was that even if double binds could not be logically solved—that is, cured—one could nevertheless learn to *endure* them in ways that kept open a space for some kind of meta-learning and meta-change. "Care" would be another name for this kind of therapeutic strategy in a situation that has no apparent end or cure, the activation of an attentiveness to the different communicative levels and their paradoxes that could hold open a space for play or creativity—or experimentation.

There is some shared sense in and around data-intensive sciences like contemporary genomics that, while scientists and in some ways science itself may not be suffering from a schizophrenia-like pathology, they can find themselves in situations that are at the very least difficult and debilitating. This sense may not be a particularly new aspect of scientific culture, but it does seem to be reaching new prominence and intensity. The data-intensive science "plateau" (to use Gregory Bateson's concept appropriated by Gilles Deleuze and Felix Guattari to signify "a continuous self-vibrating region of intensities whose development avoids any orientation toward a culmination point or external end") exhibits ever more overtly and urgently its need for ongoing care.[41] These genomics plateaus, and the situations in which they place scientists, are (infra)structured by double binds.

Or, to put it in terms of chapter 2's figuring of the genomics situation: like the "experimental systems" of which it is composed, the whole cultural endeavor of science is a labyrinth always in the process of being constructed. Walls of data, equipment, and concept-metaphors coalesce, build up, and extend forward with astonishing and increasing rapidity, in a pattern interdependent with *and yet* well beyond our intent. Genomicists—or let's say, our friends in the sciences—are guided and blinded by these emergent structures through which they feel their way forward as they secure the boundaries, oscillating between exhaustion and exhilaration, work and play, dull routine and creativity. Every opening appears only because of a closure, but every blockage promises a possible path. The sciences, genomic and otherwise, are a labyrinth of aporias, paradoxes, contradictions—and affects. Bundles of them, cat's cradles of them. A labyrinth built with care, in constant need of ongoing care, and also the very materialization of that care.

Matrix for the Diffractions of Care

Chapters 4–7 rematerialize some genomics labyrinths for readers, to re-present the affects and enactments of care that have been essential to it—stories of minding genomics, how it got better not in spite of but because of its material, conceptual, and cultural limits. It also works to rematerialize and re-present something of the experiences of careful genomic scientist-subjects in their labyrinthine *tâtonnement*, carefully muddling their way through some opening in a direction that *will have been* known as "forward" once the new walls have been assembled and their directionality emerges.[42]

To say that this labyrinth of my own devices presents genomicists as "carefully muddling their way through" is only to say, more carefully, "doing science." If my rematerializations and re-presentations work, care and science should begin to blur into each other and confound each other, much as (and here I do shift rhetorical figure, to metaphor) a catachrestic particleXwave is diffracted through an experimental setup to form a discernible pattern of differences (see figure 3.2). Like light passing through the double-slit apparatus so iconic of quantum mechanics, careXscience is textually passed here through a series of double-bind structures to manifest as the diffraction patterns, the "weak care theories" I'm calling curation, scrupulousness, solicitude, and friendship.[43]

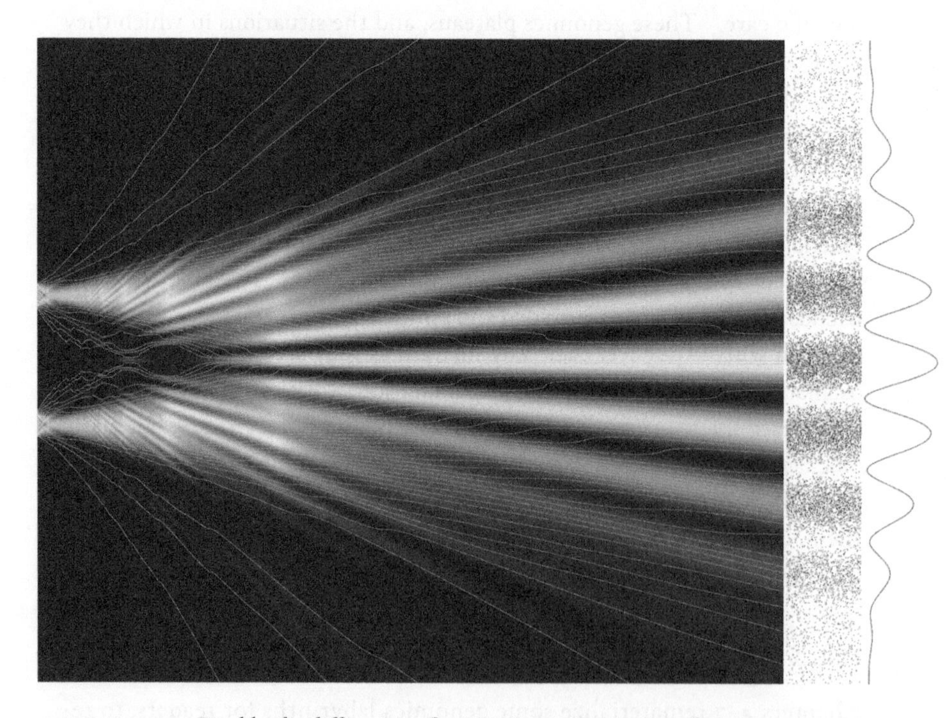

FIGURE 3.2 · Double-slit diffraction of careXscience into its interference patterns of curation, scrupulousness, solicitude, and friendship. Alexandre Gondran, own work, CC BY-SA 4.0, https://commons.wikimedia.org/w/index.php?curid=54072531.

PATTERN	INFRA/STRUCTURING DOUBLE BINDS	OBJECT OF RELATION	TEMPORAL EXPERIENCE
Curation	MetaXdata SecureXinsecure GroundedXundermined Too muchXnot enough	Data differences	Repetition (of the same)
Scrupulousness	MetaXanalysis ParanoiaXtrust ModestyXboldness DoubtXconviction	Experimental systems	Repetition (of difference)
Solicitude	MetaXscience PartXwhole ThreatXpromise ImXpossible	Systemic limits	Ever on the verge
Friendship	DifferenceXidentity DestructionXrepair CriticismXlove	Self-other relations	To come

FIGURE 3.3 · Matrix of care's patternings.

Care becomes, under diffraction, a not-bad name for these patterns of relational events that happen between scientists and their multiple objects of relation, and affection. The crucial dynamics in these different situations are double binds: impossible contradictions that necessarily elicit particular modes or styles of response that we minding systems call "care." The particular styles or patterns of relational care are ways of enduring unresolvable or impossible double binds that arise in the course of doing science. Instead of short narrative previews of my longer narrative weak care theories of curation (chapter 4), scrupulousness (chapter 5), solicitude (chapter 6), and friendship (chapter 7), they are arranged here (figure 3.3) into a matrix that telegraphs the particular double binds or impossibilities that infra/structure the conditions of care, the object of relation with which each pattern of care is in give-and-take, and the sense of time experienced in those relations. These differential patterns of relational care in/of/as science are what I read carefully—ethnogrammatologically—in the remainder of this book, as the minding of genomics.[44]

POEM-LIKE *TOLLS* 2 ·

an interlude

Human TLRs and IL-1Rs in Host Defense: Natural Insights from Evolutionary, Epidemiological, and Clinical Genetics

JEAN-LAURENT CASANOVA, LAURENT ABEL, AND LLUIS QUINTANA-MURCI

The immunological saga
of Toll-like receptors (TLRs) began with the
seminal discovery
in 1981 that antimicrobial peptides are a key
mechanism of innate host defense
in insects.

This was followed by
the observation in 1991 that the fruit fly
Drosophila melanogaster Toll
and mammalian interleukin-1 receptor
have an intracellular domain
in common.

These studies paved
the way for elucidation of
the role of Toll in controlling
the synthesis of some of
these peptides in *Drosophila*.

These discoveries soon led
to the identification of a human
TLR, followed by
the discovery of a function for
TLRs with the demonstration that
lipopolysaccharide (LPS) responses were
abolished
in mice with spontaneous TLR4 mutations.

The similarities
between the Toll and TLR
signaling pathways
in invertebrates and vertebrates
were initially interpreted

as evidence of a common
ancestry for these defense mechanisms and
subsequently of convergent evolution,
emphasizing their evolutionary
importance.

The 15 years or so following
these findings have witnessed a substantial
rise in interest in
the role of Toll
in *Drosophila* immunity,
of TLRs in mouse host defense,
and even of TLRs in diverse other animal species.

Indeed, interest
in TLRs has been such that just about
any immunological phenomenon imaginable—
ranging from host defense and tumor immunity

to allergy and autoimmunity—has been examined from
a TLR perspective.

This phenomenon
has even extended to processes only
remotely connected with immunity, such as
atherosclerosis and degenerative diseases, and
has also stimulated research into the role of human
TLRs in the pathogenesis of most,
if not all,
human diseases.

Various schools of immunological
thought have conferred different names
on pathogen receptors, including
pathogen associated molecular pattern (PAMP) recognition receptors,
pattern-recognition receptors (PRRs),
innate immune sensors,
and microbial sensors.
Whatever the terminology
used, the underlying
idea is that TLRs detect
a wide range of microorganisms,
discriminating between these microbes and
distinguishing them from self on the basis of
their type, through the detection of
specific,
conserved
microbial patterns,
molecular patterns,
or molecules.

Does this commonly expressed view of
TLRs and IL-1Rs reflect the biological
reality?
Like most immunological
knowledge, it is based mostly on experiments conducted in
the mouse model.
However rigorous, accurate, and thorough

such experiments are, can experimental
findings in mice really provide a
faithful and reliable representation of host defense
and protective immunity in other species,
in their natural setting?

There are differences between species, including several
identified differences between humans and mice, and
immunological generalizations from
a single species may be
perilous.[1]

Cnidarian-Microbe Interactions and the Origin of Innate Immunity in Metazoans

THOMAS C. G. BOSCH

Bacteria are an
important component
of the Hydra holobiont . . .

It came as a surprise
that examinations
of the microbiota in different Hydra
species kept in the laboratory for more than 20 years
under controlled conditions revealed
an epithelium colonized by a complex community of microbes, and
that individuals from different species but cultured under identical
conditions
differed greatly in
their microbiota.

Even more
astonishing
was the finding
that individuals living
in the wild were colonized

by a group of microbes similar
to those found in laboratory-grown polyps, pointing
to the maintenance of specific microbial communities over long periods.

Bacteria in Hydra, therefore, are specific for any given species.

For microbial recognition, Hydra uses two types of receptors
and signaling pathways, the Toll-like receptors
(TLRs) with MyD88
(myeloid differentiation factor 88) as signal transducer
and the nucleotide-binding and oligomerization domain (NOD)-like
receptors
(NLRs).

The patterns of
differentially regulated host genes as well
as changes in the bacterial colonization process and pathogen susceptibility
in MyD88-knockdown polyps strongly indicate
TLR signaling has a role in sensing
and managing microbes. Thus, not only
are TLRs the long-sought cell-surface receptors
that recognize common microbial
features such as bacterial cell wall components
(e.g., flagellin), but their role in
controlling the resident microbiota could date
back to the earliest multicellular organisms, as
humans and Hydra share
the molecules involved in
the TLR signaling
cascade.[2]

minding the infrastructures of genomics

PART II

4 · curation

OF DATA'S LIMIT

MJP: Can we clean up metaphor?

GAYATRI CHAKRAVORTY SPIVAK: No. It's like cleaning teeth. You know, you will never be able to clean your teeth once and for all. But cleaning one's teeth, keeping oneself in order, etc.—it's not like writing books. You don't do these things once and for all. That's why it should be persistent.

MJP: So political practice is like housework?

GCS: And who doesn't know this? Except political theorists who are opining from the academy with theological solutions once and for all.

· GAYATRI CHAKRAVORTY SPIVAK, *POST-COLONIAL CRITIC*

Data and Their Vicissitudes

The introduction of the concept-metaphor "data curation" into our scientific vocabulary can be traced to the then-emergent fields of genomics and bioinformatics in the earlier years of the Human Genome Project. Probably. It might have sprung instead from the then-established field of museum studies, in which curation had long been a necessary and valued—indeed, essential—practice. This difficulty in assigning what today's data scientists would call the term's *provenance* tells us something about the multiple and hybridized natures of both data and curation, and the underlying infrastructures and activities on which both of those terms depend.

It was at a 1993 Invitational DOE Workshop on Genome Informatics sponsored by the US Department of Energy that "curation" was first used to name *something* that was needed for the many types of data that were being generated at an ever-accelerating pace by the HGP: DNA sequence data, of course, but also protein data, genome mapping data, and, "although the Human Genome Project does not fund gene-function or related studies," the meeting report pointed out, "such data must be captured in a useful way if the genome project is to be truly successful. Genome researchers, funding agencies, and the entire biological community should ensure that an infrastructure is in place to capture this type of data."[1] This was a notably forward-looking workshop and report, convened and issued in the early years of the HGP, that clearly drove and shaped the development of bioinformatics as a new scientific field of research. It demonstrates again the pronounced emphasis on building and maintaining scientific infrastructure that was part of the culture of the DOE and its work within the infrastructure-building HGP.

Data, in turn, is the infra/structure of genomics infrastructure, what supports it and flows through it. And as chapter 3 showed, data itself has an infra/structure, a relational difference within or beyond it, a not-self in "data itself," that I signed for there as (meta)data. We might also say that (meta)data requires its own infra/structuring in order for it to be data. (Meta) data requires curation, it has to be cared for, if it is going to be data; it must be "captured in a useful way," or else it's useless—that is, not data. Curation, then, is our first interference pattern in the diffraction of care, the minding and mending of these foundational but unstable objects.

(A brief side note to those readers who may find slashed or partially parenthetical signs off-putting or showoff-y: Introducing a slash into infra/structures is meant to occasionally foreground what Brian Larkin calls their "peculiar ontology": that they are "both things and the relations between things." Infra/structures are simultaneously material structures and their *infra-*, what's beyond them and yet within them. A DNA sequencer is a structure here, infrastructure there; infra/structure is a shorter if not sweeter way to mark this divided, unfixed quality. For similar reasons I sometimes complicate "data," the infra/structure of genomics infrastructure, as (meta)data or meta/data, to foreground its instabilities and doubledness. And sometimes, respecting the limits of all of our collective patience, I just use the everyday signs.)[2]

The 1993 DOE report referred to the different "community databases" that preexisted the Human Genome Project, and that were growing in size and number as more and more data of more and more different types were generated for more and more different "model organisms," and not simply the eponymous humans of the HGP.[3] Even in these early years of the internet, such databases were frequently accessed and increasingly interconnected, and already acquiring some of the functions and importance of the published scientific literature. "As databases take on a role similar to the primary literature," the report noted, "curation will become increasingly important. Tools are needed to allow and encourage data submitters to take responsibility for the continuing quality of their submissions. Curators must be appointed to oversee long-term quality and consistency of data subsets in community databases. A new professional job category, not unlike museum curators, may develop for these databases. Professional database curators and tools for direct author curation should be supported."

If I point out here that I know more about the science, technology, politics, and people of the HGP than almost anyone else, it's not for reasons of immodesty, but only to benchmark the fact that I recognize only two or three names on the list of twenty workshop participants. They represent the truly infrastructural personnel of this infra/structure project, the scientists who are difficult to see and credit, even for those of us trying to pay close attention. They are the ones strategizing and advocating for the increased support of the mostly invisible, mostly undervalued, but utterly foundational work—that is, care; that is, science—of data curation. I can republish their names in a footnote, as a small ethnogrammatological data-curation gesture of my own that recognizes and values their work.[4] But I still can't credit the anonymous reader comments that they also appended to their report that suggest some of the data curation challenges evident in genomics in 1993; here's a partial list of those care demands:

- The report does not note that its recommendations span a wide range of difficulty, from relatively straightforward implementation using current technology to outstanding research challenges in computer science. If the more difficult challenges are to be addressed, the genome project should direct some of its informatics support toward more basic computer science research.
- The report omits any mention of knowledge representation (KR) technology, which is a very useful approach to information management.

KR technology may be the best current technology for building the type of shared data models discussed in the report.

- Current researchers in genome informatics rarely publish on the details of their methodology or on the lessons learned from their research. This reticence hinders the development of the field and efforts should be made to secure better exchange of general ideas in bioinformatics. Regular workshops that emphasize general principles and methodology would be helpful, especially in the development of improved data models and better user interfaces.
- A common misunderstanding is that the lack of exchange formats has significantly hampered database intercommunication in genome informatics. In practice, the semantic differences among the databases greatly overshadows issues of physical data *format*. Having the data expressed in the same format is of limited use if the conceptual database schemas are incompatible. Practical experience shows that the major effort is involved in understanding and cross-mapping semantic differences between databases and in developing tools to transform the data accordingly (often this cannot be done in a fully automated manner). In comparison, the effort to develop syntactic parsers is insignificant. Currently, it is not possible for groups not associated with a particular database to understand the semantics of that database well enough to cross-map semantic content accurately.
- Many community databases are significantly underfunded, given the expectations of the community. If a proper information infrastructure for biology is to be developed, appropriate levels of support must be identified and maintained.
- Direct submission of data will not entirely replace published literature as a source of data input until the bench scientist approaches the publication of a data submission with the same care as the preparation of a traditional publication.

Many of the concerns and issues expressed in these comments, and in the report overall, may be summed up by what Diana Zorich, the anthropologist and museum studies scholar who is most responsible for installing this DOE report into the history of data curation, calls "the idiosyncrasies inherent in manual recording systems."[5] Zorich took up what she called "the DOE's curation metaphor" and reissued a similar programmatic call for museums and related institutions. The DOE, she wrote, treats bioinformatics databases as collections of information which undergo a sequence of processes not dissimilar to those employed for a collection of objects housed in a museum. Information enters a database through various channels, it is stored, accessed, modified and stored again, and it may be removed permanently from the database. To ensure the integrity of that information throughout this process, policies and procedures are needed to govern each phase. Data sets need to be examined for consistency, long-term quality and relevance time, and new sources of data (i.e., from libraries, archives and other resources not necessarily within the institution) must be identified and assessed. Changes or updates to data require authentication and verification. Tools which support databases, such as authority lists, thesauri, data dictionaries and other documentation resources, need to be maintained, updated and distributed at regular intervals, while data security and access must be considered. All these concerns constitute the discipline of data curation.[6]

Rather than read the "curation" in "data curation" as a metaphor borrowed from a domain (like museums) in which it is legitimate, though, we should read again for catachresis: signaling one pattern of "care," curation is an improper name for an impossibly long and heterogeneous list of elaborate, often tedious, resource-intensive, theoretically challenging, and usually difficult practices *for which there is no manual.* Those practices are always demanding and always necessary; handling the idiosyncratic, the quotidian, the unruly differences that inevitably arise as part of the day-to-day activities of scientists is a monumental chore, like housework. These practices of curation often begin as tacit and ad hoc but can then be drawn out and turned into more explicit routines and algorithms, partly but never fully codified; they traverse all scales, from the molecular to the organizational, but can be both particularized and generalized; and they are underappreciated and undervalued efforts, as almost all infrastructural work tends to be. Curation practices can be foregrounded and advocated for as not only important but also *essential* work, where essential means necessary and requisite, as well as "constituting the very essence of." Data *is*, in essence, care.

Since all of these qualities and demands are structured as double binds, they recur interminably; that's a rule of unruly matters of care. Identified in long but never quite complete lists of dense and elaborate idiosyncrasies, they are each their own singular mixture, as the earliest Greek meanings of "idiosyncrasy" suggest. Even as the care work of curation works to organize that disorganization, it's only ever a partial and temporary win over the forces of entropy, fragility, or mutability. Data curation is like teeth cleaning and housekeeping, and similar hygienic and domestic responsibilities: the need for it never stops happening, things never stop needing care—a reality we life scientists are always needing to come to better terms with than we have.

German-speaking cultures may have already started to come to better terms with the double binding of care and data: there, the compound name for these compound practices is *Datenhygiene*, suggesting there is not stuff called "data" that then needs curatorial care, but that there is only datacurating. Data is ever prone to vicissitudes, to use what Spivak would no doubt consider a "mindy" term—or data *are* vicissitudes, fateful turns of happenstance.[7] Less mindy-ly: stuff is always happening to stuff, and since data is stuff just as teeth are, they constantly need tending to—the undervalued but invaluable work of maintenance.[8]

Exploring Curation Space

No one, it's safe to say, gets too terribly excited about doing laundry or brushing teeth, even if they are careful about attending to these essential tasks. There aren't many surprises, either. One thing that makes the curation of not only Big but Excessively Large and Avalanching Data in genomics different from oral hygiene, though, is that it is capable of generating surprise and excitement in addition to the necessary, admirable, but maybe just a tiny bit *resentful* rolling up of the digital curatorial sleeves.

I've long been partial to quoting a 2000 review article in *Nature*, "Exploring Genome Space," by molecular biologists Ognjenka Goga Vukmirovic and Shirley M. Tilghman as a marker of the "intellectual and experimental sea change" that not just genomics but all of biology was undergoing at that time, primarily as a result of the excessive amounts of genetic, protein, and other data that had begun pouring out of university, government, and corporate labs.[9] (We read about Tilghman's conscription into the planning bodies for the Human Genome Project in chapter 2; a year after this 2000 publication, she would be named president of Princeton University.) "This avalanche of data," they wrote, was unleashed by the "fortuitous confluence" of radical improvements in numerous technologies: DNA and protein sequencers, mass spectrometers, nuclear magnetic resonance (NMR) spectrometers, X-ray crystallography, and other imaging technologies, to name only a few.

This data deluge had only "whet our appetite for more," they noted—another indication of the affective forces in play here, the excitement and surprise to be found in encounters with genomic data only recently more likely to be experienced as sort of boring. Indeed, Vukmirovic and Tilghman noted that, fewer than fifteen years earlier, when the initial HGP discussions mentioned in chapter 2 occurred, "many argued that investing in genome sequencing was unwise until we had the tools in hand to understand the sequence. Funds spent acquiring the sequence would be better spent developing tools to first understand it. To physicists and engineers, in disciplines where data are very often acquired well before their utility is apparent, this seemed illogical."[10] But by the time they were writing in 2000, with the Human Genome Project still in progress, the putative logic of physicists and engineers had gotten through to biologists. Few doubted "the value of such large-scale data acquisitiveness in biology," and the idea "that data are inherently good" had become "a central philosophical tenet for biologists."[11]

How does an avalanche—an excessive, overwhelming, disordering engulfment—come to be valued as "inherently good"? More than likely, Vukmirovic and Tilghman intended their characterization of data as "inherently good" in the Levi-Straussian sense of "good to think." Their article is an elaboration of many of the cognitive reasons why data becomes good to think, some of which I return to in later chapters, on care as scrupulousness and solicitude.

But it may also be the case that such an excess registers affectively as well on genomicists—surely some of first scientists who should acknowledge they are hardly disembodied intelligences, but rather embodiments of biocultural evolution. To them, who only think because they care, to say that data are "inherently good" doesn't just mean "Data acquisition has been optimal and conducive to advancing my cognitive schemas and schemes," but also can and should be read in the sense of "*mmmmmm . . . data good! I'm totally stuffed but, if you insist . . .*" I think we don't want to lose sight of this reading, this alternative signaling pathway, this hypothesis I began advancing in chapter 3's opening segment, quoting genomicist David Smith: there's an appetite for data, part of an epistemophilic drive system.[12] But I also don't want to dwell on it here, raising it only as a reminder to read for the complex transductions, or at least their possibility, among affect, science, and care. The stated intent of their article was not to explore what pleasures might be found in excessive "avalanches" of data, but rather to step out momentarily from that space and describe "some of the challenges that biologists face as they acclimatize themselves to this change in the data landscape." What new study designs, experimental strategies, theoretical perspectives, and methods were genomicists inventing to be kept from being swept away or buried?

Vukmirovic and Tilghman describe many such data-driven trends in genomic practice and thought at the turn of the millennium; some of these I discuss in more detail in later chapters. What interests me more here is one that they do *not* discuss, even though it undergirds them all and makes them possible. There's an iconic desktop computer representing the "computer scientist" in the article's illustration of the "many scientific perspectives" from which "gene function" can be approached, along with an iconic cell for cell biology, iconic test tubes and flask for the biochemist, iconic mouse and rat for the physiologist, iconic protein model for the structural biologist, and iconic double helices for the geneticist (see figure 4.1). It illustrates the "transdisciplinary" effort that genomics was becoming—as one -omic among many -omics—and is also a healthy reminder of the enduring importance of "wet," "material" practices in such pursuits. Yet it does not illustrate, and indeed directs attention away from the fact, that although flasks, mice, and globs of organized or denatured biological materials would almost certainly not be found anywhere near a computer scientist's work space, the work space of every other professional in this symbolic circle would have one of those desktop computers in it. The computer or data scientist does not directly depend on the flask to do her work as a life scientist; every other iconic scientist around the rest of that circle interacts directly, regularly, and often for large chunks of time with data accessed electronically through a computer.

Only later would careful historian-philosophers like Sabina Leonelli and Rachel Ankeny show that a crucial part of the wetter intellectual interdisciplinary developments was the collective scientific work in the late 1990s that built the databases for model organisms such as yeast, fruit flies, and mice, and developed the technical and social protocols that rendered them interoperable.[13] As is the case with almost all such infrastructural work in/of the sciences, efforts to build, improve, interlink, and generally *mind* genomic data and databases tend to be far less visible than the flashier, more exciting intellectual discoveries that emerged on the basis of that work.[14]

It's for reasons like these that I put this kind of scientific activity under the sign of care, patterned as curation. Genomic data has to be cared for because one must dig out from any avalanche of excess, even the minor ones that occur at slow speed in any busy, lively, domestic space. And *every* space is, in matters of care, a domestic space. Genome space as it began to emerge and be explored in 2000 was indeed exciting, but it was also a labyrinthine domestic space that needed continual curation—the kind of care that rarely gets the appreciation it richly deserves. "Sequencing is boring but everyone is doing it" got transformed, and transvalued, into ubiquitous genomic data curation.

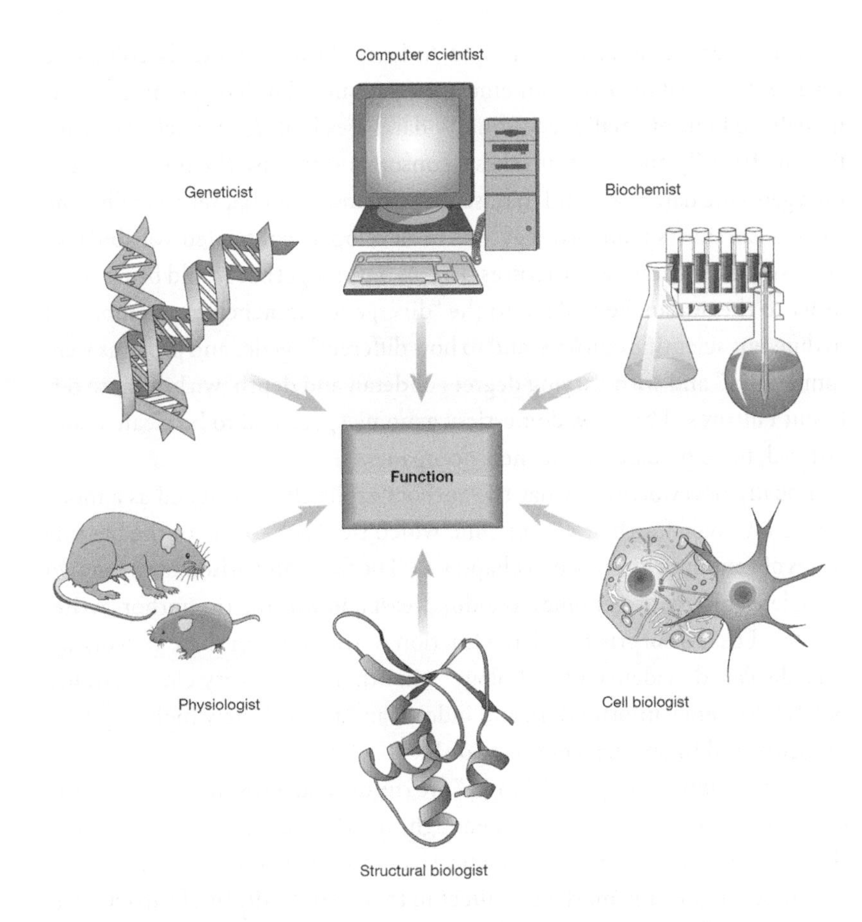

FIGURE 4.1 · Icons of transdisciplinarity. From Vukmirovic and Tilghman, "Exploring Genome Space."

An earlier 2000 publication in *Nature Genetics* had announced the Gene Ontology (GO) Consortium and described its initial work. Vukmirovic and Tilghman did not refer to this in their review article, but it was later resurfaced and analyzed by Leonelli and Ankeny. The article was signed by twenty coauthors, about half of whom were women (extrapolating from name), and half men, suggesting that at least sometimes, in some place, genomics data curation may not be so stamped by gender as other forms of care work are. *And yet,* a gender-balanced list of authors being a relative rarity in scientific publications, such a significant percentage of women might be a sign of cleanup again being shouldered unevenly. The question needs more ethnographic data.

At the very least, as Leonelli and Ankeny show, a markedly collective spirit got carried over into an emergent genomic database community as it coalesced out of smaller community databases built for model organisms like the fruit fly and yeast. It takes a consortium to raise the infrastructure for a genomic database—and to raise so much more, as chapter 7 details. The GO Consortium's main strategy was to develop a "controlled vocabulary" and associated database structures (i.e., an ontology) that would bring some order to the avalanche of data, to the "divergent" nomenclatures employed in different scientific cultures, and to how differently genes and proteins were "annotated," and with varying degrees of detail and depth, within those different cultures. This new, domestic-y genome space had to be cleaned and ordered, to be prepared for its new occupants.

Equally noteworthy is what the authors explicitly mentioned as a motivating rationale for their consortium, which ties back to one of the big surprises of genomics discussed in chapter 2. "The first comparison between two complete eukaryotic genomes (budding yeast and worm)," the authors write, "revealed that a surprisingly large fraction of the genes in these two organisms displayed evidence of orthology"—in other words, very close matches of DNA sequence information, and indeed, an "astonishingly high degree of sequence and functional conservation."[15]

"Surprisingly" and "astonishingly" are the only adverbs that appear in the entire text. I confirmed that easily enough, and the data is there for you to do the same. There are a number of adjectives that offer some embellishment, but these are the only markers of affect in this (admittedly brief) article that otherwise follows scrupulously the rhetorical conventions that have infrastructured scientific prose for centuries.[16] The text as a whole, aside from these short phrases gesturing toward an affective response associated with matching otherwise boring sequence information, concerns entirely unastonishing "matters of concern": how a "structured, precisely defined, common, controlled vocabulary" could render the many different databases that had grown up around particular organisms and other projects "interoperable," why "flat file formats" were the proper way to archive these vocabularies and the annotations employing them, and similar descriptions of or recommendations for standards, protocols, conventions, and tools that would facilitate work and queries across databases, organisms, and fields. The GO Consortium was one of the earliest expressions in genomics of "organic" collective care in the pattern of curation, those difficult-to-discern expenditures of time and effort to assemble and maintain the mundane infrastructures supporting, indirectly and often at a great distance, interest, surprise, and even, on occasion, excitement. If genomicists crave those affects, as I'm hypothesizing, most of them know, at least some of the time, that they will have to care for their data.

Biocurators

As if it were some giant heap of laundry, in genomics circa 2000, every data point, every data model, every database had to be sorted, cleaned, and finally folded neatly and put into a place prepared for it in order for it to exist as Avalanching Big Data. This collective work, like that represented by the GO Consortium, became referred to now as "data curation," a term that we saw just entering genomics' vocabulary. The root term is worth paying attention to, for both its affinities with and divergences from care.[17]

The first use of curate as a verb in a scientific context appears in the pages of *Nature* in 1972, with the news flash "mineralogist to curate meteor collection" (*OED*). In the 1980s and 1990s, curating remained predominantly a practice in the worlds of art, art performances, and museums, even if they included things like meteors; not until the mid-1990s did curation became associated with data, beginning with genomic and biological data. "From Impressionism and Pop Art to phosphorylation sites and interacting atom pairs," in the words of two biocurators writing for *PLOS Computational Biology*, "the realm of curation has been expanded."[18]

Caring is no guarantee of curing. A cure may occur over the course of care, providing a climax, finality, or dramatic transformation into another state, but this can be as much a matter of gift, luck, or chance as it is of causal logics. The need for curation is interminable, and can't be restricted to the teleology of cure. A house well cared for can in no way be said to be a house that is cured, never in need of attentive care again. The same has to be said of genomic and other scientific databases: they require constant care, and are never *curated* once and for all. The constant cleaning, ordering, and other domestic-y chores of curation, whether these are done on houses, museum objects, or DNA sequences, constitute some of the best data we have in support of that craziest of Nietzsche's hypotheses, the one about *something* always returning, *somehow*, eternally. At least these dynamics provide evidence of what Freud, who claimed to have never read Nietzsche, wrote of as the reality of, desire for, and sometimes compulsion for, repetition.

To recapitulate abstractly, then: the repetitious nature of care-as-curation is an effect of the vicissitudes of idiosyncrasies, the fact that any object or event is a complex mixture, a precarious composite that, if left unattended, falls apart or withers away. That may be pertinent to all of life, but at least all of the life sciences and their data-objects, their data-tools, and their data-bases are composites of such composites. Genomics—genomic sciences, practices, technologies, databases—became increasingly composite over the course of the 1990s and beyond, as the surprising complexity of genes and genomes became increasingly evident and increasingly pursued. Databases dedicated to singular substances, functions, and/or organisms came to require not only increasingly careful composition in their own right, but also to require increasingly careful composition as an interlinked ensemble.

By the mid-2000s, biocuration was well on its way to becoming a full-time, paid occupation, especially at the larger databases and consortia. It could then be the subject of "perspectives" or editorial-like articles in scientific journals, as with this less-than-rousingly titled "Biocurators: Contributors to the World of Science," in *PLOS Computational Biology*. This modest introduction of a few articles written by biocurators nevertheless went on to indicate not just the importance but the absolute essentialness of these "contributions" made by the care of curation: "Computational biology is a discipline built upon data (mostly free access), found in biological databases, and knowledge (mostly not free access), found in the literature. So important are these online sources of data that the discipline, and indeed this Journal, simply would not exist without them."[19]

In this 2006 issue, several scientists associated with the Research Collaboratory for Structural Bioinformatics Protein Data Bank (RCSB-PDB) wrote about biocuration in terms of the affective dimensions of "job satisfaction"; the complexity of the mix should be evident:

> There are good benefits to the biocurator job. It provides opportunities for those with analytical skills who want to remain involved in their field but tire of the uncertainties associated with academic life. There is the ability to work from home if necessary, or even from a remote location. It can certainly be fun to see the hottest, latest structures during annotation and interact with the members of the structural biology community. Curating entries for inclusion in a resource that is used by so many people (10,000 individuals per day) also instills pride. For people who need to "get something done" every day, there is a unique sense of accomplishment after processing many entries in a particular day or week. Again, contrast this to the research lab, where days or weeks can go by with much hard work and few or little results.

On the other hand, the process of annotation is essentially the same for every structure. Reviewing all of the information associated with each structure can be tedious and monotonous. The pressure involved in maintaining the data rate can be distressing. Add to this all the duties previously mentioned, and the situation becomes a concentration challenge even for an experienced multitasker.

To combat the monotony of annotation, some annotators engage in other activities such as teaching, management, outreach, programming, or structural bioinformatics research. However, the reality is that because of the high data influx, curation plus another outlet can equate to two full-time jobs. It can be overwhelming, and burnout is a distinct possibility.[20]

Stability and variability, hot fun and tedious monotony, the concentrating and dispersing forces of stress and multiplicity: these are pointers to some of the tensions and, in at least some cases, the double binds associated with the territories of data curation.

"Coordinating the efforts of biological and computer scientists," these authors went on to note, "is a complicated, time-consuming process that requires mutual respect and patience." Not to mention money, as indicated by the list of funders for this project provided only as part of the article's metadata: "The RCSB PDB is supported by funds from the National Science Foundation; the National Institute of General Medical Sciences of the National Institutes of Health; the Office of Science, Department of Energy; the National Library of Medicine; the National Cancer Institute; the National Center for Research Resources; the National Institute of Biomedical Imaging and Bioengineering; and the National Institute of Neurological Disorders and Stroke."[21] (The Protein Databank itself, I would carefully point out, is a multiple, distributed entity that came into existence in 1971, back when the Department of Energy was still the US Atomic Energy Commission; the Research Collaboratory part was added in 1999, and by 2003 the "collaboration among PDB deposition centers was formalized as the worldwide PDB (wwPDB)"—where "worldwide" refers to the RCSB-PDB in the United States, the Molecular Structure Database–European Bioinformatics Institute (MSD–EBI) in Hinxton, United Kingdom, and the Protein Data Bank–Japan (PDBj) in Osaka, Japan.) Continuing to cobble together the financial support for databases and their growth and maintenance is another item to add to the long list of the stress-producing job responsibilities of the biocurator.

But let's stay focused on the "time-consuming" aspects of curation rather than its money-consuming aspects (although the two are obviously interdependent). Remembering that care always involves relationality, let's attend more closely to the difficulties in mediating relations between biologists and computer scientists, and the relational virtues of respect and, importantly, patience that are necessary to the endurance called for by the repetitive demands of curation. The difficulties are numerous and interrelated, and so I too have to plead for the patience necessary for the reader to accompany me in my own curatorial work here.

Two other careful, standards-oriented, infrastructure-dedicated scientists, Cath Brooksbank and John Quackenbush, sum up the selective history I've barely sketched here. By the mid-2000s, "grass-roots efforts" since the late 1990s in and between each of these communities, "driven by small numbers of dedicated individuals but built on broader community consensus," had resulted in a "quiet revolution" (the transformations of care are always quiet ones) in which each of these communities, "often with little or no funding," had developed various standards (gene ontologies and the like) with a "widespread impact." There are already numerous double binds to linger over here, but for now I single out what may be the most fundamental one, concerning the relationality most central to that "widespread impact," the double bind between data and metadata sketched in chapter 3. "Access to primary data alone is of little use," wrote Brooksbank and Quackenbush, "unless those data are presented in a form that is amenable to analysis and interpretation."[22]

This seemingly simple statement harbors a knotty double bind; tracing its twists will shed light on care's absolute necessity. To reiterate it in the terms I previously sketched in chapter 3: data is not data unless it is accompanied by metadata, data about data, data *concerning* data, the data without which we cannot take what data gives. In other words, the foundational sovereignty of data does not exist without the interpretive and analytic efforts that order it and render it useful as a foundation. In still other words, data is "something given" (as the meaning of the word was given to us by Roman culture), but it is only given in a "form" that has been made "amenable" to us. Data depends on us for its independence.[23] Or we could say that datagiving is an always precarious transaction, subject to its own vicissitudes, a matter in constant need of care. We'll return to precariousness in the conclusion of this chapter, but continue now to explicate more double binds of the biocuratorial form of life.

Brooksbank and Quackenbush detail some of the "community-led standards initiatives" that together constituted the grassroots revolution in genomic data, practices, and sciences from the late 1990s to the mid-2000s: the Microarray Gene Expression Data (MGED) Society and the voluntary development of its MIAME (Minimal Information About Microarray Experiments) standards, for which the "involvement of stakeholders from all segments of the microarray community, from manufacturers, through users, to journal publishers," was "crucial" to its success;[24] the Proteomics Standards Initiative; the MIRIAM (Minimum Information Requested in the Annotation of Biochemical Models) standards group; the Reporting Structure for Biological Investigations Working Group (RSBI WG); and of course the GO Consortium, among other curation-centric efforts. From her fieldwork with the GO Consortium, Sabina Leonelli describes some of what she calls "tensions," but I would call double binds, that characterized these kinds of collective efforts. One of them involved the bind between the grassroots efforts driving the development of a standardized vocabulary or ontology and, increasingly, the more top-down authority necessary to make the ontologies and their databases work and be sustainable organizationally, financially, and conceptually:

> The curators' strategy to avoid terminological imperialism consisted in restructuring the division of labour between experimenters and bioinformaticians, advertising themselves as "mediators" between different epistemic communities, and transforming database management into a scientific task requiring appropriate expertise.... This work requires their active judgement and intervention. A remarkable activity carried out by curators is, for instance, the extraction of data from scientific publications, so that they can be classified through bio-ontology terms.... Curators cannot compile data from all available publications on any specific gene product. Often there are too many publications, some of which are more up-to-date than others, some of which are more reliable than others because they come from more reliable sources. Curators need to be able to select one or two publications that can be "representative" for any given gene product. They have to assess what data can be extracted from those papers and whether the language used within them matches the terms and definitions already contained in the bio-ontology. Does the content of a paper warrant the classification of the data therein under a new bio ontology term? Or can the contents of the publication be associated to one or more existing terms? These choices are unavoidable, *yet they are impossible* to regulate through fixed and objective standards. The reasons why the process of extraction requires manual curation are also why it cannot be divorced from subjective judgement. The choices involved are informed by a curator's ability to understand both the original context of publication and the context of bio-ontology classification.[25]

As curatorial collaborations grow, they become more professionalized in their need to standardize the standardization work of creating data about data; as a result, the curators and users, once more or less the same, diverge further and the relations between them grow "uneasy":

> All the curators I interviewed emphasized the difficulties they encountered in eliciting feedback from GO users. Interviews with users corroborated this finding. Users have little time to familiarize themselves with GO, and thus to engage in providing critical feedback: they wish to get data out of databases and go back to experimental work. Given these constraints and interests, users are happy to accept the curators' authority on how data are to be classified and distributed. According to several interviewees, it is the curators' responsibility to provide a system that matches the needs of experimental research and it is not the users' job to tell them how to do this. The result is a vicious circle: to gain users' trust, curators needed to establish themselves successfully as epistemic authorities. Yet, such authority ends up undermining their dialogue with users, with damaging consequences for GO itself. Without user feedback, GO cannot claim to build on "community involvement," nor can it claim to elude the threat of terminological imperialism.[26]

The tensions, vicious circles, or double binds apparent in these more social relations in turn set up another, "higher level" double bind, one that again throws us back to matters of precarity: "The original creators of GO, all of whom are distinguished experimental biologists, have always been aware of the changing and pluralistic nature of biological knowledge and, thus, of the tension between the instability of such knowledge and the stabilizing effect of representing it through a set of well-defined terms."[27]

The difficulty here is captured in the doubled meaning of "ontology": the ontology required to build and operate any biodatabase—the "set of well-defined terms" on which the system runs—maintains an always precarious relationship with an unstable biological ontology, the changing "file cabinet-like" things of the living world like *toll*-like receptors, that continue to be the ever-emergent outcome of experimental work. The ontology is a catachresis for the ontology.

"We Love Spreadsheets, We Hate Spreadsheets"

Thus far I've stressed the increasing senses of surprise during the period 1990–2005 arising from new high-throughput abilities to produce, store, and compare DNA sequence information from different organisms and species, and the increasing need for care (curation) that accompanied that growth. High-throughput technologies would leap still further in the period 2005–2010, prompting another phase of innovation and rededication to curatorial efforts.

But even with the elaborate, painstaking, and dedicated efforts elicited by DNA sequence information, that particular form of information still constituted, as Brooksbank and Quackenbush pointed out, a "relatively simple problem" as far as curation went, since "age, tissue, treatment, and a host of other variables have little or no effect on genomic sequence."[28] Other forms of biological data—and in this context, metadata—that are analyzed in relation to DNA sequence *are* shaped by such variables. As new databases for these forms of biological (meta)data arose and then multiplied, they too required new meta/data about that data (metadata), new metadata and data models and new annotation structures, all of which had to be rendered interoperable with the genomic sequence metadata. (That deliberately confusing sentence condenses multiple ways to write "data," to again highlight the catachrestic nature of signing for this unstable precarious entity.) As later chapters address "geneXenvironment interactions" more directly, I'll further detail how genomicists turned toward developing new research methods and experimental designs, and thus even more ways of caring for the new and more complex data types, more complex databases, and more complex analytic methods. But it's worth emphasizing here that the simplicity of DNA sequence information is indeed a relative simplicity, and the complexity of such relatively simple sequence data is perhaps best appreciated not through the curatorial work of the data archivist, but in the different but no less careful work of the genomic data manufacturer.

Let's look at the case of one of the largest, most generously funded, and most important of the large genome research and sequencing centers, the Broad Institute in Cambridge, Massachusetts, where historian and anthropologist Hallam Stevens conducted fieldwork in the late 2000s. The Broad Institute at the time was housed in two main buildings, sharply differentiated: there was the "shimmering glass and metal" of its Seven Cambridge Center (7CC) building, the largely transparent, open, light-filled public face of the research-centric side of the institute, sporting large video screens in its spacious lobby, and furnishings and design features generally "more appropriate to a California mansion than a laboratory."[29] President and director of the institute Eric Lander—who we encountered in chapter 2 pushing the trope of the Human Genome Project as "infrastructure" and who became a leader of the publicly funded HGP—had his office here.

"For most visitors, scientific and otherwise," Stevens writes, 7CC "is the Broad Institute."[30] It is where, to all appearances, the science of genomics gets done, and although I am certain that there are rich stories of care that could be told about the scientific work in that building (and indeed, Stevens tells a few of them), those are not what you are meant to *see* through all of the glass in all of the open spaces at 7CC; indeed, all that enlightened transparency may be something of a distraction, temporarily blinding us to the more dimly lit and muted shapes of care that must be transpiring in its interior. But we are headed elsewhere.

In contrast, the Broad Institute building called "320 Charles," a few blocks away, looks like just another featureless warehouse in a neighborhood of featureless warehouses, including "an outstation for an electric company and a pipe manufacturer." A thoroughly infrastructural structure, 320 Charles is a "rabbit warren" of "windowless rooms, long corridors that seem to lead to nowhere, unevenly partitioned spaces, unnecessary doorways," and "odd mezzanine levels" all fitted out with "rubber floors, white walls, and metal staircases." Far from the California vibe of 7CC, an ethnographer or some other rare visitor might be struck by "the provisional feel of the place."[31] Maybe it's the manufacturing mundanity of the place that draws care to the foreground: what else is there to see or say?

Stevens describes the goings-on at 320 Charles in exquisite ethnographic depth and detail, and what goes on at the Broad Sequencing Center is indeed a manufacturing process, one calling for the finest strokes of care at every step of its production of DNA sequence data. "The Broad's raw material is samples (it deals with thousands); its products are bases of DNA sequence (it produces billions per year); in the middle, petabytes of data are generated."[32] It's this middle that most interests me in this section, a middle where care is compounded, as manufacturing metadata is generated to guide and improve the generation of DNA sequence data.

The scientific workers and managers there practice Japanese "lean production" techniques and philosophy, imported by MIT management scholars. The emphasis is on skilled workers, "motivated" by and toward a community of knowledge and practice, dependent on coordination, communication, and teamwork. For Stevens, one set of lean production principles stood out as particularly important for this kind of biological data production, one called 5S, from the Japanese terms *seiri, seiton, seiso, seiketsu*, and *shitsuke*. These can be translated as sort, straighten, shine, standardize, and sustain, more names for care as well-suited to the Avalanching Big Data of the biocurator as they are to the domestic care of the home curator, as they are to the Japanese auto worker.[33]

Stevens argues that the Broad's overall success as a preeminent genome research center is due in no small part to the "special kind of worker" who works at 320 Charles, "who is neither an automaton in the Fordist sense, nor a lab bench scientist in the mode of a Pasteur or a Sanger"; instead, "he or she (and both genders are well represented) is what might be called a 'lean biologist,'" who "combines the individuality and creativity of the scientist with the work ethic and team orientation of the production line worker."[34] We might also call him or her a biocurator, someone in the relational in-between who minds genomic data, thoughtfully caring for its creation, use, and sustained existence while carefully thinking about its significance and potential.

In such a light, Meredith, in her lab in the Broad's "rabbit warren" building, is as much a biocurator as someone caring for a new annotation structure as part of the Gene Ontology Consortium. She is also scrupulous and solicitous, as Stevens makes evident, adept at those other patterns of care that I present in later chapters but which are always woven with curation. She monitors the whole production process that turns samples into sequence data, through bar codes stuck to every piece of equipment and every reagent and every sample—"Never remove a bar code from anything!" read signs at numerous places on this data factory floor—and through "tracking sheets" to record each and every step in the process:

> Before when I started here there was no standard tracking sheet. People would do your very common diary-type that molecular biologists do in the lab. . . . Which is great, except when you need to troubleshoot and figure out why this is so good or why this is so bad, you go back . . . and many times people didn't think that was a very important piece of information to keep. . . . There's not much reliability in the data. . . . When you do a standard tracking sheet, you know it's there, and it's always there. You can also enforce, or at least you can see, that it's been done the same way over and over again.[35]

Hence a slogan of the lab, the phrase that I used as a kind of bar code for this section, that allowed me to track evidence of the double bind in enactments of care: "We love spreadsheets, we hate spreadsheets," Meredith quipped to Stevens, and reiterated in the more condensed form: "Our life is spreadsheets."

Our own experiences with love-hate in quotidian life should help us relate to the love-hate relationship that biocurator scientists like Meredith and her coworkers have with their data, and with the bar codes and tracking sheets and all the other paraphernalia of care that *gives* that data, that renders data to be taken as given. It's the boring exciting work through which (meta)data is mended and transformed and given to itself and to us, "over and over again."

Minimum Articulation of Care in Science

"Ha!" was the clinical research biocurator's first response, when I asked her how the data for asthma research had changed over the course of the thirteen years in which she had worked caring for all kinds of data at a major university hospital's research center. I could leave it at that—all my recounting of and accountings for care have to be radically condensed anyway, because care is really incompressible, as information theorists say, and because your time, interest, and patience are finite and precious quantities, so why not take the need for metonymic selectivity to its limit? *You ask me about care? Ha!*

But having relayed a compressed account of the curation of data production in genomics, I want to give at least some minimal sense (call it the Minimum Articulation of Care in Science, or the MACS standard, that every ethnographic database such as this one should employ) of what curation at the other end of scientific throughput looks and feels like. Closer to its most finely finished form in a scientific publication, here is data curation in the research lab that, as we've seen, is structured like a labyrinth: an ever-emergent and shifting space of constraint and possibility through which scientists carefully, haltingly feel their way forward as they build it. Here I relate some bits of a conversation with Lori Hoepner, whom I interviewed in 2013 in connection with her work as part of a research group trying to understand environmental factors in asthma prevalence and severity.[36] The group worked with a large cohort of mothers and children, followed their respiratory health and its interaction with an "environment," from the uterus to the census block, pervaded by cortisol, diesel fumes, pesticides, and other substances ready to activate a body's pattern recognition receptors (PRRs).

"I see the data divided into the questionnaire data and the lab data," Lori said, beginning to unpack her "ha!." "Have you seen the questionnaire?" I had: it's a sixty-five page document, adding many of its own questions to extend other standardized questionnaires for epidemiological studies, and those used in asthma research more broadly, such as the decades-long International Study of Allergy and Asthma in Children (ISAAC) study: age, household income, ethnicity, Medicare recipient, birth weight, wheezing episodes, asthma diagnosis, roaches or rats sighted in the home, pesticides sprayed in the home (professionally or nonprofessionally), exposure to lead paint, and on and on.[37] All frequently updated.

> That one questionnaire is so rich in data, and it's informed every single paper that's come out, and every grant. So it was a good starting point.
>
> *So part of your job was to transfer this from paper to silicon?*
>
> Yeah. And then train people on it during it. There were a little over 800 questionnaires at the beginning, and then it dropped, because actually delivering and getting everything that we needed to be fully enrolled—you have the prenatal questionnaire, plus the prenatal cord blood, plus the mom blood, or both. And if you didn't get the blood at delivery, there was sort of no point in going on any further with the subject. So now, fully enrolled, there are 727, but actively about 500. But the prenatal questionnaire is the basis of it all. . . .
>
> Almost all of the demographic data used in all the different research comes from the prenatal questionnaire. Some of it might come from the medical record, like date of birth. Well, date of birth—a lot of times we have gold standards. We collect the date of birth of the child like a million times, but human error being what it is, you don't want people like, "well, I'll use the date of birth from the first year questionnaire," and somebody else uses the date of birth from the medical record, and somebody else uses the twelve year questionnaire. So early on we created a gold standard for the date of birth, which resides actually outside of the database. That's an example of a derived variable, and everybody's just expected to use that.[38]

Even birthday data are complex and need care. Curating the many different kinds of lab data (dust, allergens, polycyclic aromatic hydrocarbons [PAH], phthalates, organophosphate pesticides, pyrethroid pesticides) that different researchers would then link to the questionnaire data to analyze, say, exposure to diesel fumes over time in a particular neighborhood and its correlation with wheezing episodes or emergency room visits—that was an even more complex care-job:

> Then there's the dust data—teams go in prenatally and then a few times postnatal, at one, three, and five. They vacuumed, and the dust is collected from the filter. That's another one where they have their monitoring log, and they have the lab data, and things have to come together.

And making them come together is your job, or your job in talking with the primary researcher?

A little bit of both, but with the dust data that was a little more simple, because it really was just matching on date and time point, and I didn't have to do anything beyond that. Sometimes I get lucky.

Let me pause to make something clear: even the least experienced, barely attentive ethnographer could tell that Lori loved (her word, not mine) doing this work of data curation. I figured she was very good at it, too, but aside from the evidence of her longevity at the center and her frequent inclusion on publications, that was harder for me to assess. But the interest-excitement was all out in the open:

I'm so glad that somebody's interested! I feel like such a nerd. . . .

I'm interested in nerds! I like nerds! When you got your degree in maternal and child health, did you think you would become an asthma data nerd?

Absolutely not. After I put in my time with a municipal public health organization—which I recommend to everybody—my title was an epidemiologist, but I was actually doing very data related things: I was doing quality assessment, first with TB control, and direct laser therapy, and then in AIDS research, and then in childhood immunizations, auditing clinics to make sure they were in line with President Clinton's immunization action plan. But it's not easy to work at the city, so I applied for this job. When I started here I was a computer programmer, with my main job being the Center, but I had about 24 other projects. When I was interviewed they asked, do you know Scientific Information Retrieval—SIR, which is the name of our databasing program. I said I'd never heard of it. They said it's OK, no one's ever heard of it, and we're going to train you on it. So I took it and ran with it. I kind of fell into a lot of things; I've been here since 1999. The woman who was my quasi-supervisor left a few months after I started, and then, as I say, the Center was my baby. Of all of my projects, it was the one I really became very attuned to, and fell in love with, and got very involved with. My other projects I was very involved with, but not in the same way.

A scientist's care gets enacted in quotidian places people have never heard of. You fall into those places, or they get thrown at you; they are ubiquitous, and you learn to run in them, run with it, with care, mostly on your own but always attuned to others.

There are so many things to do with the data—what I'm calling "care of the data" for the book I'm working on. How did you learn how to do that, and to know you have to do that?

The lab data is what it is—it comes in as a bunch of numbers in an Excel file, when I'm lucky. Back in the early days they were sending .txt files and .dat files, so I'd have to find fancy new ways to read it in. But with lab data it's just a bunch of numbers, and all I need to know from the PI who's in charge of that set of data is what do I do with it? Is there anything I have to do to transform the data before getting it back out for analyses? So with urinary data, for example, you have to correct—it might be done on their end, it might be done on my end—do a specific gravity correction. . . .[39]

There has to be really good communication between me and the PI. So for instance, the PAH air data—talk about intense! I have sat down numerous times with J., and we just sit at the table and go through: all right, now, explain this algorithm to me again. So first I have to do this correction, and then I have to check against the flags that came from the lab, and then I have to do another correction, and then I have to convert the units, and then I have to check against something else. I mean, the syntax itself is . . . long, and sometimes multi-part.

And is it different for PAHs versus PM2.5 . . .

Yeah, well, sometimes—L.'s group is really good. They understand how much work goes into it, and I think that's L.; she's gotten that across. So sometimes a lot of the processing gets done on their end, but still there has to be more that gets done on my end. So for the PM2.5, they actually process most of that for me, and even though it comes over with all the PAH data, I do very little handling of it. But then all the PAH data, because half of it comes from the lab and half of it comes from their log, where it has other information, like the amount of time they collected for, you have to take that into account, and their flow rate, and then you have to take that into account versus the values that come in from the lab, and the flags that come from the lab. That they don't do, but the stuff that comes in from them they will process for me. But that's all air data. Then you compare air data to blood data, urinary data to tooth data—we have data on teeth, yes!

What do you get from teeth?

There are pollutants and chemicals that invade teeth—they're looking for phthalates, and also Tylenol consumption. I don't think there's anything to do with asthma on that.

One of the studies used two city air monitoring stations—are there corrections that have to be made for those? And do you talk to the monitor people?

Yes, no, well . . . it depends.

Do you have to understand the monitor the way you have to understand the specific gravity of urine?

Yes. Well, really I don't process that monitor air data; that data comes in to me processed, with whatever corrections they have to do. It's state monitoring data, and it used to be downloadable from the internet. But what I was doing with that data—they were looking at the data from the monitor versus symptoms from the kids, or also the air monitoring done in the kids' homes. And to deal with that, depending on what's being monitored and which monitor it is—they get it like every three days, or every five days, that the data is put on paper, or collected. So I have to go through and do a lot of temporal recoding, and algorithms, in order to match, depending on what they wanted to do—I did one that was like a week of the year match, a month of the year match, and then recently they wanted like a grace period, like a sort of bookmarking, I think it was nine days—there was one day from the kid, and then there was nine days surrounding that of the air monitoring. [Laughs]

So like the temporal recoding algorithm—is that written down somewhere?

No, I have to dream it up. My boss and I figure it out—he's much more experienced and skilled, and we sit down with a paper and pencil and think, OK, is this the right way to do it? These are the variables we have. So it really is a huge thought process, and I have to consult on that. But I write the code.

M. is the director, and we decided that we need to really start paying attention to what data was being distributed, because there are so many PIs. I'd get an email like, oh, that data that you sent me in September? Can you resend that to me? And I'm like: I've sent you like five data sets since September, which one is it that you really want? So we actually have a very strict method: now we have a form, where they have to write who the collaborators are, the main point of contact, the hypothesis, or hypotheses, the type of analysis that they propose to do. The PI has to sign off on it. And it's all on a spreadsheet where they fill in their predictors, their co-variants, whatever other variables they're looking for, and they have to go through my code book, which is huge—it's html but it's huge, and they have to find the name of the variable, describe the variable, and check off if it's a derived variable, because some variables are processed on my end, even though the raw data is in the database, I don't process everything that's going on, especially maybe from a questionnaire. Sometimes there's syntax that's written on the going-out side of things, so I have to know, even though in theory I'm supposed to have every variable memorized, I need people to point out if they know it's a derived variable. So it's a very structured process: I assign it a tracking number, it goes to the executive committee, if I don't hear any complaints—it's sort of a default that there's not a complaint, but people definitely let you know if they have a problem with it. And then, within a couple of days, depending on my schedule, I shoot them the data.

What kinds of problems would they have with it?

Well, you know, in research, people don't like to have their toes stepped on. So let's say for instance someone inadvertently puts in an asthma variable, but they didn't put the lead asthma investigator on the data request. You really need to have the people who know the data at least named, and get approval from them, in order to move forward, and ask them, "am I using the right variables?" Maybe they have better suggestions, or maybe they don't think there's any reason at all for you to be using those particular variables that are under their purview in your analysis. So there's some back and forth there, and usually it's not a problem. But it's really more when people wade into waters that they're not as familiar with, and they do a little kitchen sink stuff, and don't think about who's in charge of stuff. Sometimes I'll catch it and try to head things off at the pass and say, did you clear this with so-and-so? Because PAHs are her thing, and did she say this was good? It avoids trouble.

Despite its domestic-y aspects, Avalanching Big Data cannot be done in the kitchen sink. Despite or because of its resistance to systematization, it needs to be subjected to strict methods, tracking sheets, prosaic codifications, negotiated social protocols. The quotidian demands of data are only partly sensible in even the most wonderfully prolix of nerd narratives. Monosyllabic "care" hardly hints at the long disseminated chains of meanings and doings for which it stands in.

Data Precarity

Curation returns eternally because data, like pretty much everything else, is precarious.[40] In the sciences, every apparatus, every data point, every data set, every epistemic object, every shred of knowledge, every institution, every community, every movement, every work, every relationship, every body, *everybody* is precarious and requires constant care—anxiety, trouble, tenderness, scrupulousness, attention, solicitude, mourning, friendship, love—to keep it all together. John Dewey placed something like precarity in high prominence in his philosophy of nature, and in his pragmatist philosophy of sciences that worked and thought with and on that nature. Dewey's nature was a world evolving in time:

> A plaintive recognition of our experience as finite and temporal, as full of error, conflict and contradiction, is an acknowledgment of the precarious uncertainty of the objects and connections that constitute nature as it emerges in history. Human experience, however, has also the pathetic longing for truth, beauty and order. There is more than the longing: there are moments of achievement. Experience exhibits its ability to possess harmonious objects. It evinces an ability, within limits, to safeguard the excellent objects and to deflect and reduce the obnoxious ones.
>
> The stablest thing we can speak of is not free from conditions set to it by other things. . . . Every existence is an event.[41]

Curation names the event of a scientist's extended safeguarding, within all manner of limits, of excellent but precarious objects such as data—itself an event, something that is a stubborn ground only because it goes on and is ongoing with our constant involvement.[42] Why name this event "care" rather than something like "tacit skill," or the "work of standardization," or (a personal favorite) "muddling through"?[43] Why this particular catachresis as the improper name for these long tales of data curation, detailing the tasks and activities that otherwise don't share a scale or logic?

First, because care reminds us to keep contradictions, impossibilities, and/or double binds in mind. Curation and data alike are impossible: curation must be done, and it can never *be* done, so it must be done again—and that won't do it, either. Repeat. Love spreadsheets, hate them, love hate love. . . . The incessant repetition of reparative caring, curating data is an effect of its paradoxical doubledness, both self-sufficient and needy, a reassuringly settled ground whose precarity is a constant source of anxiety.

Similarly, Lisa Gitelman and Virginia Jackson titled their edited volume of superb essays on data's need for care *Raw Data Is an Oxymoron*, as "a friendly reminder" that there is a "self-contradiction" embedded in this conventional characterization, and to prompt us to again look more carefully into what's going on in the kitchen. *And yet*: acknowledging as "truism" the oxymoronic, self-contradictory quality of "raw data" nevertheless enacts, I think, a sense of satisfaction in the analyst, whose privileged position outside the kitchen allows them to expose claims about rawness as in error. In a more paranoid rather than friendly style (see chapter 7) the humanists and the social analysts present the culinary practices and styles that scientists have overlooked or are covering up in the name of scientific authority. The *binding force* of the contradiction is almost immediately attenuated in this way, and rather than being asked to imagine what the self-contradiction of raw data *feels* like, readers are told that the essays will allow them "to *see* its self-contradiction, to *see*, as [Geoffrey] Bowker suggests, that data are always already 'cooked' and never entirely 'raw.'"[44] Seeing things this way, anthropologists and other readers of the sciences feel their suspicions confirmed, and feel reassured that their newfound insights into the fine culinary arts of cooking data (albeit in the most admirable senses of that phrase) give them an analytic leg up on scientists still insistent on the inviolate rawness of data. I'd like to reemphasize so as to help reexperience the gnawing anxiety of the force of data's double bind, instead of resolving things too quickly with our revisions, by reminding us of the simultaneous truth in this paradox that is usually left unstated or at the very least understated and underexplored: *the data are always already raw and never entirely cooked.*

I'll reiterate that these sharper data literacies are valuable analytic tools and narratives to have in our repertoires; there are numerous claims about data, and numerous claims about behavioral or social patterns erected on the putatively unshakeable grounds of data, that demand and deserve suspicion. At the same time, any suggestion that data "is always already 'cooked,'" scare quotes notwithstanding, can be seized upon in this cultural moment by individuals and institutions eager to equate cooking, dishonestly, with dishonesty, like "cooking the books." If science is to be "pure," that (specious but common) argument goes, the data has to be "raw." These philosophemes of raw data and pure science—mythemes, actually, little amalgamated chunks of conceptual presumption and cultural condensation carried unconsciously and cathected to conscious thought—need critique and revision, and we need enhanced public literacies about the careful work that goes into composing reliable, "good" data.

But we also need another kind of "friendly reminder" from scientists and their worlds about the resistances that are part of their encounters and work with data, the *something* that remains and imposes demands but often surprises, for which "raw" is a not-bad, ready-to-hand name. No matter how convinced I am that this event-object is more accurately written as (meta)data, I've learned to respect the impropriety of "raw" and I empathize with the anxiety and dissatisfaction provoked when scientists hear the word "cooked."[45]

Remembering that "affects do not always come in tidy packages," as Tomkins said, "curation" gives us an access code to the affective dimensions of the repetitive practices of caring for data: anxiety-generating and satisfying, worrying and interesting, numbing and boring and exciting. Some people, at least, having been raised properly and placed in modestly conducive conditions, *want* to do the work of curation. It might be their day job, it might be just part of some other job, but often they can say that they enjoy curating, and are enthusiastic about it, and even love doing it, even if such "rewards" are construed as "intangible" and even if there is no end to it. Curating data, like brushing one's teeth, is something that has to be done repetitively. Detailed and perhaps prolix analyses of the many ways in which the always partial transformations of precarious data-events into stable-enough data-objects through the work of care are vital to a public culture of science.

Foregrounding care and curation better evokes the eventfulness of genomic data and begins to value the work that makes it happen and keeps it happening. The double bind of data—completely sovereign and thoroughly relational, grounding stability requiring constant stabilization—has to be continually confronted, endured, worked, and played without being fully resolved. But always resolved enough, for long enough. Curation is a not-bad name to designate the structural conditions sketched above—the ongoingness of data/care, the repetitiveness of its demands, the glorious ingloriousness of it, its ineradicable multiple relationalities, the ease with which its quotidianness can be overlooked and undervalued, the difficulty in directing social resources toward it.

I'm grateful to Lori Hoepner (and other friendly scientists) for sharing their great data/curation stories with me to share with readers, and not just because those stories surprise and interest me, not only because something like my *toll!*-like receptors *recognize* something in them and I enjoy the signaling pathways they spark into life and hope readers will, too. More importantly, I think they can begin to help us imagine scientists' complex but overall *enthusiastic* affective attachment to this kind of data work. I wish now I had thought to ask her more directly about those dimensions in our conversation eight years ago; ethnographers of science haven't been in the habit of asking about and exploring affect in the sciences, to better engage and understand both of them and their knotted ties. We need better data on this, better data that will require its own kind of curation-work.[46]

More importantly still, the stories of curatorial care I've presented here narrate what happens at the limits of genomics and the sciences with which it is increasingly interwoven or, rather, at one of those limits. The following chapters trace others. Here in the data infrastructures of the biosciences, data about data (about data about . . .) are continually made to be continually given and continually taken, a temporarily stabilized "harmonious object" achieved through safeguarding care (to re-sound Dewey). In this beyond within, at the internal limit where a grounding foundation must be refounded anew, it's once more into the breach of (meta)data for the bioscientist, who anxiously binds it together in its own continual undoing.

Lori Hoepner is an amazingly careful—thoughtful, diligent, crafty, dedicated—scientist. She's gone on, as she told me she expected to ten years ago, to earn a doctoral degree in public health and to become a noted environmental health scientist, with a research program of her own and a strong public voice on the health effects of PAHs, pesticides, and phthalates, among other awfully persistent precarities. She is rightly proud of her now more than twenty years of "organizational and analytical data management expertise involving complex health assessment and public health research datasets," that infrastructure of curatorial care that underwrites her (and others') ongoing efforts to understand "the intricacies of race/ethnicity, sex and socioeconomics as they pertain to environmental health from a global perspective, as well as from a community-wide perspective."[47]

Publicly visible and laudable as they are, these exceptional achievements are not what matter most to me here. It's that unexceptional, quotidian work of "data management" that I've wanted to foreground, as bioscience's vital infrastructural care that takes place beyond the calculable crankings-out of algorithms and reproducible techniques, and becomes evident only in the winding narratives of curation. From a wider angle, Lori Hoepner is herself infrastructural, one of hundreds of thousands of bioscientists whose careful work has been produced and supported through federal funding funneled through the US National Institutes of Health and other agencies. The biomedical research system developed during the Cold War era has been driven by this funding for "basic research," the kind with apparently no immediate applicability, although now toward the end of that period "translational research" has become more of a thing. What becomes of this system and the bioscientists in it, all so dependent on collective public resources, as the scientific social contract of that era gets renegotiated? The interest and excitement that scientific research engendered in the affective registers of the bioscientist were a central part of that social contract, maybe not as significant a driver as the monies involved, but powerful nonetheless. When these positive affects, elicited even in the midst of the most mundane and repetitive domestic work of data curation, become increasingly overshadowed as constant grant writing becomes the norm; as postdoc positions dry up or extend years longer as the time to a researcher's first NIH grant award lengthens; as more young researchers leave the field or never enter it because it gets harder and harder to have any fun, to get to shout or just quietly mutter "*Toll!*" even occasionally—what happens then? How does a public assign a value to the enjoyment that can be found in doing science, even at its most quotidian, carefully?

5 · scrupulousness

OF EXPERIMENT'S LIMIT

The man of science, in fact, simply uses with scrupulous exactness the methods which we all, habitually and at every moment, use carelessly. · T. H. HUXLEY, *T. H. HUXLEY ON EDUCATION*

From "Race" to Care

As complex conditions came to claim genomicists' attention more and more in the early 2000s, with the HGP declared completed and a first wave of genomics-driven biotech corporations quickly building capital on NASDAQ, "transdisciplinarity" was becoming a thing for anthropologists to do and, reflexivists that they are, a thing for them to analyze. From 2002 to 2004, Evelynn Hammonds, Rayna Rapp, and I represented the disciplines of history and anthropology in the transdisciplinary group Ethics Research Consortium on Smoking, Race, and Genetics organized by public health policy scholar Alexandra Shields, then at Georgetown University. The rest of the core group consisted of legal scholar Patricia King, psychologist Caryn Lerman (whose research on the genetics of smoking prompted her to help organize the transdisciplinary effort), and psychiatric geneticist Patrick Sullivan. A changing cast of additional disciplinary experts—population geneticists, epidemiologists, medical geneticists, health policy analysts—joined us in several one- or two-day-long deliberative sessions. At the broadest level, our group was trying to understand what in the discourse were called "implications"—those zones of genomics where ethical, legal, and social matters had somehow been folded (*-pli-*) into the science, like some convoluted protein structure somehow taking shape from the linear chains of amino acids extruded by a ribozyme. "Implications discourse" is an improvement over a discourse of ethical and social "impacts," suggesting a fast-moving science slamming into an immoveable, or at least slower, legal framework or social structure.

What cultural effects might unfold from somewhere within a genomics infrastructure now capable of fast, cheap, "high-throughput" production of DNA sequence information? What might emerge from genomic sciences and scientists not simply capable of, but interested in and excited by, the capacity to cluster people into multiple new subpopulations according to ever more finely grained categorizations of genetic sameness and difference, in combination with other kinds of carefully curated data avalanches? What could be said, for example, about and to the subpopulation of those who have the DRD2 allele of a dopamine receptor gene, *and* who smoke menthol cigarettes, *and* smoked them for more than five years, *and* who self-identified as "Caucasian" by checking off the appropriate box on the appropriate form, *or* were categorized as "Caucasian" or "European" or just "white" through some system that was so much a part of the infrastructure that few really paid much attention to it? The promised future in those early post-HGP years was (and still is) one of "personalized" or "individualized" (although one only becomes this kind of individual by being a member of these fine subpopulations) pharmacogenomic clinical care. In the case of smoking, the thought or hope or plan (it wasn't always clear) was that a physician would be able to match different quitting therapies (transdermal nicotine patch, nasal spray, bupropion, and so on) to an individual's genomic profile.

At the same time, however, genomics was producing another avalanche of new data about human genetic variation, resulting in new theories and numerous articles in the scientific and medical journals questioning the use of "race," "ethnicity," "race/ethnicity," or "geographic ancestry" as variables in biomedical research, particularly research aimed at elucidating the potential genetic mechanisms involved in something so involved as smoking—a complex event at once cultural, biochemical, sociological, behavioral, physiological, and political-economic. Publications began accumulating in the scientific journals (mostly in psychology and genetics) claiming correlations between members of sets like "African-American" (in the United States, a lot of this research was organized around Black-White dichotomies) and "pack-a-day smoker trying and failing to quit more than two times."

This was a dense tangle of issues, beset by differences, difficulties, and disagreements too numerous to detail here, all undergoing rapid change.[1] Developmental biologist and feminist theorist Anne Fausto-Sterling called it "one of the most confused and confusing debates in modern biology and medical science," in which actors took "a bewildering array of positions . . . on the proper use of terminology and categories both for practicing physicians and for genetic and public health researchers."[2]

The difficulty our group was trying to address, and that many other researchers were also beginning to address, was less one concerning how a technology should or should not be used—what and where its impact would be—and more one of "proper" naming and categorizing: How *should* genomics researchers name the populations in their studies of complex conditions (like smoking, but also schizophrenia, diabetes, asthma, and so on) when the new genomic technologies of the twenty-first century did not align with the socially-constructed-but-naturalized-as-biological racial categories of nineteenth- and twentieth-century anthropology and biomedicine? I did not continue to pursue such questions as part of my own research, but many anthropologists, sociologists, and historians did, for the better part of the decade, and there is now a large literature to consult.[3] A lot of trans- or at least cross- and multidisciplinary scholarly attention became focused on this bewilderment in the mid- to late 2000s, with the eventual result that most careful genomicists avoid the use of "race," agree that the category has no biological significance, and talk and write instead in terms of geographic ancestry. (I return to this claim of a more careful genomics of human difference at the end of this chapter.)

Our transdisciplinary group published an article in a 2005 special issue of the *American Psychologist*, mailed to all members of the American Society of Human Genetics, in which we defined a set of principles regarding the use of categorizations like self-identified ethnicity (pretty much useless), geographic ancestry (still an emergent practice at the time but doable), and limited genotyping through DNA sequencing (the best method then available) as different variables in genomic studies of complex behaviors. We reviewed the entire scientific literature on genetics and smoking published to that point (about sixty articles), documenting the general confusion and disagreement about categories and methods of categorization that characterized this subfield of research; in many articles, there was no explicit discussion of these matters at all, and one could only guess why researchers used the categories they did, and by what criteria they placed their research subjects into one category or another. We also reviewed the long history of changing expert definitions of race, and the frequency with which those unstable, far from natural yet completely naturalized categories were unfolded into harmful outcomes. The article recommended that researchers investigating these kinds of complex behavioral conditions like nicotine addiction should use only partial genotypes, that is, DNA sequence information, and avoid entirely any other kind of categorization scheme.[4]

In making the recommendation we were relying heavily on the expertise and authority of our collaborator and coauthor Patrick Sullivan in such matters. We all liked and trusted Patrick, who patiently explained things like population stratification to our whole group, and then explained it several more times. We all admired and respected his intelligence, professionalism, affability, and just plain decency. As I do with any scientist with whom I engage in fieldwork, although it is usually not as closely collaborative as this was, I began reading as many of his publications as I could, trying to figure out what made him trustworthy in our eyes when we did not always make that same assessment for other genomicists who attended some of our sessions. This became my first lesson in what I then started to call, long before I even thought much about catachresis at all, the "care of the data."

It's probably obvious to many that the phrase "care of the data" owes a large debt to Michel Foucault's "care of the self," the phrase he used to name "a principle of stylization of conduct for . . . giv[ing] . . . existence the most graceful and accomplished form possible." In *The History of Sexuality*, Foucault detailed how, in classical Greek and Roman cultures, the ethical strategies to cultivate "care of the self," particularly in relation to sexuality, could take a number of forms that would allow one to live well within the kind of open, complex, excessive, agonistic field that was sexual conduct:

> There are . . . possible differences in the forms of *elaboration*, of *ethical work (travail éthique)* that one performs on oneself, not only in order to bring one's conduct into compliance with a given rule, but to attempt to transform oneself into the ethical subject of one's behavior. Thus, the sexual austerity can be practiced through a long effort of learning, memorization, and assimilation of a systematic ensemble of precepts, and through a regular checking of conduct aimed at measuring the exactness with which one is applying these rules. It can be practiced in the form of a sudden, all-embracing, and

definitive renunciation of pleasures; it can also be practiced in the form of a relentless combat whose vicissitudes—including momentary setbacks—can have meaning and value in themselves; and it can be practiced through a decipherment as painstaking, continuous, and detailed as possible, of the movements of desire in all its forms, including the most obscure.[5]

The scientist's epistemic virtue I am naming "scrupulousness" most closely resembles the third strategy outlined by Foucault: a painstaking decipherment as detailed as possible, an elaborate working-through of a complex chain of interrelated elements that together make up a genomic experiment or a genomic truth-statement. A scientist's scrupulousness is directed not only toward cultivating, developing, or caring for a particular kind of self, but also toward scrupulous cultivation of objects, methods, and data, whose vicissitudes may lead him or her astray.[6] Care is always an entanglement, and in scrupulousness the self-subject and other-object, and the numerous demands and difficulties each presents, show their deep entanglements most clearly. The most extreme kinds of experimental entanglements (and maybe there's no other kind) will place a scientist in a double-binding situation in which the smallest details must be scrutinized closely—*and yet* some *must* be overlooked.[7]

Etymology again affords an entry point into the double bind that scrupulousness marks with its X. We follow Foucault back a few millennia to a Roman world: we inherit our word from the Latin *scrūpulus*, referring literally to a pebble, transformed figuratively by Cicero into a minuscule but nevertheless palpable cause of uneasiness or anxiety. "A thought or circumstance that troubles the mind or conscience," a scrupule was like a quantum of anxiety, a minimal provocation to care, "a doubt, uncertainty or hesitation in regard to right and wrong, duty, propriety, etc." The OED immediately identifies the double bind as it elaborates that "etc." with an intensifying "esp.": "esp. one which is regarded as over-refined or over-nice, or which causes a person to hesitate where others would be bolder to act." Attending to the smallest trouble or doubt threatens, always and immediately, to be indistinguishable from an obsessive paralysis of over-refinement—what you needn't and shouldn't worry or care about. To be scrupulous is to risk being . . . scrupulous. We might be tempted to say "overly scrupulous," to distinguish the matter of intensity that is in question here, but that obscures the difficulty: how scrupulous is overly scrupulous? How would one know where the line is between virtue and vice—through scrupulous attention, perhaps, but that only displaces the matter—and how would one keep from crossing it? Scrupulous care is a necessary virtue—but don't get carried away! Care, and don't care too much.

The *OED* is unambiguous about this double-binding dynamic of scrupulousness. Sometimes it appears as an imperceptible slide into a more rigid or degraded form of itself: "Troubled with doubts or scruples of conscience; over-nice or meticulous in matters of right and wrong. . . . Prone to hesitate or doubt; distrustful; cautious or meticulous in acting, deciding, etc. . . . Rigidly directed by the dictates of conscience; characterized by a strict and minute regard for what is right." If being scrupulous means being "careful to follow the dictates of conscience," there is a fine line to be followed in this "following": on the one hand, following means "giving heed to the scruples of conscience so as to avoid doing what is wrong"; on the other hand, no "heed" at all needs to be given or taken when one only needs to be "strict in matters of right and wrong." You just need to follow the rule strictly, including the rule, "Your rule-following is always subject to question."

The primary sense of virtue is always closely accompanied, the *OED* suggests, by a secondary sense "developed chiefly in contexts with a negative expressed or implied." Scrupulousness is simultaneously virtue and vice; the scrupulous person may be admirable, but may also be intolerant and insufferable. Like *toll* the gene and "*toll!*" the expression, what a "scrupule" signals is a matter of interpretive performance in response to local conditions, each so complex as to produce divergence and even contradiction.

The everyday habit of scrupulous method identified by Huxley in the epigraph to this chapter has to be amplified and refined into a scientific habitus—and this condition of scientific creativity is prone to its own hesitancies and obsessions that can lead, along some signaling pathways, to a rigid ossification. A careful scientist, then, justly occupied with the minutiae of her experimental system and the analysis of its products, may constantly find herself worrying that she has become preoccupied by what Shakespeare had Hamlet voice as "Some crauen scruple Of thinking too precisely on th'euent."[8] Precision, in thought and act and expression, is surely one of the greatest epistemic virtues a scientist could cultivate—but thinking and acting too precisely? Crauen. And because in this *toll*-ed up state there is no algorithm, no sure means to titrate precisely the dosage of precision necessary to keep scrupulousness from rigidifying into the over-scrupulousness from which it is indistinguishable, scrupulousness becomes subject to its own demand for scrupulous care.

Scrupulousness, then, like other patterns or styles of care, will be as repetitious in nature as curation, an eternal return—well, that's a little hyperbolic. I really should be more careful sometimes. Let's say: scrupulousness for the genomicist manifests as a recurring replay of careful, patient experimentation at the limits of knowledge and practice, demanding painstaking attention to the proper working of every component and every interconnection, and a bit of trial and error tinkering—which, even when successful in expanding those systemic limits, will need doing again, as new unanticipated gaps make themselves known and new unknowns arise, again and again, at the always reimposing limit.[9]

Checks: On Lists and Scientific Behaviors

Let's begin with a closer look at the forms scrupulousness took in psychiatric genetics circa 2000, in the early stages of Tilghman and Vukmirovic's "sea change" of the new "genome space" described in chapter 4, through a close reading of some of Patrick Sullivan's publications. The dominant genomics study design at this time was the "candidate gene approach," wherein researchers take genes already identified as associated with a condition like asthma or schizophrenia (in a smaller study) and try to see whether those genes can be identified in individuals with those conditions drawn from a larger population. (Genome-wide association studies, discussed later in the chapter, which proceed without having these candidate genes, would become the dominant genomics study design only five or so years later.) Sullivan and three coauthors published "Genetic Case-Control Association Studies in Neuropsychiatry" in the *Archives of General Psychiatry* in 2001. As I read and reread this text, I came to understand it less as a scientific report on the current state of psychiatric genetics and more as an ensemble of precepts that allowed a behavioral geneticist to "check" (I'll use Foucault's terms in quotation marks throughout this paragraph) his or her conduct. "Check" here should activate two signaling pathways at once, conveying a sense of consulting an enumerated, detailed checklist of practices, and a sense of arresting or reining in a tendency to which one might succumb. This scientific article advocated the "renunciation of certain pleasures" to be found in the new genome space, an ethos of "austerity" to be learned through long years of practice. Finally, the text guided psychiatric geneticists in their "relentless combat" with the "vicissitudes" of incomplete knowledge, dizzyingly complex conditions, excessive amounts of data, and the inexact rules of precision that one had to rely on while simultaneously striving to move beyond them. By establishing such guidelines for scrupulous care, and by cultivating a scientific self capable of such care, this article counts in my reading as much more a work of bioethics than the professional ethicist's statements that usually get that stamp.

What particular conditions brought the need for scrupulousness to heightened attention in genomics circa 2000? Like the ancient sexual economies analyzed by Foucault in his development of the "care of the self," the experimental systems and data economies of psychiatric genetics in 2001 were characterized by Sullivan et al. as an economy of multiplicity and excess that endangered the propriety of both genomicists and their truths. Genotypes (DNA sequence data) had already become "one of the cheapest biological markers in neuropsychiatry." The costs per base pair of DNA sequencing had been steadily decreasing, and sequencing rates had increased, two promises of the HGP that were kept. Companies like Illumina had begun mass manufacturing "chips" (microarrays containing hundreds and eventually thousands of different DNA sequence segments, representing different genotypes) and selling them to a growing number of labs. Emboldened by "notable successes in complex disorders" such as Alzheimer's and type 1 diabetes (for which a few genes had been shown to be very informative) and "hastened" by the "availability of the primary sequence of the human genome," studies using genetic markers to identify risk factors in neuropsychiatry had become "easy to conduct," and therefore "popular." These kinds of "case-control" studies, forerunners to the genome-wide association studies discussed later, compared the frequency with which certain genetic variations appeared in one group of people diagnosed with a certain complex condition such as schizophrenia—the "cases"—to the frequency with which those same variants appeared in people who had not been categorized as exhibiting the condition, the "controls." In both the case-control study design and in the GWAS design discussed later, positive and thus publishable and thus creditable results hinge on establishing a significant statistical association between having a particular variant of *toll*-like receptor 4 (TLR4), for example (in the case of asthma), and having been diagnosed with asthma. To establish a significant association, a genomicist or, more accurately, the genomics community must first establish what will count as significant—a complex process, as we'll see. While publishable positive results would often be reported in the media in a misleadingly simplified way, like "gene for asthma [or schizophrenia] found," many if not most of these associations deemed to be statistically significant by one lab would often fail to be replicated or validated, and sometimes were outright contradicted or disproven by other studies in other labs. This pattern caused controversy and consternation within the genomics community and in a broader public. It's a pattern produced by insufficient care, and the genomics community had to develop new ways of minding it.[10]

Even if you do not want to go so far as to read this cheap, easy, fast, and popular genomic data economy as an economy of profligacy and wantonness, like the sexual economies that drew Foucault's attention, it seems pretty clear that these psychiatric geneticists were calling for the cultivation of more disciplined, scrupulous relationships within and to this economy of genomic multiplicity, openness, and excess. If the scientist-subjects of genomics did not better discipline or care for themselves and their data practices, Sullivan and coauthors cautioned, the current state of "considerable confusion" would persist, as case-control association studies would continue to be published in journals with increasing frequency. And more often than not those published genomic truths would increasingly remain unvalidated or even wind up being retracted, while the whole effort would continue to be "controversial," and indeed, something of a scandal.

"In an era of increasingly high-throughput genotyping," the authors went on to note, the possibility of obtaining false-positive results simply due to chance in such studies was greatly increased. How the psychiatric geneticist could avoid false positives and other missteps at this very fundamental level (that's why they're called type 1 errors)—basic missteps like choosing "inappropriate control groups"—were the first matter of care to be listed by the authors in their long checklist of behaviors and practices to be monitored closely. Because "many genotyping methods require subjective judgments," and because "genotyping is a complex undertaking and a multitude of technical factors can lead to erroneous data," a psychiatric geneticist was obligated to exercise "considerable care" when producing data. Better minding genomics would require adopting practices such as genotyping blind with respect to case or control status, and processing cases and controls at roughly the same time, rather than in separate batches months apart.[11] "Scrupulousness" is my catachrestic translation of these quotidian, infrastructural, and rather unexciting if not downright boring practices of "considerable care."

Diagnosing patients in the era of Avalanching Data also demanded heightened degrees of scrupulousness, the authors insisted. For complex conditions, "investigators should recognize the limitations of psychiatric nosologies," and respect that "our knowledge of the etiology of these disorders is imprecise." Such acknowledgments seemed simple and obvious, but without these checks "most statistically significant candidate genes will be false positives"—that is, more type 1 errors. Another ethical injunction in this extensive discourse of *shoulds* read, "Investigators should adopt a life-long perspective," meaning that their case definitions "should allow for the lifelong (rather than current) presence of a disorder and, ideally, for its life-long absence in controls."[12]

There is much more to this rich work that is simultaneously a work of science and a work of ethics, a meshing of reason and care, but published in a journal categorized and recognized exclusively as "scientific." Genre-bending is not tolerated in the highly stylized scientific article, supposedly, but articles like this make it clear that matters of science can't help but be merged with matters of care. But we've arrived at our own nosological limit here: What do we name as "science," and what do we name with "care," so that these appear on the written page as two distinct things with their own identities, that can then be "merged"? Is there another way to write science and care, and to write about science and care, here at the limit separating them and where they join to form a paradox: the same and not the same, different and not different?

There might be, but it's not going to be easy and it's going to take time.

First thing to note: all of the concerns, all the matters of science requiring care raised by Sullivan and his colleagues would echo and re-echo over the next decade and beyond, as we'll see later in this chapter and in later chapters. These difficulties are structural features of the system, patterns of (infra)structural effects that indicate the limits that must be there, an impossibility, paradox, or double bind that can't be resolved but must be endured.

Second, note the fairly broad terms employed here, not because they are somehow inadequate or in need of finer-grained specificity and greater rigor, to be refined either by these same authors at a later date or by the ethnographer writing after the fact, but because they gesture toward the essential double binds of care. How much care is "considerable care," and how exactly does one "exercise" it? The answers "a lot . . . and not too much" and "carefully" don't seem very satisfactory, but they are in fact the most suited to matters of care. How should one "recognize the limitations" of nosological categories, the effectiveness of which, after all, hinges on their ability to establish a limit and exclude? How do you establish a limit, *and yet* at the same time keep it under question? The impossibility of an easy, succinct, codifiable answer to such essential questions is not an excuse or cause for nihilism, but only makes it necessary to raise and respond to them again and again.

Scrupulousness isn't the proper name of something or some quality, with clear limits that can be traced and filled: "Be 15 percent more careful, and you will be sufficiently and successfully scrupulous!" "Be scrupulous!" is a catachresis for something with no proper name or referent; it names its own impossibility as the condition of its creative production: "You must be considerably more careful, and then reconsider, since that will not have been careful enough. And maybe too careful."

We could also say that "carefulness" is an incessant demand, a necessary supplement to the bioinformatics algorithms and the statistical methods and the calibrated machines, *and yet* care itself has no algorithm, no reproducible method, no calibration setting that anticipates or obviates the need for its next necessary supplement. Minding genomics by being more scrupulously careful makes for better bioinformatics algorithms, and more refined equipment and methods—in short, a better science, but still with its necessary limits, no matter how expanded or well-defined, still imposing an unspecified but urgent need for care. Care and science, like writing, embody the "exorbitant" logic of the supplement analyzed by Jacques Derrida: the sciences need the supplement of an external care, *and yet*, at the same time, the sciences are themselves already thoroughly constituted by or as care.[13] Or, *There's no genomics without care, so the demand "Be more careful!" is only another way of saying "Do more genomics!"* Or finally, to re-sound the seemingly far-from-exorbitant Huxley: *Genomics, "in fact, simply uses with scrupulous exactness the methods which we all, habitually and at every moment, use carelessly."* We're all at the limits of language here, solidifying nebulous entities into stalwart words, drawing sharp borders to separate them only to try to join them back together again, find a relationality that has been defined away.

The moment of genomics circa 2000 I am analyzing here is one beginning to be plagued by the problem of what other prominent biomedical researchers and statisticians would name, over a decade later, as a problem of "excess significance."[14] The techno-political economy of genomics had become so fast, cheap, and out of control by the end of the Human Genome Project that truths—"significance," the form that truth takes in a post-Popperian era, when truth is supposed to be always and only ever-on-the-verge of being falsified—had become so easy that even those of little epistemic virtue, the scientifically careless, could do them and be suitably rewarded.

Read scrupulousness in light of this context of "excess significance." The "easiness" of genomics, its high-throughput capacity to produce the finest differentiations—an A substituted for C in an otherwise identical chain of thousands of nucleotides; to reiterate such a differentiation a thousand times, in thousands of people; and then to produce meaning out of those differences by coupling them to a growing list of multiple differences, generated by multiple -omics technologies whose data is archived in multiple linked databases for protein structure, biochemical pathways, tissue-specific expression, clinical histories, environmental conditions—could all this multiplexing and avalanching *not* heighten both excitement and anxiety, enjoyment and worry, elation and consternation in its practitioners? These doublings provoke a crisis in the scientific culture, sensed first by some few of its inhabitants, who begin to articulate a need for limitation, and the means by which genomics' excesses might be trimmed through the cultivation of a reformed scientific subject capable of more scrupulous attention to the many ways one could be misled in this new territory of excess—too much (and never enough) cheap technology, too much (and never enough) funding, too much (and never enough) opportunity and demand to put out "significant" publications, and most of all, too much (and never enough) data.

Learning to Read, Scrupulously

This is a good time to pause and remind ourselves that "care" is not only a catachresis, but a bundle of catachreses, with each of its modes nesting within others in a complex pattern of reinforcement and interference. Here in particular I should stress that scrupulous attention to the vanishingly small also characterizes the work of curation, even as it shades over into the solicitous embrace of the impossibly large that is the focus of chapter 6. The differences between these should only hold for a while, enough time to do some analytic work, and then be allowed again to become less distinct. It would be (more) unbearable to be overly scrupulous in our differentiation of care's patterns.

And yet—even if curation and scrupulousness share an attentiveness to detail, the object of relation, or where attention is directed, differs: a biocurator tends to precarious objects (including data objects), which, left to their own devices, will fall, over time, into disorder or disarray; curatorial care works, repetitively, to sustain things as they already are, in their present forms. The life scientist practicing scrupulousness, however, is attending to each of her many devices—her equipment (old and familiar, and new and difficult), her protocols, her methods, her reagents, her data, and her questions—and attending to making them work together as an experimental system, not to sustain the continuity of data/objects but to produce new ones in new arrangements, and to some degree unanticipated—a *toll* gene or a *toll*-like receptor, for example, or a robust association between a *toll*-like receptor and a set of people who have difficulty breathing. A successful experiment is a rupture, even if a modest one, in the existing order of knowledge objects and truth telling, but a carefully controlled or scrupulously produced rupture that leverages the whole game up a notch, establishing a new present and order out of a future beyond order. Curation and scrupulousness have different temporal orientations as well: curation sustains a past to endure into the ongoing present, a repetitive return of the same; scrupulousness realizes a new present with the tinkered fragments of a virtual future, a repetitive return of difference at the limit.

Scrupulousness, like the other patterned modes of care, is something that has to be learned, practiced, cultivated in a collective space, a pedagogical place that cultivates tacit and embodied knowledge along with book knowledge. Learning to be a scrupulous life scientist is a process difficult to observe; we have a few firsthand, autoethnographic accounts and analyses, like those of Ludwik Fleck, Francois Jacob, and physical chemist Michael Polanyi. But we could do with more ethnographers, historians, and other scientists turning their attention to pedagogical practices in the sciences: the teaching and learning of not only how to be a scientist, but how to be a careful one.[15]

Scientific publications, read carefully and closely, can still give us clues to understanding the changing standards, technologies, research patterns, and practices of scrupulousness in genomics, its persistent limits and failures, and the subsequent persistent efforts to learn from those limits and failures and to improve—to improve one's practices, improve one's thinking about complexity, improve one's scientific self, and improve a community of practice. And just to remind you of the impossible contradictions involved here: the scientist-subject who improves in this way is one who is, peculiarly, learning to fail better.[16]

As one example of this kind of minding of genomics by learning to fail more scrupulously, the next section presents some episodes in genomics research into one of the complex conditions that our transdisciplinary group of the early 2000s first came to focus on, asthma.

An Extremely Selective History of Asthma Genomics, from Candidate Genes to Genome-Wide Association Studies (GWAS)

Like obesity (another complex condition that got early attention in genomics), asthma in the 1990s seemed to hold just the right degree of complexity, to have a biomedical knowledge base sufficiently broad and deep to offer good starting points for genomicists, and was a rapidly growing and serious enough problem in populations around the globe that the need for and potential of genomic research seemed clear. Or at least promising, in the complex senses and enactments of that term figured out in *Promising Genomics*: asthma promised to be a tractable genomic research problem, while promising some good return on the research investment, in both knowledge and capital. Asthma, in other words, looked to be fairly low-hanging genomic fruit for which there might be a market.

Some of the earliest efforts to find "asthma genes" (not a catachresis but a misleading misnomer) were conducted by or in collaboration with two of the earliest genomics corporations, Sequana Therapeutics and Millennium Pharmaceuticals. Their strategy centered on populations defined as "isolated": families in Anqing, China, in the case of Millennium Pharmaceuticals, and the island of Tristan da Cunha for Sequana.[17] Many genomicists at the time thought that these populations would exhibit "founder effects" that would ease the identification of what was projected—hoped, really—to be a few genes associated with being diagnosed with asthma: a "founding" asthmatic ancestor far, far back in the lineages of families recruited into the studies, setting the patterns of inheritance through transmitting his or her genes down the chains with minimal admixture from other populations. These would at least provide the "candidate genes" that could then be searched for in different, larger populations of people diagnosed with asthma. The supposed isolation of these initial populations (sometimes referred to as "inbred") also allowed researchers to control for (or at least think they controlled for) "the environment." Tristan da Cunha, for example, was represented, by researchers with Sequana, as

> the most isolated inhabited location on earth. . . . All of the island's inhabitants can trace their origins to a few ancestors who settled the island in the early 19th century. Two previous studies have suggested that at least 30% of the inhabitants have asthma. In addition to the inbred nature of this population, important environmental factors make this an excellent population study. Virtually all of the current inhabitants were born on the island. . . . All inhabitants live in a relatively small settlement of 90 homes clustered at one end of a narrow plateau. . . . These homes are generally made of the same materials, and the islanders share common occupations (farming, sheep shearing, fishing) and have essentially the same diet (potatoes, fish, mutton, beef, and poultry). Furthermore, there is virtually no air pollution on the island because of the lack of industry and the strong prevailing winds. Thus, even though environmental factors may play a role in the development of asthma in these individuals, the differences in the environment alone are unlikely to explain the development of asthma in some subjects and not in others, because the environmental background is so constant.[18]

These kinds of statements were fairly typical of most genomics studies at the time: the environment was excluded from analysis, and well-characterized families increased the likelihood that probing across the genome with a set of markers (only a few hundred were available at the time) would fish out genes of potential import—although not "genes" so much as regions of a chromosome that a marker mapped to, but that could harbor many genes or regulatory elements. Two interrelated factors crucial to these study designs were designations of statistical significance and study sample size. Over the course of this chapter we will see genomicists become more scrupulous about these factors in particular.

There were only 289 people living on Tristan da Cunha in the 1990s studies, and even though the Sequana researchers worked up 282 of them, this was still a relatively small study size. Although the researchers supplemented the data from Tristan da Cunha families with families from a few other locations, primarily families recruited from Toronto, Canada, sample size remained a serious limitation. Using only about 250 markers (another limitation) to probe these genomes (the Sequana researchers also did not make any details on these markers public), they reported finding "13 possible linkages ($p < 0.05$) to the phenotypes of asthma or bronchial hyperresponsiveness" in the Tristan da Cunha families; they managed to replicate just one of those linkages in the Toronto families. The chromosomal region around that marker contained "approximately 16 genes, 13 of which are novel."[19] The limits of the study and the paucity of the published details did not prevent Sequana from issuing a press release that claimed, rather unscrupulously, they had "discovered a gene for asthma"; it was an early example of what a reporter for *Science* called "genetics by press release."[20]

But even the more scrupulous researchers at the time were subject to similar limitations. Working with 533 families and probing their genomes with 422 markers, the Millennium-affiliated researchers published results in 2000 showing "significant linkage" between two of their markers (details provided) and "airway responsiveness," and "suggestive evidence" of linkage between six other markers and particular measures of asthma like "forced expiratory volume" (FEV). Opting to study linkage with particular measures like FEV rather than with "asthma" is one indicator of their relatively higher degree of scrupulousness:

> The lack of a standardized definition of the asthma phenotype makes this phenotype sensitive to misclassification and relatively unreliable for genetic studies. This is especially true in the sample that we studied, since the asthma statuses of the subjects were initially defined by local village physicians whose criteria for diagnosis of asthma may have substantially differed from one another's. We have studied several algorithms for classification of asthma phenotypes in the Chinese population, on the basis of a combination of respiratory symptoms, increased airway responsiveness, and a physician's diagnosis of asthma. However, it is not clear which algorithm is optimal for linkage analysis of our study sample. In comparison, the quantitative intermediate phenotypes in each subject were measured by the same team or laboratory and by the same procedures and instruments; hence, they are more reliable for linkage study.[21]

These researchers recognized asthma's nosological limitations, to use the terms of Sullivan and coauthors above. The authors of this far more detailed paper and far more careful study noted in the end, however, that "inconsistency... exists between our results and those previously reported," and that even when they found "'consistent' linkages," these, "although very encouraging, should not be regarded as positive replications, since comparison was made among many genome scans with different asthma-related traits and since, more important, few linkages reached the suggested genomewide significance level."[22]

One of those previously reported studies came out of the Collaborative Study on the Genetics of Asthma, a network of researchers at various US institutions (including Johns Hopkins, Howard University, University of Chicago, the Marshfield Clinic, and the National Institutes of Health) whose collective effort marks an early instance of a trend toward larger and larger consortia in genomics—and larger and larger study sizes—discussed more in chapters 6 and 7. This early CSGA work on asthma genomics shows how even the most scrupulous researchers still had to operate within the quotidian infrastructure of the time—quotidian infrastructure that included not only DNA sequencing technology but also race-based definitions of populations, pretty much standard at the time but now long outdated.

They characterized their 1997 letter in *Nature Genetics*, "A genome-wide search for asthma susceptibility loci in ethnically diverse populations," as "the first report of a genome-wide screen for asthma genes in ethnically diverse populations." But their language was inconsistent, and "ethnically diverse" morphed into "racial groups" as the multiple authors detailed their ambitious attempt "to identify all important loci that contribute to the development of asthma" by analyzing the genomes of "individuals of three racial groups (African-Americans, Caucasians, Hispanics)." These researchers—among the most careful you could find, by my estimation—do not describe how or why individuals were assigned to these categories, categories that in a matter of years would be widely (but not universally) regarded as antiquated and, essentially, no longer scientific. In other passages in the article they shift from the language of "racial groups" back to ethnicity: "Our data provided suggestive evidence for novel asthma susceptibility genes in each of three ethnic groups." Pointing to "the difficulty of replicating any specific susceptibility locus in different samples," they announced their identification of six markers that exhibited "some evidence," "very modest evidence," or "suggestive" evidence of linkage to asthma.[23]

I'd like to emphasize: the CSGA researchers deserve credit for even attempting to study the genomics of asthma in diverse populations, even if that diversity was defined ambiguously or inconsistently. In these early years of genome-wide analyses, slipping between ethnicity and race was not so much a failure on their part as it was a symptom of widespread practices and languages of categorization that needed more collective care. It was then and remains still a difficult territory, and even the most scrupulous care was fraught with complexity. For example, the CSGA researchers scrupulously corrected their measures of "forced expiratory volume at 1 second" to adjust for differences across their populations: "Baseline spirometry was performed according to American Thoracic Society (ATS) criteria. FEV1 was increased 12% in African-Americans to adjust for racial differences." Here the slide back into "racial" becomes even more complicated, confounding what should count as scrupulousness in the process. The spirometer used to study asthma for the better part of a century, biologist and historian Lundy Braun has shown, is itself a racialized instrument, "standardized" in a deeply racial history of pulmonary medicine in which the lungs of "African-Americans" were supposed to be different than those of "Caucasians," forcing generation after generation of researchers to apply the American Thoracic Society's badly conceived "correction," believing it to be a true and accurate measure of care.[24] Being scrupulous about being scrupulous can be tricky.

But the alternative to such studies of "ethnically diverse populations" was worse: recruiting only those of European ancestry into cohorts for genomic analysis. And as we will see, this became the norm in genomics. In an article published a year later, the CSGA, now joined by a number of other prominent genomicists, reported on their parallel study of asthma in 361 individuals from the Hutterite community of South Dakota, with another 291 from different Hutterite communities for a replication group. The probability of some "founder effect," the well-characterized and cooperative families, and the presumed irrelevance of environmental factors ("the Hutterite communal lifestyle ensures that all members are exposed to a relatively uniform environment") were all thought to increase the likelihood of eventual success, as in the Sequana and Millennium studies.[25]

In these genomics studies, and in all since, "likelihood" is the key term. The statistical measure of significance in the association between a gene, marker, or region of a chromosome and the complex phenotype "asthma" is a difficult one to make, but is also probably the most important measure of scrupulousness in genomics. Scrupulous statistical analysis, it's important to remember, is the interpretive product of a community of practice and care—statisticians and, as a result of the Human Genome Project, bioinformaticians—in its interactions with the multiple complexities of a biocultural real: the complexities of genomes themselves, the complexities of genomic differences in diverse populations, and the complexities of genomic experimentation and analysis, to name only three. Let's read, briefly but closely, a frequently cited publication on statistics from this time when the genomics of complex conditions was first getting under way; we'll learn more about the double binds of scrupulousness in the process. Later we'll see these interpretive statistical practices become more stringent, as genomic technologies and study designs changed, and as at least part of the genomics community became even more scrupulously careful.

To begin: none of the dozen or so markers announced as linked to asthma in these first genome scans of the mid- to late 1990s really held up. There were suggestions of significant linkage and a number of hopeful signs, but of the many genomic linkages to asthma reported in these publications before 2000, none persisted as confirmed, replicable results. There were numerous reasons for this, some of which are considered in more detail below. But the central difficulty lay in the statistical analysis. Eric Lander (whom we met in chapter 2 with his "infrastructure" trope for the HGP, and who was part of the CSGA study discussed above) was a mathematician before he was a genomicist; in 1995 he and his colleague at the Whitehead Institute, Leonid Kruglyak, published a short overview of the problems in *Nature Genetics*:

> Genetic mapping of any trait—simple or complex—boils down to finding those chromosomal regions that tend to be shared among affected relatives and tend to differ between affecteds and unaffecteds. Conceptually, this amounts to a three-step recipe: scan the entire genome with a dense collection of genetic markers; calculate an appropriate linkage statistic S(x) at each position x along the genome; and identify the regions in which the statistic S shows a significant deviation from what would be expected under independent assortment.
>
> Yet, these deceptively simple instructions conceal a thorny question.[26]

Before venturing into the thorny complexity in the midst of rosy simplicity (and how I wish they had included an "and," to that "Yet"), we should note their passing reference to the affective dimension of the situation: "Biologists often greet statistical issues with glazed-eyed indifference." Genomics researchers tended (tendencies galore in this territory) to be indifferent to statistics; the subject and practice were so uninteresting and boring as to register physiologically, at least in the writerly imaginations of Lander and Kruglyak.[27] Indeed, long after 1995, the numerous genomics presentations I attended at conferences would almost inevitably reach the point where, upon being asked some question about the statistical analysis, the lead genomicist speaking would dissemble and demur and, with a wry smile or a shrug of the shoulders, or a dropping of their once-glazed eyes in a different physiological display Silvan Tomkins would immediately recognize as mild shame, say, *I don't really understand the biostatistics part, let me ask so-and-so, our bioinformatician, to come up and answer that question . . .*

Lander and Kruglyak summarized the thorny question for genomicists disinterested in statistics like so: "These deceptively simple instructions conceal a thorny question: since the statistic S(x) fluctuates substantially just by chance across an entire genome scan, what constitutes a 'significant' deviation? What standard should be required for declaring linkage?"[28]

Note that what is being regulated here, the target of this ethico-technical meditation on "significance," is what genomicists are permitted to declare: What can they say about significant linkage in a publication—that is, what can they say in public? Here the authors *almost* articulate the double bind of scrupulousness, but miss it by going through Greek mythology: "To reach our goal, geneticists must chart a prudent course between Scylla and Charybdis."[29]

Prudence is an admirable virtue, but it passes too quickly through the trouble without fully registering the double bind of impossibility that this episode from *The Odyssey*, when read more carefully, in truth relates: there is no "prudent" course "between" Scylla and Charybdis. The complicated Odysseus, counseled by Circe, steers his ship not between, but as far as possible from the whirling void of Charybdis and into direct encounter with a Scylla as multiple as any complex condition, whose six heads snatch six of Odysseus's finest fighters, just as Circe said would happen but Odysseus chose not to disclose to his crew—"still screaming, still reaching out to me in their death throes. That was the most heartrending sight I saw in all the time I suffered on the sea."[30]

So although Lander and Kruglyak depend on other similarly reassuring tropes, like "striking the right balance," that suggest a painless middle way charted by reason, their article runs again and again into the double-binding demands of scrupulousness for which the "deceptively simple instructions" of reason, logic, and statistics just don't cut it. They lay out two incompatible positions on standards of significance:

> Adopting too lax a standard guarantees a burgeoning literature of false positive linkage claims, each with its own gene symbol (*ASTH56, ASTH57, . . .*). Scientific disciplines erode their credibility when a substantial proportion of claims cannot be replicated—even more so when the claims reach not only the professional journals but also the evening news. Psychiatric gene-

tics provides a cautionary tale, in which a spate of non-replicable findings in the mid-1980s undermined support for such studies. It is thus essential that there be a sufficiently stringent standard that linkage is claimed only when there is a high likelihood that the assertion will stand the test of time.[31]

Here we find yet another impossibility that scrupulousness is meant to address, and is destined to fail: even if it is not impossible, it is obviously incredibly difficult to decide on, declare, and enforce standards of significance for the present, reflective of and designed toward current technologies and techniques—and be confident that the standard will also "stand the test of time," to hold in an unknown future. Yet they also refer to the very recent past of psychiatric genetics (Patrick Sullivan's purview), most of which in *its* near future of *this* article's present, had already failed the test of time. (Note in passing that the signs chosen by Lander and Kruglyak to represent the risk of false linkage announcements—"*(ASTH56, ASTH57, . . .)*"—cast asthma as the index of insufficient scrupulousness.)

Leaving aside the question of just how long a "test of time" might take, let's consider the other peril Lander and Kruglyak project, looking for that prudent middle way:

> On the other hand, adopting too high a hurdle for reporting results runs the risk that the nascent field will be stillborn. Initial genetic analyses may fall short of the strict threshold for statistical significance, but may nonetheless point to important regions deserving intensive investigation. Without channels by which investigators can report such tentative hints of linkage, the discovery of disease genes may be delayed in an overzealous attempt to avoid all error.[32]

In other words, their injunction to their fellow genomicists can be read as, have some zeal, but not too much zeal. Or again, to highlight the conflicting orders of the double bind even further: be scrupulous in your avoidance of error—and don't be overly scrupulous.

That's not a statement of how to make a difficult but prudent choice between two alternatives, or how to chart a safe path between them. That's a statement of an impossible double bind that can't be resolved by algorithm, statistical or ethical, and must instead be, somehow, endured and cared for. And *something* is going to have to be sacrificed. That actually sounds like a "test of time" to me, although I think a very different one than that implied by the authors.

And indeed, their "Conclusion" is one brief paragraph that ends, not with a statement concerning the issues of statistical analysis that dominate the previous five pages, but with a statement concerning . . . statements: "By adopting clear rules for communication, human geneticists will be well prepared for the avalanche of information about to descend."[33]

The real genomics data deluge is yet to come, they conclude, which will only exacerbate an already difficult (impossible, in my reading) scientific and statistical situation, and their most concrete, explicit recommendations—their "proposed standards"—are to clarify the language used for different levels of confidence and significance, to dust off and brush up the language practices of the good old days of the genetics of the "simplest" single gene conditions: "suggestive linkage," "significant linkage," "highly significant linkage," and "confirmed linkage," with publication in similarly hierarchically ranked venues from newsletters to "specialty journals" to, presumably, *Nature Genetics*.

And I would agree, even if for different reasons and under a different reading, that these are the only "standards" that can be proposed in a double-binding situation, in a true Scylla-and-Charybdis moment that has to be navigated not only once, but repeatedly—not so much a test of time, perhaps, as a test of the eternal return of the limits of experimental science. The whole point of the article, in my reading, is not to arrive at final assured standards, like Odysseus arriving home to Ithaca; the article is only (in the authors' own words) an "accessible treatment" meant to be read by the scrupulous scientist-subject, placing her in an impossible situation and putting her through her statistical paces, forcing her to engage and endure them without ever resolving them. That's why it is peppered throughout with phrases like "some backsliding may be countenanced," "nettlesome problems remain," and "a careful consideration of all these factors is beyond the scope of this commentary." (Note again the essential *and yet* meaningless insertion of "careful" in that last one, gesturing to a "beyond" that always remains.) That's why the statistical techniques are always embedded within an open-ended discourse of doubled difficulties that can read like head-scratching Zen koans:

> Failure to replicate does not necessarily disprove a hypothesis.
>
> When several replication studies are carried out, the results may conflict.
>
> While individual investigators cannot do anything about this problem, it offers an additional rationale for conservative standards.
>
> Formal procedures are useful for standardizing the general acceptance of linkage claims. Still, gene hunters should not be inhibited from pursuing all hints and hunches.

This prescription is too conservative in the case of closely related models (such as correlated phenotypes), but there is no general guidance for how to proceed other than simulation. Even simulation poses a challenge.[34]

And, as a final selection that ends with perhaps the most iconic and persistent of all ethic-of-care questions:

Notwithstanding our desire to avoid spurious linkages, we must always remember that regions that fall short of statistical significance may nonetheless be correct. Unfortunately, there is no way to distinguish between small peaks that represent weak true positives and peaks of the same height arising from random fluctuations, assuming that all inheritance information has been extracted. It would be irresponsible to consign such potentially valuable hints to the dustbin of laboratory history. What then is to be done?[35]

The best and most succinct answer to that question would also be, in my estimation, the most appropriately and resolutely noninformative: be careful.

Even small, specific instances of what counted as "being careful" in the statistical analysis of increasingly larger and more complex genomic data sets involved their own double binds. A technique called a Bonferroni correction came to be a routine way to better care for p-values, the near-universal standard for measuring and expressing the statistical significance of a scientific study's outcome. There are many shortcomings to p-values, too numerous and demanding to go into here, as a measure of the significance or, more colloquially, the probability that a study's outcome is regarded as true—or at least, to be a bit more careful about it, a less than a 5 percent likelihood ($p < 0.05$) of being false and a product of chance. The biggest problem with p-values, when it came to genomics, is that they are a single measure of a single relationship, and thus a poor measure of the multiple relationships researchers were trying to discern in the increasingly large data sets of genomics. The University of Chicago's Carol Ober, one of the early asthma genomics researchers who remains a leading figure in asthma genomics decades later, explained in a 2003 article how "false-positive results are more likely if multiple comparisons are made, either with multiple polymorphisms in the same gene, polymorphisms in multiple genes, or with multiple phenotypes. In these cases, the 5% false positive rate expected when the null hypothesis (of no association) is rejected at $p < 0.05$ no longer applies because there is a 5% type I error rate expected with each independent comparison."[36]

So if you were a researcher studying, say, ten single nucleotide polymorphisms (SNPs) thought to be associated with asthma, the Bonferroni correction entailed multiplying the overall target p-value of < 0.05 by ten, the number of comparisons involved. Each marker, then, would require a p-value of 0.005 to keep the overall false positive (type I) error rate below 5 percent and worth responsibly reporting. Being scrupulously careful, then, means applying a Bonferroni correction.

But there's a double bind, an echo of the one Lander and Kruglyak outlined: "However, the Bonferroni correction can be overly conservative if the multiple tests correspond to correlated variables. For example, phenotypes are often correlated (e.g., asthma and IgE levels) as are the genotypes of single nucleotide polymorphisms (SNPs) that are in LD [linkage disequilibrium]. Thus, these comparisons do not represent independent tests and the Bonferroni correction can be extreme in these circumstances. In fact, the Bonferroni correction can be conservative even for independent tests."[37]

The correction requires a correction, under certain circumstances—circumstances which a researcher does not know, or may know only once the experiment has been conducted. Be scrupulous, but don't be overly scrupulous—and however you end up cutting it, it's going to cost you something of value.

Scrupulousness and the "Reproducibility Crisis"

The scrupulous statistical analyses involved in determining p-values, then, were a core component of caring for genomic data and becoming a responsible genomicist—as they were, and are, for all sciences and scientists. The avalanche of "avalanches" of data in genomics that gathered over the early 2000s only turned up the tension and drew these conundrums into even sharper relief. There's no satisfactory and satisfying solution to these double-binding structures, but the limits and their tolls come to be more keenly felt and known, or at least sensed and sussed. Calls to reevaluate the adequacy of statistical analyses (and p-values in particular) occurred across the sciences. This was the beginning of what would quickly come to be called "the reproducibility crisis" cast (wrongly or at least carelessly, in the view of many) as pervading most of science, even if there was a focus on particular hot spots in sciences like psychology, cancer research, and genomics. The big anxiety about Big Data across the sciences in the early 2000s is maybe best summed up by the title of a widely referenced 2005 article by John Ioannidis, "Why Most Published Research Findings Are False"—followed in 2008 by the epistemically different but affectively equivalent "Why Most Discovered True Associations Are Inflated."[38] Ioannidis is a physician, epidemiologist, and statistician who became a combination gadfly/watchdog concerning almost anything methodological and meta-analytical, and his article made him a top cited and quoted figure in numerous articles in professional journals as well as more popular science media. Questions about the extent and seriousness of a "reproducibility crisis"—including questions of what "reproducibility" means, whether it was ever accomplished with any frequency, and why exactly it was so hegemonic—are too much to handle here.[39] We'll look only at Ioannidis's involvement in discussions and publications about caring for genomics data, after considering related critiques by statistician David Colquhoun.

It was becoming evident in the mid-2000s that genomics had a problem—multiple problems, in fact, which (singly and together) were a mix of technical, natural, cultural, and methodological factors. Although doing statistics, and doing it well, is probably the single most important enactment of care in its pattern of scrupulousness, increasingly close attention was directed toward every technical and cultural component of an entire research system, including publication pressures, cultures of competitiveness, and similar dynamics. Statistical analysis nevertheless remains a key entry point to these issues for us, a point from which statisticians are particularly well-situated to lead us through the cascading causes and demands for scrupulousness—causes and demands that persist, tenaciously, even as they mutate over time and intensify. Along with Ioannidis, David Colquhoun also drew attention to these entangled structures.

Colquhoun is an eminent British pharmacologist who also writes frequently about biostatistics, including on his blog, *DC's Improbable Science*. In the early 2000s, Colquhoun's blog ran on a server at University College London, his home academic institution, until complaints from the homeopathy and related communities that he was in the habit of calling "quacks" prompted the university in 2007 to make him move his blog elsewhere on the internet, for understandable if ineffectual reasons.[40]

Colquhoun's 2014 article in *Royal Society Open Science* (Colquhoun, like many scientists in fields I frequent, inveighs often against the "glamour mags" like *Science* and *Nature*, and against for-profit scientific publishing more generally) is more of a retrospective article but is similar in tone and content to Ioannidis's 2005 article. My readings here skirt its mathematical content, which Colquhoun is actually quite gifted at presenting in an understandable and friendly (and pointed) way. There's more worth to the article than just the juicy language squeezed out here.

"It has become apparent that an alarming number of published results cannot be reproduced by other people," his article states in its introduction, producing "something of a crisis in science." Colquhoun cited his own earlier

1971 statistics textbook to indicate that he has long been trying "to prevent you from making a fool of yourself" by making "unpublishable results publishable" through careless statistical analysis—analysis that nevertheless was perfectly in accord with the professional standards on which almost all science runs.[41] More than thirty years later, he found himself still engaged in the same effort, highlighting the same problems, and calling for renewal of the same practices of care (science). More evidence, I'd say, that the impossible double binds that necessitate care return continually if not eternally as a limit-structure (although simple and at least partially correctable human idiocy cannot be completely ruled out).

Most of his article was devoted to demonstrating the pitfalls of p-values, and their inadequacy as a measure of "significance" ("*never* use the word 'significant,'" he admonished emphatically in his abstract). He acknowledged that his paper was almost entirely taken up by this "very simplest ideal case," but such cases were nevertheless a big part of the ongoing "crisis." His paper "is not about multiple comparisons," as most experiments in genomics are, and for which "it is well known that high false discovery rates occur when many outcomes of a single intervention are tested. This has been satirized as the 'jelly bean' problem (http://xkcd.com/882/)." The link is to the always intelligent, wryly hilarious, and enormously popular (in both of C. P. Snow's "two cultures") comic *xkcd*, "a webcomic of romance, sarcasm, math, and language" created by former physicist and roboticist Randall Munroe. Since, in a shared spirit of open access, Munroe invites the reproduction of his webcomic, and since it is such a fantastic illustration of the "jelly bean" problem prevalent in postgenomics and thus of the double binds of scrupulousness, Duke University Press reproduces it here (see figure 5.1). And then came a statement that says something quite . . . significant about the state of the "culture of scrupulousness" in the biosciences today: "Despite its [the 'jelly bean' problem's] notoriety it is still widely ignored."[42]

How can it be that, in a genomic culture so reasonable in so many ways, a readily available remedy (Colquhoun uses the word "simple" over and over in his article to describe almost all of his techniques) to such a "notorious" phenomenon is so unreasonably ignored? It must be for reasons of a different kind of reason, the cultural logics and structures that Ioannidis and others were also identifying at this time. After carefully and kindly walking readers through more such statistical problems—false discovery rates, underpowered studies, inflation effects, and the like—Colquhoun concludes by enumerating some of those logics and structures that keep a scientist in a time of Avalanching Data willing and even content to "make a fool of yourself at least 30% of the time":

FIGURE 5.1 · The Jelly Bean Problem as sketched by *xkcd*.

WE FOUND NO LINK BETWEEN GREY JELLY BEANS AND ACNE (P > 0.05).

WE FOUND NO LINK BETWEEN TAN JELLY BEANS AND ACNE (P > 0.05).

WE FOUND NO LINK BETWEEN CYAN JELLY BEANS AND ACNE (P > 0.05).

WE FOUND A LINK BETWEEN GREEN JELLY BEANS AND ACNE (P < 0.05).
WHOA!

WE FOUND NO LINK BETWEEN MAUVE JELLY BEANS AND ACNE (P > 0.05).

WE FOUND NO LINK BETWEEN BEIGE JELLY BEANS AND ACNE (P > 0.05).

WE FOUND NO LINK BETWEEN LILAC JELLY BEANS AND ACNE (P > 0.05).

WE FOUND NO LINK BETWEEN BLACK JELLY BEANS AND ACNE (P > 0.05).

WE FOUND NO LINK BETWEEN PEACH JELLY BEANS AND ACNE (P > 0.05).

WE FOUND NO LINK BETWEEN ORANGE JELLY BEANS AND ACNE (P > 0.05).

NEWS
GREEN JELLY BEANS LINKED TO ACNE!
95% CONFIDENCE
ONLY 5% CHANCE OF COINCIDENCE!
SCIENTISTS...

The blame for the crisis in reproducibility has several sources.

One of them is the self-imposed publish-or-perish culture, which values quantity over quality, and which has done enormous harm to science.

The mis-assessment of individuals by silly bibliometric methods has contributed to this harm. Of all the proposed methods, "altmetrics" is demonstrably the most idiotic. Yet some vice-chancellors have failed to understand that.

Another cause of problems is scientists' own vanity, which leads to the public relations department issuing disgracefully hyped up press releases. In some cases, the abstract of a paper even states that a discovery has been made when the data say the opposite. This sort of spin is common in the quack world. Yet referees and editors get taken in by the ruse. . . .

The reluctance of many journals (and many authors) to publish negative results biases the whole literature in favour of positive results. This is so disastrous in clinical work that a pressure group has been started: altrials.net. . . .

Yet another problem is that it has become very hard to get grants without putting your name on publications to which you have made little contribution. This leads to exploitation of young scientists by older ones (who fail to set a good example). It has led to a slave culture in which armies of post-doctoral assistants are pushed into producing more and more papers for the glory of the boss and the university, so they do not have time to learn the basics of their subject (including statistics). . . .

And, most pertinent to this paper, a widespread failure to understand properly what a significance test means must contribute to the problem.[43]

These more systemic-level matters calling for a new enactments of care are discussed further in chapter 6, under the sign of solicitude. For now only note that, alongside great growth and progress in bioinformatics and biostatistics in the last twenty years, many of the technical and cultural difficulties that careful statistical/informatic practices are intended to combat nevertheless persist. And are persistently railed about.

And continue to beg for scrupulous attention to detail. Let's return briefly to 2002 and to an article in the *American Journal of Epidemiology*, by John Ioannidis and six other authors representing the International Meta-analysis of HIV Host Genetics. Meta-analysis itself had become a regular feature of the biomedical research landscape (although originating in research on educational outcomes), and as with the "metadata" discussed in the previous chapter, the "meta-" prefix indicates a shift in logical level: just as metadata is data "about" data, meta-analysis is analysis "about" analyses—where "about" means simultaneously "concerning," "approximate," and "proximate" ("somewhere close in the vicinity"). The difference between each instance on either side of the "about" is more a matter of the movement of relationality. And just as meta/data harbors a double bind—secure and insecure, grounded and undermined—meta/analysis is also a sign that, despite an apparently identical "analysis" we should expect some contradictions in these relations, as researchers change from analyzing a set to analyzing a set of sets. Scrupulousness is another name for the double bind of meta/analysis.

The original studies Ioannidis's group analyzed had tried to find whether there was a difference in outcomes (dying of AIDS being the main indicator of outcome) in groups differentiated according to whether its members had either of just two genes in their genomes coding for two different receptors (like *toll*-like receptors, but here chemokines were the "signaling" agents being "recognized," bridging the membrane differentiating inside from out) which HIV used to enter human cells. Despite dealing with this small number of genetic factors, the studies showed inconsistent results and therefore "generated controversy"—metadifferences about differences. The authors had undertaken a standard meta-analysis of the published literature (MPL) several years prior, but it suffered from problems that plagued most meta-analyses but were particularly acute in epidemiological studies like this one: "variability in study designs, poor data quality, insufficient confounder adjustment, publication bias, and spuriously narrow confidence intervals." Now they were back to do a meta-analysis of individual participants' data (MIPD), one of the first applications in human genetics of this technique that drilled further down into the individual data on all the participants, cases as well as controls, of each study.[44]

Without going into the full details of this study, we can use it to index (1) the greater degree of scrupulousness required as genomic analysis became more complex and genomic data accrued and avalanched, and (2) the increased effort and costs necessary to achieve these new levels of detailed attention and care. The more scrupulous a researcher was, the more systemic elements there were demanding more scrupulous work. In this genre, lists are a prominent feature, enumerating all the factors that threaten reproducible outcomes. The long list excerpted here summarizes the even longer and more detailed list of the authors, outlining where they thought new kinds and degrees of scrupulousness had to be directed:

- Including additional unpublished data in its analysis, compensating for "publication bias" resulting from the valuing and reporting only positive results (the well-known "file-drawer problem" plaguing all biomedical research).

- Using "consistent definitions to categorize eligible participants as seroconverters or seroprevalent subjects," which affects generalizability and "influences the magnitude of the observed effect of genetic markers that affect disease progression differentially among rapid and slow progressors."
- Standardizing for three different clinical outcomes in different studies: progression to AIDS (according to a 1987 CDC definition), progression to death, and progression to death after the development of AIDS.
- Applying "Cox models separately to each study and synthesized study-specific hazard ratios with fixed and random effects general variance models using the *Q* statistic for heterogeneity. MIPDs of randomized trials have typically used an extension of the nonparametric Mantel-Haenszel-Peto method. This method is biased when the allocation ratio in the compared groups deviates substantially from one, a frequent situation in genetic epidemiology."
- Adjusting for different covariates and confounders in different studies, "for example, . . . CD4+ lymphocyte count and HIV-1 RNA levels, two strong predictors of disease progression whose values vary substantially depending upon when they are measured in the disease course."
- Adjusting for linkage disequilibrium, since comparing "carriers against noncarriers may be biased because participants with the *CCR5 Δ32* allele preferentially cluster in the latter group."
- Adjusting for population stratification through better accounting of "variability in racial categorization and inconsistent reporting of racial subgroups across studies." [*Note that "racial categories" would themselves be radically challenged and changed within a few years, as mentioned at the beginning of this chapter.*]
- "Eliminating discrepancies in data definitions and analytic approaches."
- "Assessing sampling bias."[45]

Note that each of these "adjustments" and "assessments" are marks of scrupulousness that themselves deserve, indeed *demand* further scrupulous attention: If our methods of assessing assessments need to be reassessed, what's the protocol for that? How does one, in practice, adjust the adjustments and account "better" for the accountings that had seemed so well accounted for so recently? These are the double-binding effects of scrupulousness: its continually escalating demand for greater scrupulousness, prevented from becoming overly scrupulous only by the limits of time, money, and other collective resources, and no part of it reducible to an algorithm or amenable to routine formalization.

Ioannidis and coauthors recognized the costs that attached to these new levels of care, and at the end of their article stated them frankly:

> The time and effort required to perform an MIPD must be seriously considered. Our MIPD used a professional data-management company contracted by the National Cancer Institute. A total of 2,088 hours of data management were used in the MIPD.... The four coordinating investigators each invested between 5 and 20 percent of their full-time effort during the project. More than 1,000 e-mails were exchanged between the coordinating investigators and with the data managers. International coordination was also demanding and included several general mailings and extensive telephone, fax, and electronic mail communications.... While more conclusive answers were obtained with the MIPD, extensive resources were required.[46]

These kinds of careful meta-analyses for the genomics of complex conditions, they concluded, were "a costly and time-consuming enterprise and should not be undertaken lightly." "Given the large number of putative genetic associations" that were promised to be discovered through the recently concluded Human Genome Project, the authors concluded, "the time and resources for an MIPD cannot be justified routinely."[47]

Maybe you were expecting that sentence to end differently? When I first read it, I figured a sentence beginning "Given the large number of putative genetic associations" would continue along the lines of *"it is imperative we invest in the most scrupulous meta-analyses of this most difficult research."* Maybe the authors, having just completed such a meta-analysis of research on just two fairly well-characterized genetic variations, were exhausted and jaded by the experience, willing to let hundreds of other individual research groups continue to work with less-than-fully-scrupulous levels of attention and analysis that could nevertheless still get them published? Did they figure that scrupulousness may be justified routinely, but that was no reason to be *overly* scrupulous about it? Or was it that the genomics status quo circa 2005 was less careful than it could or should have been, but it would have taken too much time and money to do much about that on a systemic level?

Testing the "Non-hypothesis Testing" Hypothesis

Or was it that they sensed the double binds, and recognized that the only workable response was a kind of crowd-sourcing of meta-analysis, building it into the infra/structure of the genomics infrastructure, including scientists cultivating their "care of the data" selves? For it was just these kinds of increased investments in scrupulous genomic care that were beginning to be made and, most importantly, made in a distributed, decentralized, collective manner. A key component to these developments was the growth of research consortia and similar kinds of collaborative (more or less) networks to both carry out the research and to better mind its increasingly complex analyses.

These kinds of networks increased in number, and grew in size, in the genomics of complex conditions like asthma, diabetes, or schizophrenia. Every month there seemed to be a new team, or a new consortium, or a new network, or in the case of the Centers for Disease Control's HUGENet—the Human Genome Epidemiology Network and the Network of Investigator Networks—a new network of networks.

Established in 1998, HUGENet became widely known by the early 2000s not for its own genomics analysis, but for the kind of meta-analysis outlined above: HUGENet existed to test the testing protocols of genomics researchers. By May 2004, this "network of networks" had more than seven hundred collaborators from forty countries.[48] In HUGENet studies, scientific discourse became ever more tightly bound to ethical discourse, an ever more hybrid genre that articulated evolving conceptions of scrupulous care for a field undergoing rapid growth and change. In one typical article, for example, meta-analysis was targeted on the published results from genomics studies of complex conditions that thirty-eight HUGENet authors from thirty-one different institutions characterized as "difficult to assess," where evidence was "fragmented," and the necessary multidisciplinary links were "poorly developed." These uncertainties, inconsistencies, and irreproducibilities—all peer-reviewed, all validated as "significant," and all published in highly regarded journals—made it unclear how to "keep track of this fragmented, changing evidence," let alone how to "rate its credibility." Individually and collectively, scientists, argued HUGENet researchers, must create "new data synthesis methodologies," and "rules for assessing evidence" and causality, and had to do this all within an ethics of "transparency" so that "methods can be judged and significance assessed."[49] Rating ratings, assessing assessments, synthesizing syntheses: same old same old, just more meta-, and just more carefully.

To reduce such sprawling meta-developments in genomics to a more manageable narrative, let's take up again the thread of asthma research. Genomic approaches to asthma diversified and multiplied in the late 1990s and into the first decade of the new millennium—as did the number of "asthma genes," as we'll see. The previous chapter highlighted early studies that searched for genes in families and "isolated" populations—really a kind of genetics approach. When that proved unproductive, genomicists complemented those studies (and eventually mostly replaced them) with candidate gene studies, in which they searched in nonisolated, nonrelated populations for variants of known genes that had some plausible biological connection to asthma. In terms of gene discovery and our understanding of asthma, relatively little of enduring value came from these early studies.

But in organizational and methodological terms, asthma genomics was where consortia-driven research began, with the Collaborative Study on the Genetics of Asthma discussed in chapter 4. The CSGA represented a noteworthy change in genomics: in the late 1980s, as chapter 2 described, many geneticists who had worked in a tradition of what were often called "mom and pop" genetics labs funded through individual R01 research grants from NIH were wary of, and sometimes outright hostile toward, a Human Genome Project concentrated in a few large centers. That more centralized group model of research was basically imposed on the field from the top. Barely ten years later, geneticists-becoming-genomicists were self-organizing to work in consortia, collaborations, and research networks that would continue to grow in size over the next decade.

Much of that trend was numbers-driven: uncovering unknown genes, with unclear and almost certainly small effects, across the vast molecular territory of a genome, in populations just beginning to be organized, characterized, and understood in genomic terms, required larger studies that could provide the required statistical analytic power. In chapter 6 we'll look in more detail at the research of Erica von Mutius and her collaborators in the ALEX Study. The ALEX Study worked with several hundred individuals from European farming families, along with a few hundred more participants from nonfarming families as controls. We'll see how their 2004 paper tied asthma diagnosis to *toll*-like receptors, and to whether one spent one's childhood with farm animals. One of the first studies to show these kinds of "gene-environment interactions," it depended on the analysis of 609 genomes. Since half of those 609 people had asthma, the study also depended on the extended network of care and tracking, over years and even decades, that comes with the maintenance of such populations as biomedical experimental cohorts.

That was a candidate gene study, the efficacy of which is contingent on a number of conditions. A candidate gene research design begins with a gene known to be implicated in asthma, or at least hypothesized as implicated. The left side of a flow chart from a different 2004 paper, reviewing genomic research in asthma, illustrates the candidate gene design (figure 5.2). In this kind of hypothesis-driven approach, 609 samples is a respectably large number of samples, but only *just* large enough to provide the statistical power the

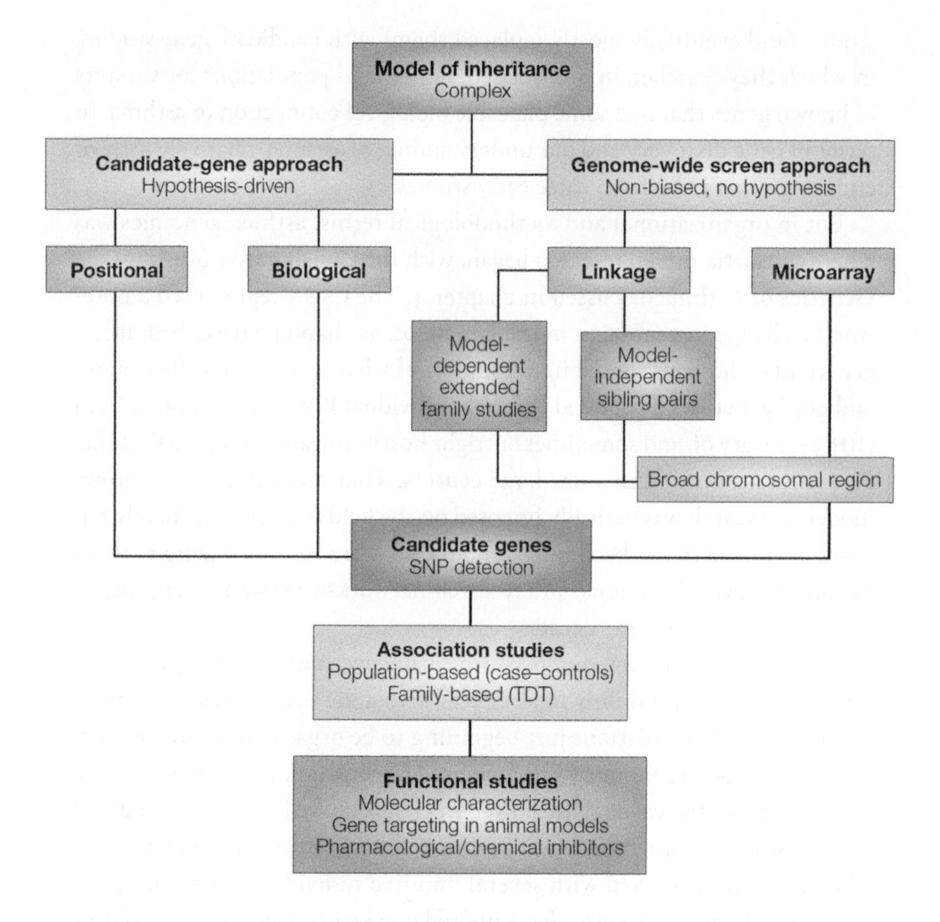

FIGURE 5.2 · Chart of genomics research designs. From Wills-Karp and Ewart, "Time to Draw Breath."

study needed. Even just a few years earlier, achieving such numbers would have been technically and/or economically and/or practically and/or analytically much more difficult, if not impossible. So the ALEX Study paper of 2004 was something of a landmark.

But on the right side of that 2004 chart from the review article, we see a new set of possibilities on the horizon: the possibility of scanning entire genomes, and not just known genes, in what is called there the "nonbiased, no hypothesis" approach. This was also sometimes called the agnostic approach, often greeted with a snarky dismissal by science studies scholars who, since at least Thomas Kuhn, can't believe that scientists can't believe. We could also say that many science studies scholars are of the belief that science and scientists are always believers in something. We (I share the symptom) are skeptical that something like a "no hypothesis" approach is even possible.

But even if none of us, scientists or otherwise, cannot *not* believe (as Gayatri Spivak's construction of deconstruction would put it), we can still engage in "persistent critique of what we cannot 'not' want"—a double move for the double bind of precarious believing and secure knowing.[50] None of us are *only* compulsive believers in something; we have other capacities and habits as well. So to counteract our habits of disbelief and the predictable predicaments to which they lead, and to acknowledge the recursive logics and double binds involved, and to supplement a language of believing and constructing with a language of testing and discovering that is more in tune with care as friendship with the sciences, my suggestion is to double down and name what genomicists were doing here as "testing of the 'no-hypothesis-to-test' hypothesis." *Now that we can cheaply genotype several hundred people in a go*, this genomic cultural logic goes, *let's just see if we can assign significance to some correlations between some areas in some several avalanches of DNA sequences and the embodied lives of some people who struggle at times for breath—and sometimes even die in that struggle.*

But before we get fully into that territory, let's take a quick look at the situation created by candidate gene approaches fueled by genomic technologies, the avalanche before the avalanche of genome-wide association studies that were just beginning to take off. A 2007 review article arrayed the more than 120 genes shown in at least one published study to have a positive association with asthma around a wheel subdivided according to physiological systems or processes like immunoregulation, lipid mediation, mucus production, bronchoconstriction, and inflammation (see figure 5.3).[51] The *toll*-like receptors that so interest-excite me almost get lost on this dartboard dense with targets (in the "innate immunity" sector on the right, between three o'clock and four o'clock). I should emphasize: most of these "targets" would not be replicated in later studies on different populations, as changing methodologies and stricter statistical measures of significance made genomicists more careful and genomics better minded. But I should also emphasize: many of the articles and review articles from this time would make exactly such cautionary statements, such as, "even the most replicated genes, such as *IL13*, *IL4*, *IL4RA*, *CD14*, *ADRB2*, *FCERIB*, *TNF* and *ADAM33*, failed to show an association with asthma or allergy in a substantial number of studies, and when successfully replicated, they typically have small effects and explain a negligible proportion of the phenotypic variance."[52] This exemplifies that situation of fragmentation, murky associations and judgments, and opaque significance detailed by the HUGENet article discussed earlier—the state of genomics that HUGENet called for new levels of scrupulousness to address.

To make matters even more complicated, many of these genes implicated partially (if at all) in asthma may in fact exert opposite effects under different conditions. Genomics here does not provide fundamental explanations so much as particularly productive entry points for improving understandings of the knotted specificities of the condition.[53] Thus, a leading asthma researcher like Fernando Martinez (more about him in chapter 6 as well) thought that the complexities and specificities of asthma becoming evident

FIGURE 5.3 · Genomic Dartboard circa 2007: genes associated with asthma arranged.

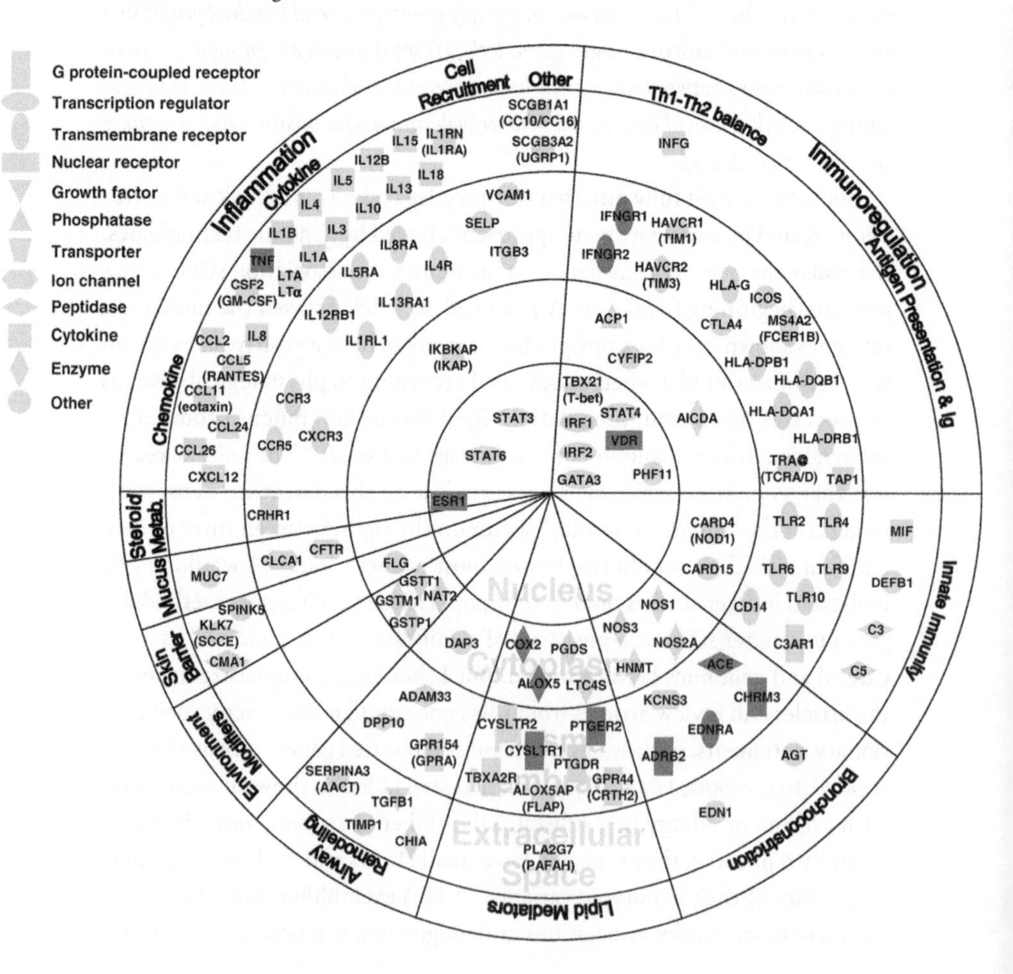

in 2007 compelled an apology to William of Ockham for abandoning his razor of simplicity. Martinez highlighted how the "weak linkages" among "flexible," "indirect, undemanding, low-information" knots of complexly interacting biological response systems produce a heterogeneous condition like asthma, where

> a specific protein may exert opposite effects when participating in coordinated responses to different external stimuli, and therefore, a genetic variant that increases transcription of that protein may enhance an "asthmatic response" to one exposure and hinder an "asthmatic" response to a different exposure. The specific role of any element of the response system is thus determined not only by its intrinsic characteristics but also by the biological context in which it is expressed.[54]

Giving up on genetic explanatory parsimony also meant giving up on "the original hope that genetic tests would allow us to identify who is at risk of which complex disease." But Martinez felt that the more complex view of asthma that genomics had helped to construct was at least "more in tune with the degree of heterogeneity and unpredictability of the expression of the disease that is evident in any asthma clinic."[55]

To continue tracking the growth of genomic collectives, their continued coupling to a cautionary discourse, and the continued invention of a more scrupulous, careful genomics, this section concludes with a quick look at one of the first genome-wide association studies of asthma published in 2007, one still noted for its non-hypothesis-driven discovery of an unexpected "asthma gene"—and one that was not even on that dartboard of just a few years previous.

Erika von Mutius is at the center of this research network again, grown out of her 2004 ALEX Study and now called the GABRIEL Project, "a multidisciplinary study to identify the genetic and environmental causes of asthma in the European Community, an EC funded collective of over 150 scientists from 14 European countries."[56] Some members of the new collective brought cohorts of asthmatics that they for years had cultivated, characterized, and cared for; others did the high-throughput genotyping; others brought new bioinformatic functions and other analytic methods that many other members of the collective probably did not fully understand. Funding sources for these kinds of efforts had also proliferated—care, remember, is expensive—with funding for this 2007 paper coming not only from the EC Framework 6 but also the UK's Wellcome Trust, the Medical Research Council, the US National Heart Lung and Blood Institute, and other individual institutions like the two children's hospitals in Vienna and Dresden, all bringing something to the work, the care, the science.

The crucial number *n*, the number of research subjects, had also grown dramatically in the three years and change of research design from the ALEX Study of variations in one candidate gene under different environmental conditions to this GABRIEL genome-wide association study (GWAS) in the "non biased, non-hypothesis" mold on the right-hand side of figure 5.2. The 609 total subjects needed for the 2004 candidate gene approach had ballooned to more than 990 asthmatic subjects on the "cases" side, assembled from two different cohorts, and another 1,200 controls from three different sources for a total *n* of more than 2,600 for the GWAS. These are the numbers of bodies necessary for the statistical power to test the non-hypothesis-testing hypothesis on the millions of (meta)data points curated carefully from their genomes.

Two tightly coupled technoscientific achievements made such an expanded research design possible in 2007 where it hadn't been just three years earlier—and not only possible, but exciting and popular. The biotech company Illumina made a "chip" commercially available that allowed for the rapid and economical genotyping of those 2,600 individuals. The chips are plates with hundreds of thousands of single nucleotide polymorphisms (SNPs) deposited onto them in a thin molecular layer; these are not genes, but just chunks of DNA sequence that differ from each other in one otherwise meaningless way. Illumina could make and market these commercial chips, part of a "next-generation" of sequencing technologies, because of another key technoscientific development in genomics since 2004: the availability of data and materials from the International Hap Map Project, the collective federally funded effort to collect and characterize the genomic variation (in the form of single nucleotide polymorphisms, SNPs) from sampled populations around the planet.[57]

The multiplexed functions and experiments connecting 2,600 individuals to these "next-gen" sequencing technologies generate data on a scale unimaginable and unmanageable a few years earlier. The avalanche elicits ever more elaborate and creative forms of scrupulousness from all involved but especially the bioinformatician, now a key node in this multidisciplinary network, who establishes significance from multiple series of minute differences and their groupings (see figure 5.4).

Quite sketchily, moving clockwise from the bottom left in figure 5.4, almost all of the thousands of differences across the genomes of asthmatics and non-asthmatics alike (box a) would be ignored, and only the highest degree of difference would be assigned significance, from a region of a few

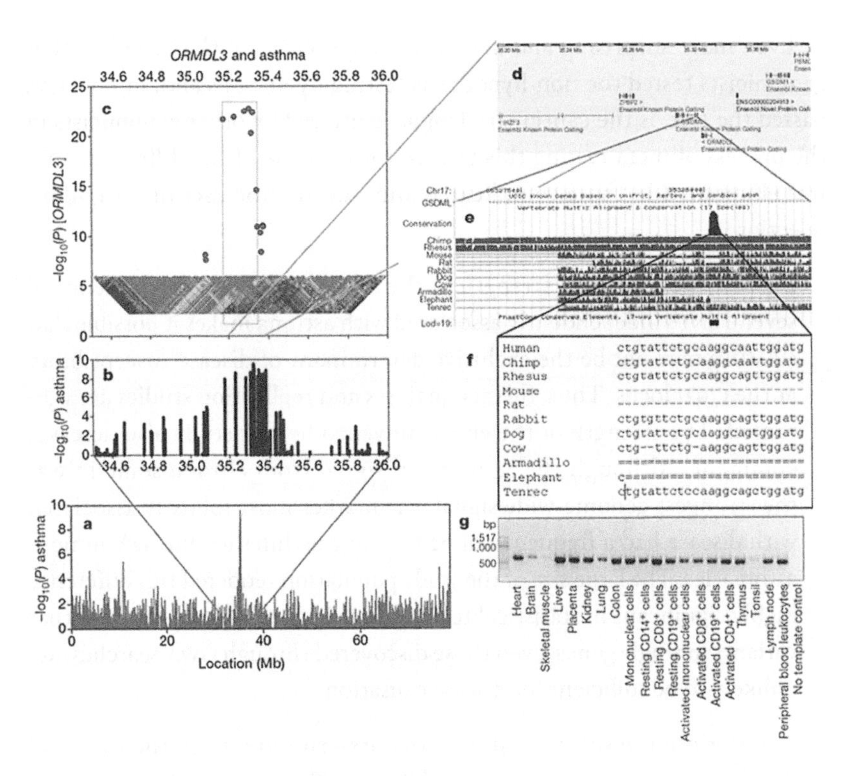

FIGURE 5.4 · One infra/structural sifting of one genomic (meta)data avalanche.

million DNA base pairs on the short arm of chromosome 17 (box b). From all of those differences in box b, only one set of differences rose above another stringent statistical threshold and was selected for further analysis (box c). These SNPs were determined (box d) to be tightly associated with a gene already in the public databases called ORMDL3, whose kinship and thus evolutionary structure were then developed (boxes e and f). Humans share ORMDL3 with most every creature between chimps and hedgehogs, and it is expressed in most every tissue—from the heart to the colon, except for the skeletal muscles and the tonsils.

And that's how we can say these asthma genomicists passed the test of the no-hypothesis-to-test hypothesis: ORMDL3 was never a candidate gene, and no one had ever before associated ORMDL3 with asthma in any way, shape, or form. It is a gene involved in making part of the endoplasmic reticulum, a basic structure of every eukaryotic cell; the genealogy of ORMDL3 extends back to even more distant kin than our *toll*-like receptors do: to the important genomic model organism, yeast.[58]

Even more surprising and deserving of appreciation than the fact that genomicists tested the non-hypothesis-testing hypothesis and, in my view, passed the test, is the cautionary language uttered by other genomicists in the process of meta-testing this test, restoring some of the difference that was eliminated in these complicated and careful processes of scrupulous selection:

> Despite the statistical strength of the association data, the presence of several SNPs independently associated with asthma makes it possible that *ORMDL3* may not be the exclusive determinant of disease susceptibility at the 17q21 locus. Thus, further analyses and replication studies are warranted. The strength of the effects unveiled by this study also deserves comment. Although the association signal from chromosome 17 was the strongest genome-wide signal, the marker most robustly associated with disease had a frequency of 62% among asthmatics and 52% in non-asthmatics. The large size of the study population rendered this difference highly significant in statistical terms, but this distribution reiterates that variants in single genes, even those discovered through GWA searches, are unlikely to be sufficient for disease causation.[59]

Translation: the substantial investments—running from the financial to the affective—made to isolate this difference that makes a difference may not translate into a useful therapeutic substance, or even just into the kind of knowledge claim that would be useful in the clinic, let alone inform at the scale of public health measures. And although that kind of translation, and the differences that make a difference to them, are the ones that are coming to matter most in this era of translational science, are they the only kinds of translation that should matter?[60]

The world got a new surprising and (mildly) exciting feature: a new knot that binds a gene, conserved over a long multispecies evolution and coding for a transmembrane protein on the endoplasmic reticulum, to asthma—in some people. Quite probably. Granted, this knot joins the massive tangle of knots that tie more than a hundred other genes to asthma, to say nothing of asthma's ties to particulate matter, ozone, the stresses of poverty, and other environmental, epigenetic, and developmental forces that have been shown to be involved in producing the tear in our biocultural fabric to which we assign the catachresis "asthma."[61]

Donato Vercelli, the asthma researcher who earlier brought us to read the ORMDL3 gene falling out of the asthma GWAS more scrupulously, also offered cautions about GWAS more generally. The "thousands of cases and controls" needed for a GWAS require not only new statistical methods and new practices of care (and about $10 million and elaborate organizational structures and movements), but "the study populations must be carefully characterized," a demand that next directs a researcher's attention to "heterogeneity in environmental exposures," a "critical" variable. Critical, but almost entirely unexamined by genomicists, certainly at the time and only slightly less so now. "More generally, even at this early stage it is clear that not all GWAS results replicate consistently. In this respect, GWA studies are reminiscent of earlier candidate gene studies, and conclusions should be similarly cautiously made."[62] The same demands for scrupulous care return, with a difference, because the knotted double binds of scrupulousness are enduring ones.

Over the next decade, GWAS attracted a lot of this kind of criticism. The *American Journal of Human Genetics* published an evaluation, "Five Years of GWAS Studies," in 2012 that opened with the question, "Have GWAS been a failure?"; the first full three columns summarized both popular and scientific literature documenting the "perceived failure or disappointment of GWASs":

1 · GWASs are founded on a flawed assumption that genetics plays an important role in the risk to common diseases;
2 · GWASs have been disappointing in not explaining more genetic variation in the population;
3 · GWASs have not delivered meaningful, biologically relevant knowledge or results of clinical or any other utility; and
4 · GWAS results are spurious.[63]

In the Postscript, I promise to return to what I think is the important difference between "failure" and "disappointment." Here I just note that these authors went on to respond to each of these criticisms in detail and delivered an overall positive evaluation of GWASs.

Five years later, the same set of authors did it all over: "Ten Years of GWAS Discovery: Biology, Function, and Translation" did not begin with that defensive acknowledgment of criticisms, however, but simply summarized at length a "remarkable" set of achievements, including in our understanding of type 2 diabetes, schizophrenia, and several autoimmune disorders.[64] Many scientists remained unpersuaded. Asthma was not mentioned at all.

I think all these differences are interesting, as are their recurring nonresolution by the genomics community. Many in that community are maddened by the inconclusiveness of so much if not all of it, or by the meager "return on investment" for the hundreds of millions of dollars and person-hours poured into GWAS to produce those floods and avalanches of genomic data. I fully sympathize, for whatever that's worth, and couldn't agree more; it's why I've gone to some length to, in effect, translate their cautions and careful checks of scrupulousness into my own—the ethnographer's obligation (and pleasure). But even if GWAS turn out in the even longer run to be judged by the majority of the community to have been a poor investment, for its lack of "translatability," I don't think that should completely overshadow the other ethnographic fact that, in minding genomics with care, the community taught themselves to become more scrupulous in multiple ways, to think about and conduct GWAS more carefully.

Genomics by the late 2000s had indeed become "better science," as Joan Fujimura and Ramya Rajagopalan assess, than genomics had been in the early 2000s: more subtle, less determinist, more attuned to the flexibilities and limits of its own categories, techniques, and analytic concepts. I read Fujimura and Rajagopalan's article-ending evaluation of how the developing theories and practices of genotype variation in large-scale populations have, over time, become "better science" as saying genomics has become "more scrupulously careful," with these terms amalgamating the cognitive with the ethico-pragmatic with the affective.

And yet, getting better also made things worse. Needing and wanting to be ever more scrupulously careful, genomicists also may be beginning to recognize that they can never be scrupulous enough. That doubled impossibility is both good—the play in and between the elements in the system keep it experimental, open to the unprecedented—and bad, undermining reliable reproducibility and trust. Beyond good and bad, if you like. Attending to the tiniest of myriad details, checking and tightening every intricate connection between them strengthens the system, heightens its capacities, and makes it more trustworthy. At the limit, however, where scrupulousness verges into being overly scrupulous, the whole thing threatens to lock up, immobilized into repeating the same patterns. Tightened connections here pull the distant connections over there out of joint, heightening existing tensions until something threatens to snap. And at some point, now when it is too late and too soon to really know what must be done, you have to step back, get outside the vast, finely tuned machinery—if there is an outside, and if you could step to it—and reconsider the whole thing.

6 · OF SCIENCE'S LIMIT *solicitude*

The simplest expression that I know of the scientist's obligation can be stated in terms of the Christian paradox, that man is called upon to try the impossible but not expected to achieve it. As scientists we must seek a truth which is unambiguous and universal, even though at the same time we must recognize that this is impossible and, indeed, strictly speaking, meaningless. In science, just as in the jury box or in the voting booth or in the recruiting office, you must commit yourself on grounds which, on reflection, must necessarily appear deficient. · MICHAEL POLANYI, *PERSONAL KNOWLEDGE*

A research experiment . . . belongs to the future of its elaboration and, being wed to novelty, cannot be said to fall on this or that side of a divide determining good or evil, usefulness or harm, and so on. Ever on the verge of becoming something, and ready to assume an identity that could be tagged, taxonomized, sectioned, it looms in the hybrid form of threat-and-promise. · AVITAL RONELL, *THE TEST DRIVE*

GxE

Chapter 5 analyzed some newly emerging practices of scrupulous care in the genomics research community, as its collective general focus moved from the iconic single-gene, more-or-less Mendelian genetic conditions such as cystic fibrosis or Huntington disease to more complex conditions like asthma, accompanied and driven by changes in (among other things) DNA sequencing technologies and genomics study design post–Human Genome Project in the early 2000s. As candidate gene studies approached their limits and genome-wide association studies (GWAS) became more paradigmatic, genomics experiments and their statistical analyses took on a greater complexity of their own, characterized by larger study populations harboring smaller gene effect sizes, pursued through gene chips and SNPs that allowed an escalating number of comparisons within avalanching data sets—a profligate genomic technological and data economy that eased experimentation but also could easily lead researchers astray into a territory of false positives, irreproducible results, and a general carelessness that nevertheless had its social rewards. Genomicists invented new practices of scrupulousness, and reinvented them, and reinvented again as these researchers—more and more bound together into consortia and networks—worked to think and operate more carefully in this new genome space of complexity and excess.

In almost all of that research, however, "the environment" remained a variable to be controlled for or excluded from analysis entirely. Whole genomes, and comparisons of whole genomes within populations that ballooned from extended families to a few hundred to several hundred and eventually to thousands of individuals, kin and non-kin, were on their own more than enough complexity to try to manage, practically and analytically, without making matters worse by opening the research door to "the environment." Even in 2011, years after the GWAS wave first began to build in 2007, asthma researchers Carole Ober and Donata Vercelli would pose the anxious question in the title of their *Trends in Genetics* article: "Gene-Environment Interactions in Human Disease: Nuisance or Opportunity?"[1]

This chapter follows some of the attempts to answer that question, as genomicists exposed their research protocols to the new confounding complexity of environments and "interactions." A techno-bureaucratic fondness for acronyms sometimes translates that to the sign GEI, which seems to strip the situation of all complexity, relationality, and difficulty, and doesn't help us appreciate where genomicists actually were in this time, in which studying such interactions were not just difficult, but verging on impossible. The reasons supporting a literal reading of this phrase, "verging on the impossible," will emerge over the course of this chapter.

As we'll see, sometimes this territory is often signed by genomicists not as GEI, but as GxE. I admit that I'm partial to the chiasmic X, and feel validated when I find it being used like this; I used it in *Promising Genomics* to mark volatile in-betweens, where apparent opposites folded into each other, constructed and deconstructed each other—the lavaXland called Iceland, for example, which issues occasional reminders of its precarious stability, or inXstability. Here I also read the chiasmic X in GxE as a sign of imX-possibility, a double-binding X that marks the spot—the "Here it is!" of possibility, where opposites entangle—and an X that marks the "not!" an impossibility. The chiasmic X is a telltale sign of an imXpossible relationship—and the solicitous care it elicits from scientists in response.

What follows is not a history of GxE research. Readers wanting a broader narrative of the emergence of gene-environment research will do better reading Sarah Shostak's *Exposed Science*, about scientists situated in the National Institute of Environmental Health Sciences (NIEHS).[2] Here I read selected episodes in GxE research, filtered again through asthma, in a way that draws out its simultaneous truthfulness and questionability, its "ever-on-the-verge" status, its weird, interesting, *toll!* relationship to both possibility and impossibility. Unformalizable care, diffracted into a new pattern of solicitude, is again cast as essential for minding genomics within these double binds. Where scrupulousness took as its object(s) the thousand different individual material, conceptual, and technical components that had to be made right in genomics, and couldn't be made right enough, solicitude takes as its object a whole—a whole experimental design, a whole research program, even a whole field. And a whole, as almost any genomicist or anthropologist can tell you, is a fine thing to aspire to and to orient your thinking around, but grasping or realizing it?[3] That's a different story.

"I Can Never Remember the Rule"

Solicitude is the art of doing impossible genomics carefully.

In a book containing at least a few difficult and, indeed, *toll!* sentences, that one may take pride of place. You sense that it maybe doesn't quite make sense. Let's let a scientist, that paragon of logic and sense, ease us into this odd territory.

"Science," reflected immunologist Douglas Green for readers of the journal *Molecular Cell*, "that very creative human endeavor to understand the nature of the reality that exists independently of us, is impossible."[4]

It's a seemingly nonsensical statement, so easily and immediately falsified by simple empirical observation, beginning with a glance at any of the research articles published in just that February 2010 issue of *Molecular Cell*. It nevertheless deserves to be taken seriously. In the process, we'll need to reckon with what "creativity" means to scientists, and to our ways of making sense of the sciences, and how it might make more sense to re-trope creativity in terms of solicitude.[5]

I'm encouraged when scientists like Green, or physical chemist Michael Polanyi, whose words mark the opening of this chapter, hit us with these counterintuitive or at least unanticipated characterizations of a science drawn from their own immersive experience within it. Or noted geneticist and exquisite writer Francois Jacob, who insisted that public-facing, rational and methodical "day science" existed only in conjunction with its opposite, a "night science" that "hesitates, stumbles, recoils, sweats. . . . Doubting everything, it is forever trying to find itself, question itself, pull itself back together."[6] These scientists know something about the sciences to which they are so connected and cathected—devoted, we often say—that we should attend to and can learn from. They sense something nonsensical about a science that in other ways is the epitome of sensibility, and reading them can tell us something interesting, something *toll!* about sense, nonsense, science, culture, wholes, and impossibility—and how *care* could be a good, or at least a not-bad term for how all of those might be bound together . . .

. . . tensely. Even the genesis of Green's short essay, "Stress in Biomedical Research: Six Impossible Things," written while he was "traveling between international meetings," is marked by the tension of being in between: no longer in a there, not yet in a here, speeding motionlessly in an X. The six impossible things Green describes as characteristic of biomedical research might as well have been marked with that chiasmic X, although he can't seem to decide—I'm not really sure I can, either—if they are in fact impossible double binds or just very difficult stresses that result from being suspended between two points. Green, by his own confession, is "not a philosopher" and not "a clinical psychologist (nor a patient)," and so he references not Jacques Derrida or Gregory Bateson but rather Lewis Carroll's White Queen and her boast that, "with practice," it's possible for all us Alices to believe six impossible things before breakfast. But Green *is* an immunologist, so he can nevertheless claim to be "something of an expert" on the kinds of stress "which are unique to our profession." He offers some suggestions on how a fellow biomedical researcher might deal with six stresses, each sourced in an impossible in-between.[7]

I've already hinted at some of the slippages here and the difficulties they raise, beginning with the difference between the "impossible" invoked in Green's title and the merely (!) "very, very difficult." Is this a difference that really makes a difference? What's to be gained by casting GxE research as impossible rather than very, very difficult? Doesn't that risk elevating or even *transcendentalizing* science and scientists (a philosopher's move) when it's far more urgent that one play the anthropologist and simply describe in detail the careful practices by which people, deserving of our admiration but not idolization and idealization, accomplish very difficult things? "Stressful things" do not have to be impossible things begging to be believed before breakfast in order to be noteworthy. Why elevate painful but manageable stress to impossible double bind?

We can read these slippages in Green's truly fine essay, placed in the out-of-place "Forum" section of a premier scientific journal, not as effects of a careless imprecision but as careful (although unconsciously so) undecidables. In other words, the difference between "impossible" and "very, very difficult," for both scientists at work and for my analysis of them here, will be one *of* solicitude, as well as a difference that can only be decided or made *through* acts of soliciting.

In other words, let's continue reading. Out of order.

"Thing #4: Stress Can Be Good for You" suggests the double bind inhabiting the concept and phenomenon of stress generally: two (or more) differences—forces—will create tension or stress unless it is possible to dissolve or reconcile them. Resolution is often impossible, however, and the persistent stress appears to be neither good nor bad, but *both* good *and* bad. Stress is an evolutionary *pharmakon*, poison and gift, and, according to Green, "the trick"—a reliable sign of a double bind, a sleight-of-word to distract from the sleight-of-hand necessitated by impossible situations, and another catachresis for an act of care that has no proper formula—"the trick is not to eliminate stress, but to master it, bending this evolutionary gift to our needs, those times when we need it." That, frankly, sounds even more stressful—and illustrates well the tightening of the double bind in play here. *You need to master this trick—a trick I can't teach you. It's tricky—don't fall for that old trick!*

Things #5 ("Be an Athlete") and #6 ("You Are Your Support Group") liken scientists to athletes and artists, working the impossibility of "same *and* different" that animates any likening. Biomedical research "is a creative enterprise that has this in common with all other creative enterprises" like artistry or athleticism, says Green: "You do it not because it provides you with security and a stable career ladder, but because you can't bring yourself to do anything else." More than a little romantic and problematic, but let's keep moving. The fact that biomedical researchers "do our best, and our work, when it *does* work, is savaged by reviewers," or that once you "finally manage to publish" the work that works, you "feel that nobody has noticed"—well, those are just the normal stresses of the impossible position of the artist-scientist. Similarly for the scientist-athlete: the best athletes know that they will always hit a "wall" in their professional performance, and when they do, they will "dig deep and press through the wall," an impossibility they have learned to do "by training, experience, and sheer will." That's "not dissimilar," Green doubly negates, for scientists: "We have to be mental athletes who struggle with difficult concepts until we hit a wall and keep thinking. And when our ideas are wrong, dashed by experimental evidence, we *keep thinking*."

(Noted in passing: another name for impossibility is aporia, and another name for aporia is blockage, and another name for aporia is path. *Keep reading*. . .)[8]

Something like a wall, or maybe just the outside that would be beyond it, also appears in impossible Thing #2, "Ideas Come from the Eighth Dimension." How does one tap into the eighth dimension? Green's answer is the seemingly simple "reading": "Contrary to [the] popular notion that good ideas are a dime a dozen (or a gazillion per euro at the current exchange rate), they are very valuable. . . . So how do you get hold of such an idea? Or a dozen? You are going to hate the answer: Read. A lot. Read everything you find interesting, inside and outside your field, and then read everything else."[9] Reading "everything you find interesting" would be very, very difficult; "then read[ing] everything else" surely qualifies as impossible. It's a recognition of an absent totality that is constitutive of any of the -omic sciences, a whole no one ever seriously expects to actually reach but is nevertheless sensed to be, somehow, necessary. It's hyperbole, that time-honored rhetorical trope employed often by scientists (we'll see another example later in the chapter) but which needs a literary theorist or anthropologist to take seriously and *read*.

But Green doesn't linger with this casually hyperbolized but still seriously referenced layer of impossibility; he doubles down and adds another impossible whole, a neural one, to explain the need for the literary one:

> Here's why this is so crucial. Creativity, as near as we can tell (I've read about this), emerges from a combinatorial process in which bits of information are rearranged and extrapolated at a subconscious level—think of it as a conceptual smoothie sloshing around in your brain. Then, when you happen to think about something you have noticed in the lab, wondered about in the literature, or worried about late at night (you do this, right?), there emerges an "aha" that might explain something that has never been explained before. (How do you know it hasn't been explained before? Because you did the reading!) This only works if there is a lot of information oozing around in the blended brain smoothie.[10]

I've already cautioned readers (in chapter 1) to care about the differences between "aha," "Eureka!," and "*toll!*," but they all serve to mark some unnarratable event, some "trick." Any way you cut it, Green seems to be telling us that there's something funny going on when a science that is conceptualized and publicly characterized as fully self-sufficient, and fully transparent and explicable in its mechanical chains of logics, actually depends on the oozing and sloshing of a thick slurry. Green names it "creativity," which, like "imagination," is a respectable albeit hand-wavingly vague ingredient theorized as essential to the scientific mind. At least since Kant, in the European philosophical systems imagination is an acknowledged human faculty, even if it is one walled off from reason. Another name for such an undifferentiated but potent substance might be "the unconscious." "Care" in my schema does similar rhetorical-conceptual work, and the double bind of scienceXcare can then be mapped onto the other double bind this chapter is driving at, possibleXimpossible.

For now, let's liken creative care to a more colloquial "fun-loving, irreverent mind," and a combination of "ingenuity and sense of humour." These are phrases used to explain the otherwise inexplicable creativity of an old-school geneticist who investigated the combination of genes and environments in the 1930s, long before anyone could identify "genes" as material entities: Lionel Penrose.

Figuring Impossible Wholes

For many years, the National Institute for Environmental Health Sciences webpage dedicated to "Gene and Environment Interaction" held an often-reproduced illustration figuring their research into the interactions between genes or genomes and different environments with (White) hands preparing to fit two jigsaw puzzle pieces together; figure 6.1 is a re-creation of that illustration, which has since been taken down. The left-hand piece held an image of the now-ubiquitous double helix icon of DNA within its sharp, well-defined contours to represent genes or genetics. Its sharp, well-defined borders and knobs and holes suggesting obvious places for matches and fits project confidence in a clear solution to this not very complicated puzzle. Indeed, given decades of careful effort by thousands of scientists, underwritten by generous and by some measures lavish institutional support and financing, we're justified in saying that the gene piece of GxE research is, as it were, in pretty good shape. Genes, genetics, and genomics have all been bountifully infrastructured. And for the sake of argument and narrative flow, although the shape of

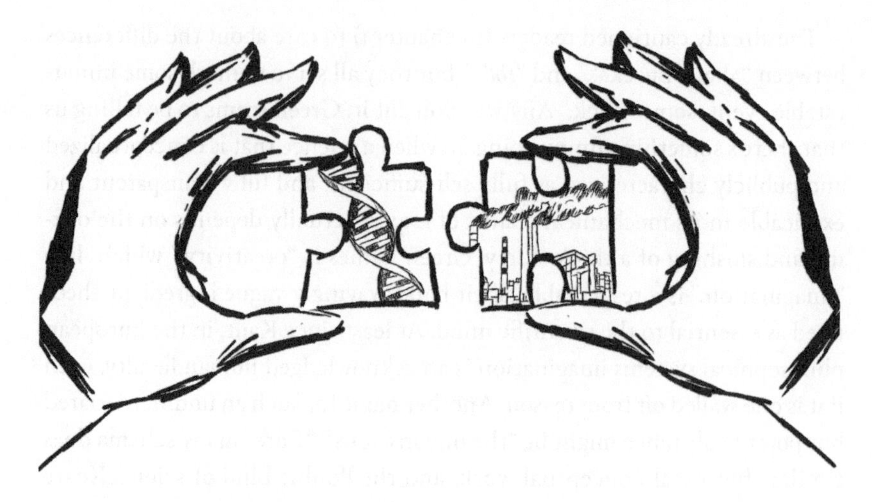

FIGURE 6.1 · Puzzling GxE. Drawing by Kora Fortun based off NIEHS image.

our environmental health research infrastructure is harder to delineate and assess, even when it's a matter of an industrial facility spewing pollution, as depicted in the right-hand piece, let's just go with the broader reading the image was meant to provoke: we have a good grasp of DNA, we have a handle on environments, they obviously go together somehow, so if we just keep fiddling with the pieces hard enough and long enough, eventually it will become clear how they go together and "the big picture" will start to emerge. Within this kind of imaginary, research on gene- or genome-environment interaction might be very, very difficult, but eventually it will take a recognizable shape. Puzzle-solving might be hard, especially with more than two pieces, and it may not be glamorous or attention-getting—there's a reason why Thomas Kuhn opposed "normal," puzzle-solving science to "revolutionary," world-changing science—but at least we know what needs to be done, and have clear reason to think it is doable.

It's a not-bad image, and it seems easy enough to improve: there's nothing but a blank space in between "the gene" and "the environment," but we can imagine researchers filling that space with things like, say, *toll*-like receptors that will bridge or signal across the gap. Or if molecules are not your thing, insert "the body." In one of his many critiques of oversimplified but widespread genetic determinisms and ideologies, population geneticist Richard Lewontin countered with *The Triple Helix: Gene, Organism, and Environment* as the necessary frame for thinking about large-scale phenomena like health, illness, or evolution. Adding more pieces to the puzzle, even poorly defined big ones like "the organism" or "the body," is certainly an improvement and may best represent what happens in practice—it's doing what's *possible*—but it doesn't adequately represent the conceptual and pragmatic difficulties that are involved.

Learning to think under or through a different sign-image might help us better appreciate the complex challenges of GxE research, and the iconic image I'd like to offer as an alternative is not a triple helix, but genomes, organisms, and environments in an impossible triangle, a figure first imagined/invented by the geneticist Lionel Penrose and his mathematician son Roger (see figure 6.2).[11]

A quick encompassing glance suggests that genes, bodies, and environments *look* as though they go together quite nicely to make a whole—at least, that's how our mind makes sense of it. But when you take a closer, more *careful* look at the representation, coupled with some appreciation of how our perceptual systems work, things don't quite match up and don't fully make sense.

Before continuing with that more careful look at geneXenvironment research, let's quickly ask how substituting an impossible triangle for a triple helix (or two, three, or even a thousand puzzle pieces) might make a difference, supplementing the concept of possibility with that of impossibility. In such a refiguring, the question framing this chapter is not, "Is it possible or impossible to research geneXenvironment (or geneXbodyXenvironment) interactions?" but rather, "What is it about our *conceptual* (rather than perceptual) systems that renders geneXenvironment research simultaneously and paradoxically both impossible and necessary?" Or, in a different form, "How is the pattern of care called *solicitude* a response to this paradox of imXpossibility, a name for another way of acting and another style of thinking through which scientists come to better mind genomicsXenvironments?"

"Contradictory Perceptual Interpretations"

The Penroses introduced the impossible triangle figured above (sometimes called a tribar) into the field of psychology and not genetics; it may very well have been inspired by the art work of Marcel Duchamp.[12] They intended their short paper for the December 1958 issue of the *British Journal of Psychology*, "Impossible Objects: A Special Type of Visual Illusion," to be a kind of holiday diversion. The tribar exemplified the "fact" that two-dimensional drawings, although capable of faithfully "convey[ing] the impression of three-dimensional objects," could also "be used to induce contradictory perceptual

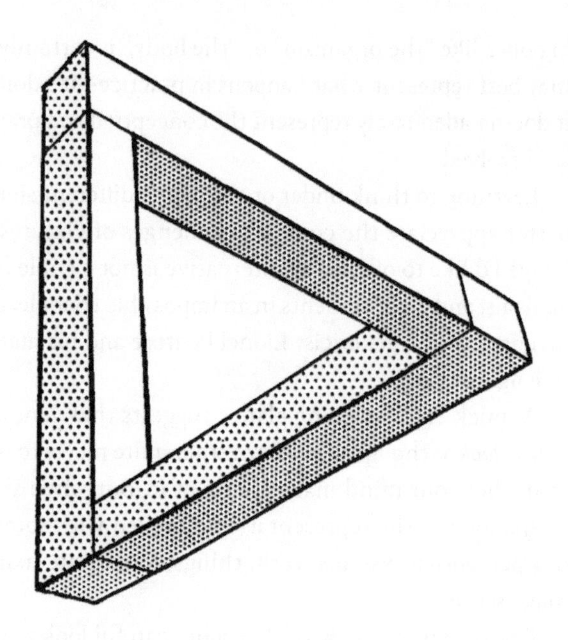

FIGURE 6.2 · Impossible triangle by Penrose and Penrose ("Impossible Objects," 1958).

interpretations." Each individual part was "acceptable as a representation of an object," but "acceptance of the whole figure . . . leads to the illusory effect of an impossible structure." "As the eye pursues the lines of the figure," they note, "sudden changes" in "interpretation" are necessary, and "reappraisal has to be made very frequently."[13]

Neither Penrose thought of their impossible object in terms of genetics, let alone in terms of geneXenvironment research. My joining them together here is not meant as a representation of historical continuity, but it does mean to introduce a contradiction that will force, I hope, reappraisal and a sudden change of interpretation. As a figure for geneXbodyXenvironment "interactions," I want the tribar to signify the simultaneous possibility and impossibility of researching such interactions, an imXpossibility that calls for the constant reappraisal of the whole structure—what I'm calling solicitude.

Solicitude names something close to what sociologist Anselm Strauss named "articulation work." In hospitals, with many actors doing many jobs on many organizational levels in many different temporal frames to accomplish many different but related tasks and achieve many different but related goals, Strauss conceptualized articulation work as "a kind of supra-type of work" that "meshed" these disparate tasks and their doers, coordinating and directing them toward the achievement of certain ends.[14] Joan Fujimura (Strauss's student and my teacher) has analyzed how this kind of articulation work was essential to conducting research into oncogenes and their role in cancer: articulation work made the research do-able, that is, possible.[15]

Solicitude, too, happens in a "supra-" zone, operating at a different logical level than the mesh "kind of" below it, and also connected to it. But solicitude is less about coordination of meshy organizations for pragmatic purposes, and more about a mindy encompassing of an intricately articulated structure—a conceptual and material infraXstructure that is a kind of supra-, kind of infra-, kind of structure—to get a better sense of the whole and its holes, a sense of the play of resistance and play, and a sense of its supra- (sort of) and (kind of) infra-: What holds the structure together without being fully part of it? Where and how does the system both hold tight and threaten to collapse? In short, where are a material-semiotic structure's limits, where do they hold, and where do they move? And, in addition to helping us understand the double binds and miscommunications that occur as one crosses and recrosses the conceptual levels of part and whole, solicitude also prompts the question, What, if anything, is outside, beyond, *supra* the supra-, and *infra* the infra-?

Structure, Sign, and Play in the Genomic Sciences

As near-synonyms for "care" go, "solicitude" is probably the one with the most archaic ring to it; we just don't encounter it much anymore in speech or text. Like care more generally, solicitude carries connotations of anxiety and worry that tend to get pushed to the background. "What we desire with impatience, being possessed, brings Care and Solicitude," wrote Jeremy Taylor in his 1684 *Contemplations of the State of Man in This Life and in That Which Is to Come*. The *OED* cites this as one of the earliest uses of "solicitude," in its sense of "the state of being solicitous or uneasy in mind; disquietude, anxiety; care, concern." Taylor's contemplation would not seem out of place, though, inscribed above the entrance to the National Human Genome Research Institute, or the NIEHS, or the Broad Institute, or any other genomics facility—or any science lab(yrinth) at all, for that matter.

If you continue reading the numerous usages provided by the *OED*, you get the sense that solicitude is a kind of heightening or magnifying of care—"a kind of supra-type" of care. In solicitude's sense of "anxious, special, or particular care or attention," the *OED* gives us Thomas More's *Works* of 1557: "What diligence can here suffyse vs? What solicitude can we thynke here ynough? agaynste the cummyng of thys almightye king." That one may not scan so well as a lab motto, but it does suggest the kind of grand, even apocalyptic scale that may be at stake—as well as the feeling that one might just not be adequately equipped for the imminent event.

I read and use solicitude here to indicate a kind of *scaling up* of care, less in terms of magnification or intensification and more in terms of what it takes as its object: solicitude takes a whole as its object of concern.[16] Solicitude warms to and worries over an entirety, verging on the abstract or ideal. You care for loved ones; Others elicit your solicitude. If curation keeps keeping a precarious thing together, and scrupulousness attends to each node in a proliferating articulated networking of precarious things, solicitude is more infrasuprastructural in scope and intent.

To further characterize the diffraction patterns of solicitude in the worlds of the sciences, consider how Jacques Derrida made solicitude practically synonymous with the entire operation called "deconstructing." In an early essay, "Force and Signification," Derrida scoped out what it means to *solicit* a structure:

> Structure is perceived through the incidence of menace, at the moment when imminent danger concentrates our vision on the keystone of an institution, the stone which encapsulates both the possibility and the fragility of its existence. Structure then can be methodically threatened in order to be comprehended more clearly and to reveal not only its supports but also that secret place in which it is neither construction nor ruin but lability. This operation is called (from the Latin) *soliciting*. In other words, *shaking* in a way related to the whole (from *sollus*, in archaic Latin "the whole," and from *citare*, "to put in motion").[17]

Solicitude attends, then, to entire structures: experimental systems, research programs, databases and their curation, institutions and their historical trajectories, emerging systems of thought and practice, standards of scrupulousness, cultural patterns of habits and desires. And "once a system has been 'shaken,'" Gayatri Spivak adds in her commentary on this crucial term, "one finds an excess which cannot be construed within the rules of logic, for the excess can only be construed as *neither* this *nor* that—or both at the same time."[18]

As a kind of care, then, solicitude is patterned activity that puts everything in motion, puts everything up in the air; by "menacing" entire structures, worrying them and even trying to break them, solicitous care comprehends them, embraces them to shake them—not to ruin, destroy, or debunk those structures (as is often and wrongly said about deconstruction), but to comprehend their "secret" movements, to sound out the loose connections, frictions, breakages, and omissions, and to sense some excess *beyond* that still escapes their logics yet, funnily, makes them possible and *structured.*

I am quite sure that deconstruction and the life sciences share a great deal, and can benefit from mutual analysis and critique. In fact, there are few things of which I am more confident and to which I am more committed. It is because both living systems and semiotic systems (and biosemioticians at least, if not all life scientists, think living systems *are* semiotic systems) are systems of differential reproduction.[19] So, too, are the sciences. To recap how life scientist, historian, and deconstructionist Hans-Jorg Rheinberger has collapsed the differences between science and deconstruction in his analysis of experimental systems,

> The coherence over time of an experimental system is granted by the reproduction of its components. The development of such a system depends upon eliciting differences without destroying its reproductive coherence. Together, this makes up its differential reproduction. The articulation, dislocation, and reorientation of an experimental system appears to be governed by a movement that has been described as a play of possibilities (*jeu des possibles*). With Derrida, we might also speak of a "game" of difference. It is precisely the characteristic of "fall(ing) prey to its own work" that brings the scientific enterprise to what Derrida calls "the enterprise of deconstruction."[20]

This structural pattern occurring in living systems, semiotic systems, and experimental systems becomes especially apparent when trying to understand complex systems and conditions, such as asthma, asthma genetics, and the complex research systems that try to elucidate the GxE interactions that constitute them—the interactions not that they undergo or undertake, but that they *are.*

In their survey of emerging genome space circa 2000 (discussed in chapter 3), Ognjenka Vukmirovic and Shirley Tilghman described structural changes not only in genome research space, but in financial, disciplinary, and campus spaces as well: "It is hardly a coincidence that many universities and research institutes, including our own, are making major investments in multidisciplinary life-science initiatives to explore the complexity of living things."[21] The new appreciation for biological complexity in genomic science—that "funny thing" that happened to the grail-like quest for a mechanical, deterministic genome harboring a "secret of life" needing only to be "decoded," the encounter with an excess that the system couldn't construe—was the dominant theme of their article.[22] But they also understood that these more complex understandings were the product of the major capital investment the US government had made in the HGP, and these now needed to be sustained and furthered by additional major private investments in genomic structures and infrastructures. Their article concluded with this summary flourish:

> Organisms are networks of genes, which make networks of proteins, which regulate genes, and so on *ad infinitum*. The amount of complex data that will be generated, and the need for modeling to understand the way networks function, will ensure that disciplines outside of biology will be required to collaborate in this problem, if the ultimate goal to deconstruct such networks is to come to fruition.[23]

"Ad infinitum" is surely hyperbole, but hyperbole may be the most appropriate trope when writing about a whole and its holistic understanding when it is clear they have not arrived, and may *never* arrive. Yet somehow from this inaccessible future, the whole provides the context, the promise from which present biology takes its meaning. Can we say, exactly, what wholeness is and how it acts on its constituent parts? I'm betting the answer to that has been and will continue to be "no"—ad infinitum. And I am equally confident that the question, because of its impossibility, is a necessary one, and that it is wise, or at least not stupid, to invest in its asking.

There's another funny thing to add to our list of genomic jokes and surprises: Vukmirovic and Tilghman characterize their "ultimate goal" as "to deconstruct such networks"—this is my ultimate goal, too! And even though we come from different disciplinary spaces with different understandings of what "to deconstruct" means, and even though *deconstruction* has the potential to lead to all kinds of disagreements and shutdowns between genomicists and ethnogrammatologists like myself, I am going to argue that we are all of us doing our best work when our goal is "to deconstruct," that genomicists and ethnogrammatologists truly share ways of thinking and reading that each qualify as deconstruction—and that "solicitude" is another not-bad word for these shared deconstructive tendencies.

Solicitude (and deconstruction) embraces and comprehends structures, rattling and stressing them to better sense how they resist and hold up, where they are prone to slip, where the necessary play in the system is located, and what exceeds the system and remains unencompassed by it. To give one more example of the play and the work of solicitude, and of the affinities between the practices of biology and deconstruction, and of how they each depend on structures that are simultaneously closed and open, possible and impossible—that is, structures that are double binds—let's consider briefly an important computer program in genomics called, aptly and simply, STRUCTURE.

We stay around the year 2000, before the HGP was marked "completed," that is, whole. But the "avalanche," "flood," "deluge," "firehose" or "other catachresis of excess" of genomics data had already arrived, producing the new "genome space" that in turn provoked the anxieties, excitements, and the scrupulous care described in previous chapters. Another line of care work at this time involved the writing of new computer programs that "clustered" genomic data excesses into meaningful groups. STRUCTURE, created by Jonathan Pritchard, Matthew Stephens, and Peter Donnelly of the University of Oxford's statistics department, was one of the first of these, and became widely used in a variety of genomic research contexts in which "a crucial first step is to define a set of populations."[24] The first crux, then, the first decisive X marked across the genomic data avalanche, is a definitional limit placed on the whole of a data excess.

The data might concern any organism (birds, Angora goats, Norwegian spruce trees), but human genomic data was their focus in that paper, as it is for us here; it required special care. Researchers might put STRUCTURE directly toward the (population geneticists') end of categorizing or defining human populations—that is, assigning people to categories of "race." The authors described this as a "typically subjective" process, based on "linguistic, cultural, or physical characters, as well as the geographic location of sampled individuals." The problem, in their view, was to decide whether "populations based on these subjective criteria represents a natural assignment in genetic terms," and "to confirm that subjective classifications are consistent with genetic information and hence appropriate for studying the questions of interest."

For some researchers, then, STRUCTURE promised what was called a "race-neutral approach" to the study of population differences.[25] Physical anthropologist Deborah Bolnick provides us with an important critical reading of STRUCTURE's limitations in such contexts, and how "race" tends to become reified even as it supposedly becomes neutralized.[26] My focus here, however, is less on the use of STRUCTURE to assign individuals to large-scale categories of human difference such as "race" (not good) or geographic ancestry (less bad), but more on how the program is used in the context of other types of genomic research. STRUCTURE's limits, as critiqued by Bolnick, remain a concern in these other uses, too, but as we will see, these are limits that are recognized (to again use that mindy word) by STRUCTURE's creators. These limits confound the easy distinction the authors sometimes make between "subjective" and "natural," double binding those together in an experimental place that calls for solicitude.

As a first cut we can say that STRUCTURE is structured by two sign systems or languages: a mathematical/statistical semiotic system, and the semiotic system you are making sense of now, displayed on the page or screen before you. Using that too-simple but useful cut, we can place the Monte Carlo methods, Gibbs samplers, nested equations for inferring prior probabilities, and other mathematical and technical dimensions of STRUCTURE on the "nature" side of the conventional nature/culture binary, while *this* semiotic system (even if it is sometimes called "natural language") can be placed decisively on the side of "culture." This cut, which also corresponds to the science/care double bind structuring my text here (and the structure/play binary, too) is a good enough cut, despite *and* because of its limitations. But in actuality the technical-mathematical semiotic operations are always entwined with a rhetorical-pragmatic semiosis that is more accessible to most readers and more open to deconstruction, always needing to be read. In other words, the program STRUCTURE is a double bind of these two systems, simultaneously a structure of "play and structure" and a play of "structure and play." Or, we could say, both statistics and rhetoric concern, and are constituted by, the structure, sign, and play of a flood of differences, a flood or avalanche that has to be rendered meaningful, that is, limited, that is, needing our solicitude.

Or, as the writers of the program STRUCTURE put it more succinctly in the *User's Guide* published several years and several versions later: "While the computational approaches implemented here are fairly powerful, some care is needed in running the program in order to ensure sensible answers."[27]

Something beyond calculation becomes equally necessary, and "care" in that sentence as in so many others in the scientific literature, is best construed as signifying the beyond of the calculable. The formal had to be crossed with the informal, methods crossed with a catachrestic care, if sense was to be made of data.

In its first decisive cut, then, STRUCTURE takes a vast set of unorganized genomic differences and clusters that data into *K* groups—one, two, five, or more populations that share some sort of sameness as established by the software's algorithms. Sequence the same small portions of the genomes of several thousand Norwegian spruce trees, for example (and by the way, conifer genomes are sometimes called giga-genomes because, at twenty to thirty billion base pairs [bp] each, they are seven to ten times larger than the three billion bp genome of a human)—and run the resulting data through

STRUCTURE, which will tell a researcher how many groups (K=2,3,4,5 . . .) those thousands of trees can be organized into, that is, the structure of the population.

Well, it will "tell" researchers a number like that if the whole procedure has been "treated with care":

> In our paper describing this program, we pointed out that this issue should be treated with care for two reasons: (1) it is computationally difficult to obtain accurate estimates of Pr(X|K), and our method merely provides an ad hoc approximation, and (2) the biological interpretation of K may not be straightforward. In our experience we find that the real difficulty lies with the second issue. Our procedure for estimating K generally works well in data sets with a small number of discrete populations. However, many real-world data sets do not conform precisely to the structure model (e.g., due to isolation by distance or inbreeding). In those cases there may not be a natural answer to what is the "correct" value of K. Perhaps for this kind of reason, it is not infrequent that in real data the value of our model choice criterion continues to increase with increasing K. Then it usually makes sense to focus on values of K that capture most of the structure in the data and that seem biologically sensible.[28]

Invoking the "ad hoc" is a sure signal that scrupulous solicitous care is needed, especially in conjunction with a rhetoric of negative qualification and hesitancy ("not . . . straightforward," "not conform precisely," "not infrequent," "usually," "seem"). When poststructuralists enclose key terms like "correct" in scare quotes, this textual device often elicits dismissive scorn; postgenomicists drop it into their texts as a matter of course, and as a matter of care, without incident.

These rhetorical devices characterize not only the less formal documentation, but the more formal published scientific article:

> In fact, the assumptions underlying (12) are dubious at best, and we do not claim (or believe) that our procedure provides a quantitatively accurate estimate of the posterior distribution of K. We see it merely as an ad hoc guide to which models are most consistent with the data, with the main justification being that it seems to give sensible answers in practice (see next section examples). Notwithstanding this, for convenience we continue to refer to "estimating" Pr(K**j**X) and Pr(X**j**K).[29]

The documentation that guides STRUCTURE's users in these technoethics of care is replete with this dissembling rhetoric, constituting a genre of "informal pointers" that is a necessary supplement to the "firm rules" of pointed, formal mathematics:

> There are a couple of informal pointers which might be helpful in selecting K. The first is that it's often the situation that Pr(K) is very small for K less than the appropriate value (effectively zero), and then more-or-less plateaus for larger K, as in the example of Data Set 2A shown above. In this sort of situation where several values of K give similar estimates of log Pr(X|K), it seems that the smallest of these is often "correct."
>
> It is a bit difficult to provide a firm rule for what we mean by a "more-or-less plateaus." For small data sets, this might mean that the values of log Pr(X|K) are within 5–10, but Daniel Falush writes that "in very big datasets, the difference between K=3 and K=4 may be 50, but if the difference between K=3 and K=2 is 5,000, then I would definitely choose K=3." Readers who want to use a more formal criterion that takes this into account may be interested in the method of Evanno et al. (2005).
>
> We think that a sensible way to think about this is in terms of model choice. That is, we may not always be able to know the TRUE value of K, but we should aim for the smallest value of K that captures the major structure in the data.[30]

The STRUCTURE program, despite what its name might suggest, does not supply rigid, unyielding, clearly delineated results. Its use can produce, in the language of its authors, "anomalous outcomes" and require supplemental "assessments" that involve "statistical difficulties" that are "not rigorous" and may lead to results that are "quite difficult" to "interpret."

My analysis does not reveal how the truths of genomics are "socially constructed" by statistical programs like STRUCTURE. Demolishing STRUCTURE or deriding its scientificity is the last thing I want to do, or am capable of doing. Instead—and I apologize in advance for this sentence, but I am trying hard to be careful—I have deconstructed the structure of STRUCTURE to comprehend how it structures data to capture the major structure in data that is structured like an unstructured avalanche. And I've learned this in no small part from genomicists themselves.

The genomicist's STRUCTURE shares more than the same name with the "structures" that Derrida and other so-called deconstructionists solicit in order to comprehend them "more clearly." They all strive to "recognize the limits" (re-sounding one of the demands for care issued by Patrick Sullivan and his psychiatric geneticist coauthors in chapter 5) of any system, experimental and/or textual, to comprehend how those limits produce both insight *and yet* blindness, simultaneously. They share an appreciation of the fragility and the limits of a system that call for and solicit our attentive, vigilant care. They share structural logics; they share a spirit of deconstruction that drives an effort to comprehend the whole, through shaking and disturbing it in its entirety, to "more clearly" (I hope I have reiterated this Enlightenment value sufficiently) comprehend the limits of the structure, how those limits are productive, and how something always exceeds those limits, a remainder.

So when Vukmirovic and Tilghman, or any other genomicists, speak or write of powerful yet precarious networks within networks constituting a systems biology that has yet to be fully comprehended: I hear them. Reading Vukmirovic and Tilghman, Pritchard, Sullivan, and many other genomicists has helped me comprehend more clearly—has helped me "to deconstruct"—biology and biological systems, as well as so-called deconstruction itself. I try to hear and read them carefully, not just what I think I want to hear and read, but the entirety of their work, the fullness of their thoughtfulness and carefulness. In other words I, too, practice solicitude.

GxE Research in Asthma

Solicitude, then, is a kind of care that seems to come from beyond a whole structure *and yet* permeates that whole structure; when we solicit a structure we worry it and attend to it from its *supra-*, its *infra-*, its outside inside, inside out. Solicitude's objects are not the myriad parts of a system and their connections, as taken by scrupulousness, but a whole and/at its limits. That whole structure can be a complex bioinformatics program like STRUCTURE; it can also be the structure of an entire research program, like genome-wide association studies, in which a program like STRUCTURE is one part among many others.

But to comprehend a whole presumes an analyst outside it, in which case the whole was not actually whole. It may be necessary to take a step back to see, if not the whole then at least "the big picture"—and now you've made a bigger picture encompassing the now smaller, incomplete whole. And where or what is that place you stepped back to?

Here is another structuring double bind of an experimental science like genomics: ideas (i.e., differences) come from the eighth dimension; there is only an eighth dimension once you've made it to the ninth dimension, and then . . . ad infinitum.[31] Minding genomics happens at the limit, where a scientist encounters the contradictory orders: comprehend a whole; there is, as yet, no outside from which a whole can be comprehended. *And yet* it happens that the limit . . . recedes? shifts? wavers? . . . and remains, and reimposes itself as a limit.

To further our understanding of the extreme-difficulties-to-the-verge-of-impossibility of GxE research, and the structure and play of solicitude in genomics research as it grappled with GWAS's limits, this section zooms in on the research trajectory of Erika von Mutius. Following her investigations, spanning decades, into the genomics, biology, and epidemiology of asthma that were introduced in chapter 5 leads us again to *toll*-like receptors (TLRs). These biosemiotic "pattern recognition molecules," or sign-things protruding from cell membranes, recognize molecules in some place's air—a Leipzig day-care center, a neighborhood in Munich, or a small farm somewhere in rural Austria—to transduce signals to a cellular interior.

Although I am running this narrative through von Mutius, let me make clear: I haven't interviewed her or even spoken to her informally, nor have I spent time in any of the labs she's been associated with, as one might expect of a proper ethnographer. But as my ethnogrammatologist's interest in TLRs developed, so did my interest in von Mutius, and as her many publications accrued in my digital folders, I began to better understand what most interested me about her, what took me a long time to be able to name as her solicitude. She has a strong sense, in my reading of her and her work, that the real complexities of a condition like asthma are such that it will always force a requestioning of it—its symptoms and causes, its etiologies and epidemiologies, its conceptualization, its harms—along with one's knowledge of it. That sense of solicitous care forces a continual reexamination and redesign of methods, approaches, and structures.

Maybe because she was first a clinician and a pediatrician, and not a geneticist, von Mutius seems particularly adept at the kind of whole-genomics-structure-shaking that constitutes solicitude. Or it may be that, in the words of Fernando Martinez, her mentor early in her career and later her colleague and coauthor, "In stereotypical terms, Erika has Latin passion and German practicality. In science that's a winner as you can develop creative new approaches then work with precision to ensure their fruition."[32] That's undoubtedly hyperbole, as Martinez acknowledges with his prefacing qualifier of "stereotypical." But since my "solicitude" is itself hardly unaffected by hyperbole, I ask the reader to read through the stereotypes and retain the patterning of conjoined opposites that both Martinez and I sense are necessarily at work in good, "winning" science: the precision that is so dominant (and imagined as exclusively so) in the familiar characterizations of sciences like genomics, tightly assembled with the more elusive or tacit dimensions that we commonly gesture toward with signs like "creativity," "night science," "care," "passion," and "solicitude." Only by being solicitous, in my view, by caring for wholes and shaking entire structures of thought and practice, can one be credited with such a large-scale feat as (in the words of a *Lancet* profile of von Mutius) "reshaping the landscape of asthma research."[33]

Early in her research trajectory, and an early sign of solicitous care for us, von Mutius expected to find one kind of environmental effect shaping asthma incidence and experience, but found almost exactly the opposite. In the mid-1980s, before the word "genomics" had been coined, she was studying the effects of air pollution on croup in children at University Children's Hospital in Munich, and "felt it had just all been a mess." Trying to avoid being in charge of another project and producing another mess, she cast her next grant proposal in such large and ambitious terms that she expected it to go unfunded. It was funded. That surprising event was accompanied by another: the fall of the Berlin Wall and the reunification of Germany in 1989. Those events together provided an opportunity to collect health and environmental data in ways that had not been possible for decades; von Mutius designed a study that would compare asthma rates and prevalence in Munich and Leipzig, populations that for all practical purposes could be considered bio-ethnically identical, but had experienced strikingly different environments. "I'd long been thinking that air pollution levels were very low in Munich and had to be very high in industrial East Germany," she recalled to the *Lancet*, "but I knew that for propaganda reasons the [Communist] regime would never allow a comparative study." She struck up a collaboration with two other doctors in Leipzig and Halle, "where the river Saale had run violet with toxic waste," and prepared to document what she anticipated to be strikingly higher asthma rates in the former East Germany than those in West Germany. But that would be the third surprise: "When her team analysed their results they were left speechless: the prevalence of hay fever and allergic sensitisation was about 50% lower in these areas than in and around Munich."[34]

Or as von Mutius and her coauthors reported it, in more scrupulous statistical terms,

> The prevalence of hay fever diagnosed by a doctor was significantly lower in Leipzig than in Munich (2.4% (24) v 8.6% (410); p<001). Typical symptoms of rhinitis—runny, stuffy, or itching nose during the previous 12 months—were also reported less often in Leipzig than in Munich (16.6% (171) v 19.7% (961); p < 0 05). Allergic triggers of rhinitis symptoms such as grass, pets, or dust were reported also less often by parents from Leipzig than from Munich.[35]

By design, there's no place for speechlessness in the scientific journal article's speech. We have to trust that, as with Nüsslein-Volhard's "*toll!*" moment, von Mutius said something (*sprachlos*? *stumm*?) to that *Lancet* writer that he honestly recounted while probably embellishing an affective moment of surprise-startle that reset her expectations and thought-patterns. (See chapter 3 again for Silvan Tomkins's analysis of this affect.) So, like any solicitous scientist presented with such small but not insignificant differences, von Mutius and her coauthors immediately engaged in some shaking of their own explanatory cage and study design.

Solicitude always has under- or overtones of scrupulousness, a piling up of the nagging, detailed questions into a mass of worries and auto-critiques, even if these are mostly there to be rhetorically parried and dismissed. "There may be some methodological limitations," they write:

> Questions concerning the lifetime prevalence are obviously subject to recall bias, which could have operated differently in the two populations. Until recently people in Leipzig and Munich have lived in a different social and political environment which could have influenced their attitude towards physical complaints as well as their responses to the questionnaire. Differences in the observed prevalence of doctor diagnosed asthma and bronchitis could be a reflection of different medical practice and diagnostic labelling in the formerly separated states. It is possible that some of the children with bronchitis in Leipzig would have had asthma diagnosed if they were in Munich. Parents of children with bronchitis, however, reported wheezing, attacks of shortness of breath, and nocturnal cough with similar frequency in both cities. . . .
>
> Differences in diagnostic labeling between the two cities could have influenced our findings. Nevertheless, not only were typical symptoms of rhinitis less common in Leipzig than in Munich, but parents in Munich also indicated allergic triggers and the typical seasonal pattern of rhinitis symptoms more often than those in Leipzig.[36]

What might explain these surprising, maybe even *toll!* or *stumm!* results? The authors preferred to end by deferring any attempt at an answer, beginning with a marking of limits that scientific articles do routinely, and that is another hallmark of solicitous, scrupulous care:

> Our study based on only two centres does not allow us to attribute disease differentials to specific causes. We aimed at giving a descriptive comparison of the prevalence of asthma and allergic disorders among children

> living in western and eastern Germany. Further studies are currently underway to address aetiological questions in more detail. The apparently lower prevalence of allergic disorders in Leipzig was particularly interesting. These findings could point toward aetiological factors for allergic disease that are associated with the lifestyle and living conditions in Western industrialised countries.[37]

How do you get from a descriptive and statistical analysis to an etiological, causal explanation? "Further studies," of course, necessitating another round of unexciting grant-writing. But something more than that must have seemed necessary, too. Here again the *Lancet* article and Martinez ascribe von Mutius's difference to "creativity," but I encourage readers to think in terms of solicitude, that reaching for or comprehending of an entire structure so that it can be turned over, shaken up, and rattled:

> During her fellowship at the Respiratory Sciences Center at the University of Arizona, in 1992, Martinez encouraged von Mutius to think creatively when researching instead of just collecting and analysing statistics. "Fernando's reflective, philosophical approach helped me make more sense of my own findings. It was only when we discussed a 1989 article by David Strachan that had shown how children with many siblings had higher resilience to allergies, that I could tie this in with how very young infants in East German state-run childcare facilities also spent lots of time surrounded by other young children. He was a real inspiration."[38]

Initially expecting to find a linear effect in which a polluted environment causes increased asthma and allergy rates in industrialized East German cities, von Mutius ended up affirming, or at least adopting as an explanation to explore, a version of David Strachan's "hygiene hypothesis," in which early childhood exposure (later research would extend it to prenatal exposures as well) to microbes, or microbial products, or *something* in an environment confers some small but not insignificant protective effect against allergy or asthma, in at least some individuals in a local population.

As the term hypothesis might suggest, this was not a broadly accepted, widely acclaimed, thoroughly supported, or, to put it simply, very popular position to adopt.[39] Which in turn suggests that some degree of careful solicitude is a necessary virtue for the adoption of "maverick" positions in the sciences—a highly developed but not fully articulate sense of a whole explanatory system, its limitations, its points of play, and its uncertainties, gaps, and omissions. Solicitude—and this is where the impossible double bind kicks in again—requires both a thorough comprehension of a whole, and also exposure to the blank incomprehensibility of an outside. You must be thoroughly composed, and out ahead of yourself.

Strachan recollected, ten years after he first proposed a "hygiene hypothesis,"

> At first this hypothesis was received with scepticism because the prevailing immunological thinking considered infection as a potential trigger of allergic sensitisation rather than as a protective influence. However, during the early 1990s a plausible mechanism arose from the distinction of Th1 and Th2 lymphocyte populations in laboratory animals and the recognition that "natural immunity" to bacterial and viral infections induces a Th1 pattern of cytokine release, potentially suppressing the Th2 immune responses involved in IgE mediated allergy.[40]

Strachan's hypothesis was out ahead of developments in immunology, where the signaling pathways led in one direction only, from infection to allergy. It would take a while (too long to narrate here) for immunological science to elaborate these more complex mechanisms, leveraging numerous new biotechnological tools, including monoclonal antibodies, and healthy infusions of capital.[41] But eventually these infectious signals were understood to be recognized in more ways than the officially recognized one, diverging to include the "natural immunity" associated with the different T-lymphocytes, as Strachan describes, as well as the newly discovered *toll*-like receptors.[42]

Von Mutius's research focus shifted then from urban settings to more rural ones, and from day-care centers to farms. With growing cohort sizes and the growing number of collaborators going hand in hand—solicitude needs a powerful social engine—a series of publications in the late 1990s substantiated the hygiene hypothesis. By 2001 Von Mutius and (by then) nine coauthors, plus another seven members of "the ALEX [Allergy and Endotoxin] Study Team," scattered across Munich, Basel, and Salzburg, could confidently state that "growing up on a farm protects against allergic sensitisation and development of childhood allergic diseases," including asthma, and that "regular contact with farm animals confers an important protective effect in such an environment."[43] The large number of participants gave their studies the statistical power to validate such claims, but even though the statistical correlations were robust, they were still in the dark when it came to actual mechanisms and molecules, including genes:

> The mechanism by which time spent in a stable and consumption of farm milk protect against development of asthma and atopic sensitisation is not known. . . . Farm children are exposed to higher inhaled concentrations of endotoxin in stables and in dust from kitchens and mattresses than non-farming children. However, swallowing could be another route of exposure to bacterial products. This theory is lent support by the independent protective effect on atopic sensitisation of . . . farm milk consumption and exposure to stables in the first year of life. Farm milk, which is usually raw, contains more gram-negative bacteria and thus lipopolysaccharide, than pasteurised milk. Therefore, the protective factor associated with consumption of farm milk could be associated with ingestion of noninfectious microbial components, with resultant changes to the commensal gut flora, or both.[44]

Indeed, the group was somewhat dismissive of any genetic component:

> We have previously shown that children who did not grow up on a farm but who had regular contact with farm animals had low frequencies of allergic diseases and atopic sensitization similar to those of farmers' children. Thus, genetic background probably does not explain the protective effect of early exposure to farming on development of asthma and allergy.[45]

But that would soon change, in concert with the changes in immunology brought not only by the discovery of *toll*-like receptors as a mechanism of innate immunity, but by the fairly rapid elucidation of population differences, that is, different people harboring different genomic polymorphisms. So, only a few years later, a 2004 publication, still grounded in the ALEX Study team and its 609 rural Austrian and Bavarian children, but now including von Mutius's former mentor Martinez, could add another hypothetical layer to the hygiene hypothesis:

> There have been numerous reports of an increase in the prevalence of asthma and allergies over the past decades in countries with a high socioeconomic status that cannot be explained by a change of the genetic background.
>
> Of interest in this regard are recent studies in Europe and North America pointing to the key role of a strong environmental influence because children and adults raised on animal farms were consistently found to have a lower prevalence of asthma, hay fever, and IgE-mediated reactivity

to local allergens than those living away from farms. Because a large variety of microbes is detectable on animal farms, these findings are in line with the hypothesis that exposure to microbial products in early life modifies immune responses away from asthma and allergies.

Many bacteria have pattern molecules on their surfaces, and these molecules interact with pattern-recognition receptors, which are part of the innate immune system. Among the known pattern-recognition receptors for microbial products, toll-like receptors (TLRs) are an evolutionarily conserved group of molecules expressed in antigen-presenting cells and epithelial cells. . . . . The genes encoding TLRs show a high variability in human populations, but whether these genetic variations modify the interaction with microbial molecules and thereby modify the frequency of asthma and allergies is not yet known.

We therefore reasoned that if microbial products are responsible for the lower prevalence of asthma and allergies among farmers' children and in children heavily exposed to endotoxin, then polymorphisms in TLR2 and TLR4 might modulate these effects.[46]

This was a much more complex and ambitious project and study design than that behind the 2001 publication, and became one of the first demonstrations of geneXenvironment interactions in asthma and allergy. It was also a much more expensive project; the involvement of Martinez's lab added NIH funding to the effort, to cover the genotyping of those 609 people (229 from farms, 380 from nearby nonfarming towns). To reiterate yet again: minding genomics is resource-intensive. It's also important to recall that, for all this effort and expense, you only ever get tentative results, concerning partial effects. Less than ideal—but not bad.

Let's take a closer look at the effort, and the small—but still *toll!*—effects observed. The platform (employing Applied Biosystems technologies) basically worked like this: the team prepared sixteen different oligonucleotide probes, short segments of DNA constructed to match up with and bind to the eight different polymorphisms of the TLR2 and TLR4 genes being studied, and fifteen PCR (polymerase chain reaction) primers to go along with them. The different probes and primers were placed into different wells in a 384-well plate, and then DNA prepared from each of the different blood samples is placed in each well. If there's a match—unknown sample DNA to known probe, a more straightforward kind of double bind—the DNA segment gets amplified by PCR, and that particular sample (say, nine-year-old farm-dwelling child #178) and that particular genome variant (say, TLR2/-16934)

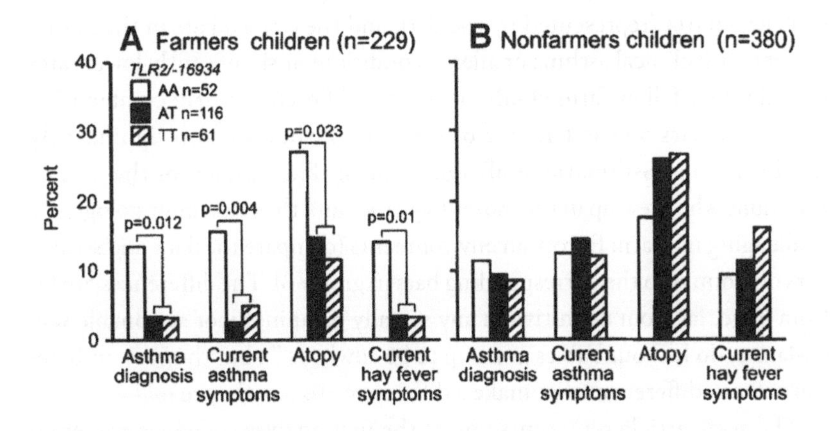

FIGURE 6.3 · Modest differences in and from TLRs. From Eder et al., "Toll-Like Receptor 2 as a Major Gene for Asthma in Children of European Farmers," 2004.

then form a signal, one bright spot in a matrix of 384 other bright and blank spots. Many layers and forms of scrupulous attention are glossed over here, but this is basically how you genotype a (small, but just big enough for statistical power) population for a small number of genetic variations in a few select genomic locations.

And then you run all manner of statistical analysis—also scrupulously, and also glossed over here to keep an already glacial-paced narrative from freezing over completely. There were really only two significant positive results from this set of experiments; one of them was the graph shown in figure 6.3.

The first thing to read here is that the differences in allergy or asthma prevalence (the y-axis represents the percent of the population diagnosed with asthma or atopy) between farmers' children (A, the left half of the graph) and nonfarmers' children (B, the right half) is not particularly striking overall; as a group, whether you grew up in the company of dirty, smelly, shedding animals of multiple species and their multiple insect, worm, and microbial symbionts or whether you grew up in a village, town, or city largely devoid of these kinds of cross-species interactions makes some difference but not a huge difference in terms of your likelihood to suffer from asthma.

But when the researchers subdivided the groups according to which versions of the *toll*-like receptor 2 gene their cells harbored, the pattern gets more interesting: children who grew up on a European farm *and* who by chance were born with either of the different polymorphisms of AT or TT of TLR2/−16934 (represented by the dark and the striped bars in the graph) present with clinical asthma or allergy conditions at significantly lower rates than do their fellow farm children with the different AA SNP combination (the white bars on the left side of the graph); they are also significantly less likely to be asthmatic or allergic than *any* kid with *any* of these TLR2 variations who grew up in the more "hygienic" and thus less immunologically challenging nonfarm European environments (compare the dark and striped bars of group A to the corresponding bars in group B). The differences are far from huge, far from definitive in any socially meaningful or actionable way ("Mamas, do let your babies grow up to be cowboys!"), but they are, in Bateson's terms, differences that make a difference. To me, they're *toll!*

The 2004 article represents one of the first analyses of genes and environments acting together in/on organisms—in this case, several hundred German *Kinder*. Wouldn't this contradict my claim, or at least challenge my hypothesis, that geneXenvironment research is impossible? To begin to respond to that question, let's look at the self-described limits of that experiment, the authors' solicitous attempt to mark out the actual and potential limitations on their own claims.

"It is not possible," they write, "to deduce from our study the molecular mechanisms involved in the interaction between polymorphisms in TLR2 and exposures present in children raised on farms because the functional significance of this polymorphism is not known."[47] So, saying that this article demonstrates the possibility of studying gene-environment interactions also has to acknowledge that it's "not possible" to actually say what the gene does, let alone what difference the different polymorphisms of that gene make for that gene function, and that it is impossible to actually say very much at all about the actual mechanism of that interaction.

And if it was possible for the authors to write that "In our study we found that a genetic variation in TLR2 (TLR2/−16934) is a major determinant of susceptibility to asthma and atopy in farmers' children. That microbial products present in a farming environment might interact with TLR2 is supported by the finding of an increased expression of this receptor in blood cells obtained from children of farmers compared with children not raised on farms," then solicitude demanded that they write how that might also *not* be possible:

> Other polymorphisms that are in linkage disequilibrium with TLR2/−16934 could be responsible for the observed associations. . . . It is not known whether a specific microbial molecule is essential for the observed associa-

tion between asthma and the TLR2 polymorphism in farmers' children or if activation of a common intracellular signaling pathway by any of the TLR2 ligands is sufficient. TLR2 interacts with components of a variety of gram-positive and gram-negative bacteria, mycobacteria, yeast, and parasites, which are likely to be abundant in an animal-farming environment, but the design of this study does not allow us to identify the specific pattern-recognition molecules underlying our observed association. . . .

The observed associations in farmers' children are unlikely due to population stratification. We found the associations in Austrian and in German farmers' children, and in both countries there were no such associations among children of nonfarmers. However, replication of our results in an independent study is clearly needed. We cannot exclude the possibility, for example, that avoidance of exposure to TLR2 ligands and other microbial products might have occurred in homozygotes for the A allele in TLR2/−16934, thus making them more susceptible to asthma and allergies among farmers' children. Further studies are needed to identify the TLR2 ligands responsible for the observed association to exclude the possibility that preferential avoidance of certain exposures might explain our findings.[48]

Tagging the unknown riddling the newly known, marking out what is not allowed within the current allowances, binding the possible sense to the simultaneous impossibility of excluding other possibilities—these are the hallmarks of whole-shaking solicitude, the signs of careful scientists minding their data, craft, and knowledge. To be solicitous is to be "ever on the verge," an excited-worried, interested-anxious state that is the dominant form of scientific experimental life: always at the limit where it may work and may also fail, experienced as that uneasy double bind of threat and promise.

Repetition of Impossibility, Step-by-Step

Let's pause here to restate and (maybe) clarify a few things. This is not a narrative built to convey what exceptional scientists Erika von Mutius and her colleagues are, even though they may be. This is also not a narrative about what exceptionally important molecular structures *toll*-like receptors are, or the polymorphic genes associated with them, even though they may be. And this is not a narrative of how GxE research came into existence in the first decades of the twenty-first century and how it grew as a field of research.

My narrative, based on a particular form of close reading rather than any semblance of historical or ethnographic comprehensiveness, has instead tried to convey the unexceptional qualities of these scientists and their sciences—their quotidian and under-attended-to practices of solicitous care, their continued minding of experimental, conceptual, and organizational whole-systems. It's a narrative produced by experimenting with the question: what can scientific writing be made to do? What other signaling pathways can be opened for the supposedly dull and strictly denotative prose of the scientific article? In that sense mine is an infra-narrative of the infrastructures of mindful care that make genomic infrastructures and genomic sciences possible, although I think they would be better written as imXpossible. Those infrastructures of care endure through time, across particular genomic "eras" like the candidate gene era (1980s–early 2000s), the GWAS era (mid-2000s onward), and into the GxE or the GWIS era (genome-wide interaction studies) as different kinds of experimental study designs become the dominant "whole" through which genomic truths are produced.

And undone: the not-insignificant linkage between a TLR2 gene and decreased risk of asthma in a population with exposure to *something* on farms, established through the candidate-gene study detailed above, could *not* be reproduced in a GWAS study barely six years later—a larger, more comprehensive study led again by Erica von Mutius and other researchers as well. Genotyping an even larger study population (for greater power) of 1,708 kids, and using an upgraded Illumina Human610 chip, von Mutius and colleagues, in what was now called the GABRIEL consortium, found that the "TLR2 SNP did not significantly interact with farming for asthma or atopy." Although this GWAS study did support linkages between a few other previously identified genes and asthma, there were really no new discoveries. "The failure to detect interactions for SNPs with interacting allele frequencies in the range 30% to 70%, for which our study was adequately powered, suggests that common polymorphisms are unlikely to moderate the protective influence of the farming environment on childhood asthma and atopy in Central Europe."[49]

The "failure" here, in the dominant conventional understanding of science, is the failure of reproducibility that, cumulatively across many scientific fields, constituted the "reproducibility crisis" discussed in chapter 5. A failure such as this is often read, metonymically, as a more general failure of a genomics whose reach for a whole continued to exceed its grasp as it pursued elusive and ever smaller genetic effects in ever more extensive and complex populations in ever more diverse and complex environments—all while continuing to attract a lion's share of public funding and public acclaim. Such failure is also linked in the conventional narrative to the "self-correcting" nature of science: the errors of a genomicists' ways that led down a path to unreproducible results were discovered, exposed, and later corrected so that genomics, like all proper sciences, as a whole progressed incrementally "closer to the truth."

But no matter how often or how seriously it is said, there's something funny about the customary phrase "closer to the truth." Saying it depends on a presumption of what that truth is and therefore what distance—even if incremental—has been covered in approaching it. This future truth, a fully whole truth given in "the fullness of time," coincides completely with the "real"—again, in the conventional conceptualization (substitute "metaphysics" if you like) of science. And this is precisely what it is impossible to access, or even view from a distance that is ever diminishing. How can you say you are closer to something when that something is not there and remains, by definition, unknown?

It's worth resisting that conventional reading, and its packaging with a conventional notion of (scientific) progress. ImXpossibility lets us read outside these bound conventional narratives of failure, self-correction, and successful progress toward completion, wholeness, full presence. In the narrative of possibility, genomics progresses by either failing or succeeding in the test of reproducibility. The motion is always upward because, in this set of conventions, up is always better, one step nearer the transcendent "God's-eye view." In a narrative of imXpossibility, a genomic science doubly bound to care places itself and needs to be understood at the limit, only "ever on the verge" of a success-and/or-failure-to-come. Reproducibility is a limit-function, an ability, capacity, or potentiality that always remains to be actualized. The truths of a more careful, better-minded genomics can't be said to be closer to the Whole Truth of the future, but genomics with care has solicited its limits diligently to arrive at some partial truth better, or at least less bad, than yesterday's.

More than the previous diffractions of care taken up here, solicitude may require us to think more with images. Maybe this is because solicitude takes shape from and as a relationship to a *whole*, a *gestalt*, a total system, or something else that, temporarily at least but perhaps permanently, exceeds our grasp. Scrupulousness, even when it is enacted through the form of a meta-analysis, can be imagined as directed at a series of identifiable procedures, attending to each of the components and connections of an experimental system. Solicitude would seem to be directed more up, out, or beyond. If scrupulousness requires scientists to get down on their hands and knees, actively and even aggressively zeroing in with exhaustive checklist and magnifying glass, solicitude finds us scientists leaning back and looking up or out into some indefinite distance, receptive, patiently attempting to approach an entirety.[50]

Here again the Penroses help us imagine this limit of imXpossibility visually. Their 1958 paper with the impossible triangle that figured earlier in this chapter ended with another impossible figure, an impossible staircase. Figure 6.4 was not published with their paper, which features two versions of the stairs, but can be found along with the paper's illustrations in a folder in the Lionel Penrose archive.

"Each part of the structure is acceptable," the Penroses wrote, "but the connexions are such that the picture, *as a whole,* is inconsistent."[51] I prefer this version of the staircase (I am not sure why the Penroses did not) with the man and his dog because the stairs alone in the published paper read as static and timeless; both members of these companion species are pictured suspended, ever on the verge, poised within the "connexions," facing forward and upward. All must seem perfectly acceptable to them. Were he able to

FIGURE 6.4 · Impossible staircase by Penrose and Penrose. Lionel Penrose, UCL Special Collection provided by the Wellcome Library, Shelfmark PENROSE/1/7/1, reference number b2021313x; persistent URL http://wellcomelibrary.org/player/b2021313x.

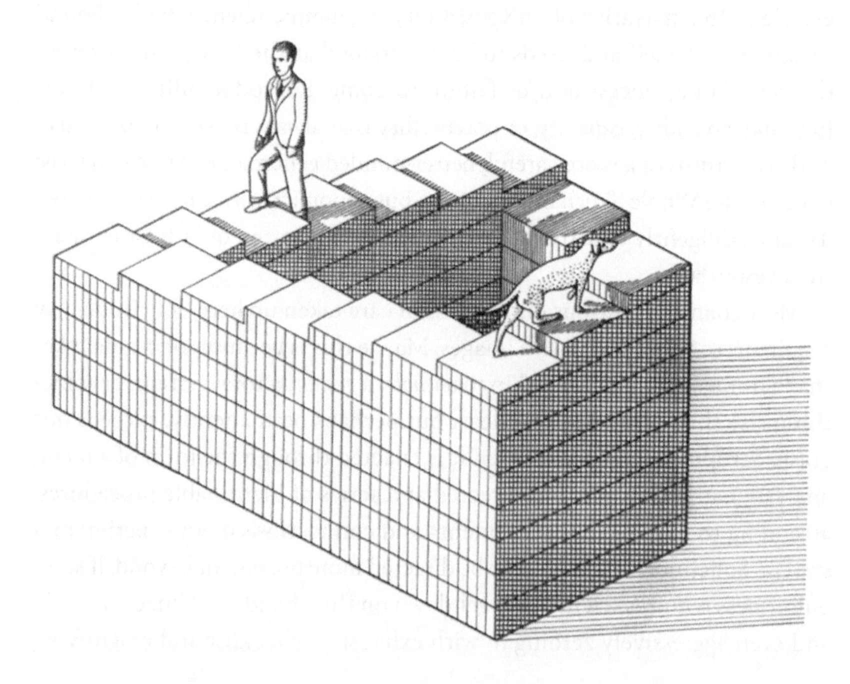

move, not to the eighth dimension but just the next one out, over, or up—*our* dimension—he could see the impossibility of his situation *as a whole.* As it is, he would just take the next possible step that presents itself. And the next, and the next, never closer and always further.[52]

It's an imperfect figural analogy for imXpossibility; what figure wouldn't be? I've shown it in formal presentations and informal contexts to scientists; almost all "get it," and as you might guess I would describe their most frequent response as: "*Toll!*" Cool, weird, interesting . . . and anxiety-producing. But I hope that its imXpossibility forces, as the Penroses put it, "sudden changes in the interpretation of distance"—here, the distance between the partial truths we always have and the full truth we like and maybe need to imagine we are getting "closer to." In the labyrinth life of genomics (and *all* experimental sciences), "closer to the truth" is not a particularly meaningful phrase, even if it's good for one's public or self-image. "Successfully stepping into new zones of truth blocked and opened up by the walls you have carefully albeit semi-blindly constructed" is more felicitous even if . . . no, *because* it's wordier and less congratulatory, and takes some effort and previous preparation to read. What complicates (to say the least) both statements is a limit that marks both a (possible) whole and the (impossible) beyond which, strangely, it both blocks and promises. It's that conjunction I'm marking as imXpossibility, a double bind of possible and impossible, part and whole, system and beyond.

There's another way to articulate the impossible double bind structuring the limits of experimental science, where formal logics offer no resolution and to which the appropriate response is solicitous care. In the "classic" form of the double bind the Bateson group outlined (discussed in chapter 3), a scientist-subject recognizes two conflicting messages, one at a "higher" logical order, that cannot be eluded or resolved but only endured: "Get outside the whole system to grasp it, to comprehend it, to understand how it all operates and where it and why it is no longer working"; "there is nothing outside the system, *il n'y a pas de hors-systeme*, at least nothing sensible."

Imagine yourself back in 2004, reading von Mutius's newly published article claiming significantly less risk of developing asthma for farm-raised children who harbor in their genomes a particular version of a *toll*-like receptor gene. Is it reproducible scientific progress, or unreproducible scientific failure? Is the experiment a promise or a threat? It's imXpossible to tell in 2004; the consortium's authors, having ascended many steps with scrupulous care, are poised once again "on the verge." Given time and money (same difference), the NIH could have funded efforts to directly attempt to reproduce these results, and we could have had an answer, eventually and hypothetically.

Although that happens slightly more frequently since the "reproducibility crisis," it is still unusual. Unless the stakes are fairly high, it's usually not worth anyone's time and money to conduct such boring, uninteresting, *affectless* studies. It's more likely that, over time, the "self-correcting" scenario comes into play as the larger scientific community either builds on or is unable to confirm previous work. All these outcomes are in the realm of possibility.

Reproducibility is not the mark of success in science, and missing that mark is not the sign of science's failure it is usually made out to be, often with a moralizing edge. Reproducibility is the limit of the system, a test less about science in its formally understood sense and more about a science-care double bind whose main quality is its iterability, the necessarily continual reconstruction and retesting of a limit. That's not to say there isn't failed and even pathological science; there most certainly is, and it is most certainly a social problem. It's a problem we should be interested in and should care about, and more and more people and institutions are. But it's a different problem that interests me here.

What is it possible to say, with no funny language business? Erica von Mutius and her colleagues were scrupulous and solicitous. They cannot be fully confident in advance that their significant findings will be reproduced—and indeed, the larger future version of their consortium and its larger future study and more encompassing study design, one step further. But they carefully minded genomics. What's continuous across these changing step-structures of genomic truths are the infrastructures of care that underwrote them—the solicitous, scrupulous, curatorial minding of genomics that made genome scientists (and sciences) better over time.

Again, my point is less about the repetitive *failures* of genomic analyses, although they are real, and more about the repetitive and at least partially successful or *heeded calls* to do and think genomics with care in new ways. Both failure and heedful solicitude are real; the former are more visible and draw more critical attention, but here I've tried to give the latter the ethnogrammatological attention I think they deserve as well.

Ever on the Verge

The "GWAS era" is generally cast as opening around 2006, driven in part by "next-generation" sequencing platforms capable of faster and cheaper genotyping, and the public release of data from the HapMap project that provided more detailed views into the patterns of genomic variation in more and more populations (although still almost entirely Eurocentric, another severe limitation).[53] Somewhat prematurely, perhaps, asthma researchers Carole Ober and Donata Vercelli in 2011 augured the arrival of a "post-GWAS era" in which study of gene-environment interactions would become more important—but, as they suggested in the title of their *Trends in Genetics* article, it was not clear if this was a "nuisance or opportunity." With the advantage of hindsight and a retroping under the sign of imXpossibility, we can focus at the limits of GxE research as nuisance *and* opportunity, the threat *and* promise of a science "ever on the verge."

An illustration: in the later stages of the GWAS period, in 2013, the NIH issued a new Funding Opportunity Announcement, soliciting proposals for "Analysis of Genome-Wide Gene-Environment (G x E) Interactions (R21)." (And let's pause to bask once again in the awesomeness of conservative institutions adopting the sign of radical deconstructive chiasmus.) It noted how "genome-wide association studies (GWAS) have contributed substantially to our understanding of the genetic variants associated with many complex human diseases and phenotypes," identifying "many genetic loci of interest" and providing "insights into biological mechanisms." But "clinical applications to date have been extremely limited," it continued, and one reason for this was "the complexity of gene-gene (G x G) and gene-environment (G x E) interactions. To date, many GWAS have focused on characterizing marginal genetic effects and have not fully explored the potential role environmental factors play in modifying genetic risk by including environmental components into genome-wide analyses."[54]

Curiously, though, the proposal solicitation also suggested that there was already a lot of activity in exactly this area: "Despite current enthusiasm for potential investigation of gene-environment (*G* x *E*) interactions, the most effective way to apply and detect interactions in the context of GWAS or other large scale population studies remains unresolved." So again, as at the opening of the GWAS era, the popularity and technical ease of doing this type of work was outpacing the development of the qualities of care that could make it "most effective." But the community "enthusiasm" was nevertheless real, noted by science administrators, and acknowledged, in exceeding formal methods and propriety, as a source of anxious difficulty.

The first step in minding the ease of these new GxE possibilities was to note their difficulties; this amounted to reiterating the old GWAS difficulties that demanded their own new forms of care: "Establishing the existence of and interpreting gene-environment interactions is difficult for many reasons including: the selection of appropriate theoretical, statistical, or bioinformatics models, inadequate statistical power, true genetic and disease heterogeneity, and the limited ability to measure accurately both the genetic and environmental components."[55] The call for proposals was also noteworthy in that it did not support the recruitment of new subjects or new "collection of phenotypic, environmental, or other medical data." Through this call the NIH would support only the reanalysis of the existing "wealth of genome-wide data that could be more extensively explored for gene-environment interaction effects," as well as "data harmonization" efforts. The NIH was calling for meta-analyses, in effect, soliciting studies to solicit GxE studies.

Other writings, internal to the scientific establishment, about this particular moment also support this narrative of widespread enthusiastic activity in need of more careful minding; they may even bind to an appreciation of impossibility. "The study of gene-environment interaction (GxE), has become an important area of study across multiple disciplines," wrote a group of prominent psychology researchers in 2015. "That said, few research topics have generated more controversy and less clarity than the study of candidate gene-environment interaction (cGxE)."[56] The previous decade had seen "an explosion of interest in this area," accompanied by "a growing skepticism about the replicability of many of these findings . . . and increasing concern about the quality of this rapidly expanding literature." Their review went on to summarize "why the existing GxE literature should be considered with a degree of caution."[57] Reading that article, one could easily get a sense of déjà vu all over again, since the long and detailed litany of problems, concerns, and questions recapitulates many of the demands for increased scrupulousness discussed concerning GWAS in chapter 5.

And then you could have easily gotten a sense of déjà vu all over again all over again when, two years later, you were reading Muin Khoury's 2017 editorial introducing an issue of the *American Journal of Epidemiology* focused on GxE research. Khoury surveyed the territory from his leadership position in public health genomics at the Centers for Disease Control:

> As shown in these papers, in spite of the enthusiasm for and interest in GxE analysis, there are only a few genuine success stories from the study of GxE. Uncovering interactions is fraught with potential biases and methodological issues. Inherently, there is the problem of low statistical power when testing for GxE in studies designed to uncover main effects of variables. There is also the problem of the complexity of measuring environmental exposures and the difficulty in assigning temporality, especially in case-control studies. The problems of false positivity, data dredging, and selec-

tive reporting of positive interactions further complicate the field. Other problems include the limited range of genetic and/or environmental variation, scale dependence in the definition of statistical interaction, and a lack of biological data on the health impact of many genetic variants.[58]

In this complex territory, the double binds double down, or up; if it was difficult to reproduce results from GxE research, it was also difficult to reproduce the reproducibility measures and protocols, as another article from this special journal issue made clear:

> Replication is an essential component of genetic association studies, and the requirement for independent replication contributed to the success of GWASs. However, replication and meta-analysis become challenging as GxE studies become sophisticated in analytical methods, exposure assessment, and incorporation of functional information. Differences in the underlying distribution of environmental exposures, patterns of linkage disequilibrium, and genetic modifiers can reduce the power to detect the same level of interaction in independent studies. Moreover, an appropriate human replication study might not (yet) exist in studies of a rare disease, genetic variant, or environmental exposure; where exposures are unique to particular populations; or where the initial finding was obtained within a large consortium comprising all known studies of a specific outcome. . . .
>
> Moreover, as the field considers gene- and pathway-based approaches to study G×E, replication may become further complicated as different combinations of genes in different data sets may be observed in the interaction. . . . More consideration is needed of standards for replication, definitions of replication, and alternative approaches for replication and verification of G×E results.[59]

The concerns were real, the solicitation of new care of/in/for genomics was necessary and admirable. But why so little enthusiasm for enthusiasm? Why trope enthusiasm, as these scientists have (and as I too have, to some degree beyond ethnographic fidelity), as an affective event that *only* manifests as in need of damping, if not extirpation? Isn't it the impossibility, or at least the inadequacy or inefficacy, of previously cared for, established, tested methods of genomics respectfully within the limits of the science, that is also the welcome driver of scientific enthusiasm *at* the limit? Shouldn't there be a way and a will for pushing the enthusiasm further while binding it more tightly to anxiety and care? Chapter 7 and the postscript go deeper into this increasingly knotty territory.

Being solicitous doesn't guarantee that you will be able to separate in advance the threat of failure from the promise of success, as many who have bemoaned the "reproducibility crisis" seem to think is desirable and, more importantly, possible. Solicitude in science recognizes that a whole or a "real" isn't here now; it doesn't present itself. *And yet*, scientists recognize that care's solicitude is an absolute demand at the limit of that whole that is forever on its way, not so that threat might be separated from promise, but so that the current undecidability of their doubly bound state can be endured for some unknown time. It's a tedious space of bored routine and mundane infrastructural work and an interesting-exciting space of potentially startling surprises, enervating and energizing. As the experimental limits encroached over and over, as genomicists found themselves on the verge of being on the verge yet again, at the same time enthusiastically beyond themselves and anxiously aware of their limits at every step and stage—so too did the need for solicitude iterate. It's these carefully plodding steps—which are, obviously, always limited, by which genomicists today have been soliciting asthma and other complex conditions over and over, from one plateau of science to the next, and the next—that ask to be understood ethnogrammatologically.[60]

7 · friendship

OF COMMUNITY'S LIMIT

Friendship is not an easy concept to define precisely. · JOAN SILK, "USING THE 'F'-WORD IN PRIMATOLOGY"

There is no thought, there is no thinking being, at least if thought has to be thought of the other, except in friendship. Thought, insofar as it has to be thought of the other—and this is what it must be for man—does not happen without *philia*. Translated into the logic of a human and finite *cogito*, this results in the formula: I think, therefore I think the other; I think, therefore I need the other (to think); I think, therefore the possibility of friendship lodges itself in the movement of my thought insofar as it requires, calls, desires the other, the necessity of the other, the cause of the other at the heart of the cogito. · JACQUES DERRIDA, *POLITICS OF FRIENDSHIP*

"O my friends there is no friend!" · DERRIDA QUOTING NIETZSCHE QUOTING MONTAIGNE QUOTING DIOGENES QUOTING ARISTOTLE QUOTING

Patterns That Connect

Friendship is my last diffraction pattern of care, and the one that most resisted its writing here. Friendship seems out of place, doesn't it? Despite their quotidian qualities, curation, scrupulousness, and solicitude sound as somewhat exotic and slightly exalted, appropriate to the actual doing of sciences that we still, after all these years (or centuries—choose your time scale), have to work to understand. But friendship seems more familiar, something that readers have some actual experience of or with, and that might seem an unusual choice of analytic concept for writing about science and scientists.

But beneath or within this familiarity—in the infrafamiliar, let's say, to echo the similar pattern of relationships in play in infrastructure—lies something that is not easy to define, as the epigraph from anthropologist Joan Silk above suggests, at least with the kind of precision that scrupulousness would ask of us. But this uneasiness, this difficulty in establishing formal criteria or protocols for friendship, should be the first sign that it has something to do with excessive, not-easy-to-name care.

Silk was among the first primatologists to ask if the concept of friendship would be appropriate for characterizing nonhuman primate relationships. The title of the article from which that quote comes, "Using the 'F'-Word in Primatology," wryly suggests the profane and even taboo terrain she was negotiating. Could this f-word be appropriate for characterizing relationships among genomicists as well? Could "friendship" help us appreciate and understand something that "network" or "consortium" or "working group," say, can't touch? Might friendship be a bond that holds (enough) scientists tightly (enough) together that they would refer to themselves as a "scientific community" without thinking twice about it, as if it were the most natural thing in the world that scientists form communities—without it occurring to them to wonder what this might mean?

Sarah Hrdy is another primatologist who can help us rethink the relationalities of care as shaped by evolutionary forces acting on populations of organisms. Hrdy opened her book *Mothers and Others* with a thought-experiment scene of complex social relations and interactions that many of us have experienced: boarding a plane with diverse others, including noisy children and their apologetic caregivers, semi-rude and/or semi-clueless people struggling to place bulky items in the overhead bin, and stolid suit units. Smiles, forced or genuine, and the meeting of eyes in some kind of recognition are the pivotal facial gestures. After a long flight all passengers should deboard intact, with only minimal emotional harm at the other end. Hrdy asks us to then imagine, in the chapter titled "Apes on a Plane," loading a similarly close-quartered metallic tube with chimpanzees, and opening the door hours later . . . to a bloody mess.

This capacity for positive relationality—a dense amalgam of affects, cognitive habits, and social technologies—that allows humans to negotiate air travel without lacerating flesh is the same capacity, let's hypothesize, that would allow genomicists to come together as a "community" and endure, nonviolently, a two-day NIH workshop in a windowless and nearly featureless conference room to discuss "implicating sequence variants in human disease."[1] The screenshot in figure 7.1 from one of the YouTube videos archiving the event sets the scene.

In this and other videos from the 2012 workshop, you can watch abstract care played out in all its concrete diffraction patterns around a u-shaped table, and on screens personal and shared. More than a few people have watched it; the video had 317 views in June 2016, three and a half years after the National Human Genome Research Institute posted it and when I watched hour after hour of it.[2] (Ethnogrammatological fieldwork, like genomic sequencing and other careful labwork, is not always exciting.) It's one of thirteen videos documenting this workshop on GenomeTV; one has nearly a thousand views, but most have three hundred to six hundred views by some smaller number of careful, scrupulous, solicitous genomicists, and some far, *far* smaller number of careful, solicitous ethnogrammatologists of genomics. I am grateful to the NIH's archival impulses for making it possible not only to do fieldwork at some distance in time and space, but also to be able to visit the same scene repetitively—for in all of care's diffractions, repetition is an essential dynamic of double binds, of doing the impossible.

Endurance and patience are essential to care in all its patterns, including friendship. If you watch even a few of these hours of video, you'll see numerous Starbucks cups and Diet Pepsi bottles and granola bar wrappers peeking

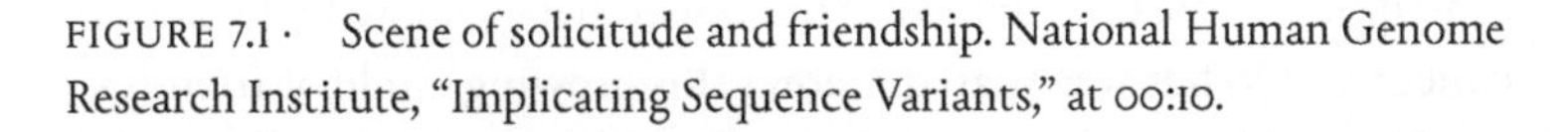

FIGURE 7.1 · Scene of solicitude and friendship. National Human Genome Research Institute, "Implicating Sequence Variants," at 00:10.

out between every participant's laptop in their places around the u-shaped table. Everyone stays seated behind their bifolded name placards, dressed snappy-casual, as my Texan in-laws would say, with another name tag (for good measure) affixed to their shirt or blouse or jacket. They're genomic-elite enough so that most everyone knows who most everyone else is, but they are not completely beyond the need for name tags, I suspect, even if they are mostly for the cameras and posterity. It's an "accomplished lab director" age cohort, but there are a few outliers on both the younger and the elder ends. But they've all left their labs and their offices for several days, giving their time in this windowless conference room to each other and to the questions and problems—especially the impossible ones—that they share as a scientific community.

A scientific community: a name unlikely to inspire ire in even the most staunchly positivist of scientist readers. Perfectly common and even banal. But what does it actually mean, this catachresis? What are the bonds that hold a scientific community in place if not together? What are the limits of community? And why put this community workshop under the name of the f-word, friendship?

Well, its originating, organizing impulse came of friendship—at least as it manifests on social media. In a post on an autism research blog promoting the guidelines that later came out of the workshop, Chris Gunter and Daniel MacArthur (the young genomicist who was one of its coconveners) wrote that the "entire effort began as a conversation between the two of us on Twitter, and quickly spread to involve other researchers who we knew shared similar concerns."[3] What was shared among this growing circle of friends—"colleagues" would be more neutral, but I don't think that does justice to the dynamics involved—were the worries, anxieties, and other matters of care that are indelibly part of the sciences of genomics, ones that can't be resolved through formal channels, strictures of logic and statistics foremost among them. MacArthur and Gunter expressed their hope that the guidelines published in *Nature* would serve as "a starting point for a far broader conversation." (More about the workshop's conversational quality—born of close conversation, carried out through extended conversation, and aspiring toward an even broader conversation—later in the chapter.)

The tone of the conversations throughout the NHGRI workshop was of course calm, polite, studious, measured, and engaged, but with an undertone of troubledness, dissatisfaction, even crisis. There's no shouting at any point or interruptions or arguments, though there were mild disagreements or at least respectful differences. In the discussions after each presentation, people wishing to speak were first recognized by one of the cochairs (MacArthur and Teri Manolio, an epidemiologist working at the National Human Genome Research Institute), then pushed the button on the mic stand on the table in front of them, speaking as the red ring around the mic illuminated and one of the camera operators staffing this entire workshop panned over and zoomed in. This scene of care, then, is a thoroughly collective and conversational one, its rapid but measured movements coordinated and technologically mediated.

The numerous presentations were framed by the most basic of questions, like, "Study Design: What Sample Selection and Data Processing Procedures Maximize Power and Minimize False Positives in Identifying Causal Variants?" Others session titles were equally quotidian: "How Can We Investigate whether Candidate Causal Variants Have a Biological Effect on Disease Risk?" "How Can We Sum across Different Classes of Evidence to Assess Overall Confidence in Variant Causality?" The questions seem at once flat, nondescript, matter-of-fact, and borderline boring—and yet so large, fundamental, important, and consequential that, on reflection, it's odd that they are being asked in the midst of an ongoing, active, and even (always the case with genomics) avalanching research arena. These orienting fundamental questions are not the mundane but necessary prequel to a contemplated scientific initiative; they've imposed themselves long after they should have been settled. It's the conjunction of these paradoxical qualities, their out-of-jointness, that makes this scene of care, this worried conversation among friends, such an interesting and urgent one: the questions come both too soon and too late, and are both numbingly simple and overwhelmingly complex. We'll see how they necessitate a prolix discussion of all care's patterns presented in the previous chapters, weaving in and out of each other. We'll hear talk of data curation as an ongoing inventive art, a recurrent worry, a cause for excitement, and an escalating expense. We'll learn about the tuning of new experimental equipment, evolving innovative techniques, calcified conceptual habits, and many of the other dimensions of doing genomics happening back in the labs the people in this room have left for a few days, that require the kind of scrupulous attention discussed in chapter 5. But above all, perhaps, we might sense how they have left their labs and come to this more featureless territory to discuss at some remove how all those fairly precarious components get assembled into an entire collective socio-technical system of scientific knowledge production that requires the solicitation of that very system—the painstaking yet unsure reach for *what's just not right* in all the rectitude they have helped build and in which they take such pride.

The care considered here shuttles through and across all of care's diffracted patterns, from curating data to scrupulous detailed attention to solicitous embracing rattle.

This is a long, boring, and exciting conversation, in other words, about the boring and exciting infrastructures of genomics—it's the technical discursive *infra-* to the technical discursive structures of genomics through which its truths will be established, the care that minds the science as it forms that science from its beyond. It's a long and demanding conversation that, as more than one participant would say toward the end, leaves their brains fried—a common but still revealing expression that gestures toward the limits of cognitive capacities. It's a conversation taking place at multiple limits, about how to better endure those limits since they will continue to recur, and a sign of the kind of open, honest, difficult conversation that needs friendship. Or something like it.

Friendship-Like Receptors

A lot of life sciences research has been collective and even collaborative for a long time, and in genomics it has become even more so.[4] But what would it mean to understand this phenomenon or event of collectivization as an instance or effect of growing friendship? What difference would the difference between collaboration and friendship make? For starters, asking about and inquiring through friendship could disclose the affective dimensions of life in science: in actively seeking out collaboration, desiring it, scientists might be figured as building something more than some Latourian "network of allies." If what's being built or woven in the minding of genomics are networks and fabrics of care, in the different relational patterns analyzed in previous chapters, these may all be tied together by my last of care's patterns, friendship.

Even if we don't have biological receptors for "relationality" in the same way that we have *toll*-like receptors for the lipopolysaccharides extruded by bacteria, my immodest hypothesis here is that our long evolutionary history as ex-ape *Homo sapiens* has somehow predisposed us to be both biologically and culturally receptive to, able to produce and amplify and indeed to be dependent and thriving on, affectively rich relationality. People have, to put it in the terms archaeologist John Terrell chose for the title of his 2014 book, a "talent for friendship." "Evolution has equipped us," he writes, "to be social in intimate and often largely subconscious face-to-face ways."[5] Terrell asks why anthropologists rarely use the term "friendship" to describe human relations, but instead prefer to discourse on things such as "exchange networks" or "trading systems." In Margaret Mead's earliest ethnographic writings, for example, Terrell thinks she was "willing to write that people locally obtained things like net bags, wooden plates, shell rings, and clay pots by 'haggling,' 'bartering,' and the like," but she "dismissed the fact that those involved spoke about their acquisitions in terms of affection, friendship, and gratitude." In Terrell's reading, Mead "evidently believed this was just a ruse to cover up the actual business being done under the guise of civility and friendship."[6] It's not that concepts like vending and trading are inaccurate or unimportant or unproductive; they just don't convey the enjoyment and well-being that people obtain from these experiences involving other people. And the headaches they get, the anxieties they bear, and the general pains-in-the-asses they try but sometimes fail to endure.

Humans evolved to be cooperative more than competitive, the rough revised story of evolution goes, and Terrell follows a number of evolutionary biologists, anthropologists, and primatologists to argue that what some would call "facultative mutualism" is clearly evident in human life—although Terrell finds that term somewhat unsatisfactory, since it "makes it sound like human beings cooperate with one another largely because they can see intellectually that they will somehow benefit in a direct and tangible fashion." Those cognitive factors are important, but Terrell opts for the affective: "People collaborate with one another because they can and like to do so. Simply put, because it feels right, and it feels good."[7]

In Terrell's view, enactments of friendship connect people with one another through "weak ties" that nevertheless result in amazingly robust and far-reaching social networks capable of transmitting vitally useful information, mobilizing people to action, and buffering individuals, families, and communities against the trials and tribulations of life.[8] What, though, are the specific qualities that would allow us to recognize friendship, and respond to and characterize it in ways that differentiate it from, say, love?

Relations without Form

While the new "genome space" was reshaping the experience of genomicists in the early 2000s, Joan Silk was arguing that her field had to transvalue "friendship" from being primatology's f-word—used often in social interactions but never in print, and with awareness of crossing norms of polite and proper discourse—to a valid and productive part of its conceptual apparatus and, equally important, of its methodological tool kit. "Although few of us take up primatology because we are fascinated by methodological issues," she wrote in 2002, "we will make little progress if we don't attend to this problem. The methods that we have developed are useful for studying social behavior, but not very useful for analyzing social relationships."[9] Primatologists tally up grooming incidents, diagram patterns of proximity, time the frequencies and duration of numerous interactions, and code a long list of behaviors that various primate collectives engage in, but assigning an interpretive name to such events seemed to make them a bit anxious. They might cloak a word like friendship in "protective italics," Silk notes, to avoid allegations of anthropomorphism. And here is where an appreciation of catachresis is again helpful.

Like protective italics, catachresis recognizes the dissociation between word and thing or event, but rather than settle for an easy irony, it installs into discourse a relentless demand for careful engagement and attention. Silk recognized that the act of naming a behavior, relationship, or pattern in primatology is something of a double double bind: the name comes too early and too late, and as a result supplements a relationship that has too little meaning with a name that has too much. "We ought to have a fairly good understanding of the function of a behavior before we label it," cautions Silk, while at the same time "we often find it useful to name a behavior well before we fully understand it."[10] Meanwhile, using "neutral" terms like "infant handling" or "allomaternal care" to name the approaches, contacts, and other actions a non-kin monkey might engage in with an infant monkey and its biological mother seem uninformative or unsatisfying, or both, while names like "aunting," "infant grabbing," and "kidnapping" evoked too many connotations already in place. Moreover, such terms were "strangely persistent," continuing to be used and to shape experience and theory "long after we had good reasons to suspect that the connotations of these words were not appropriate for the phenomena that we observe."[11] Damned if you do name, damned if you don't. It's impossible for primates to do primatology.

In a different way, Michel Foucault also understood the catachrestic nature of the name of friendship. In a 1981 interview published under the title "Friendship as a Way of Life," Foucault discussed friendship as a kind of problem, related to the problem of a homosexual identity. The latter was not a problem of "discover[ing] in oneself the truth of one's sex," discovering a certain "form of desire"; one had "to work at becoming homosexuals," not as a form of desire but instead "something desirable," and to work that "something" in order "to arrive at a multiplicity of relationships." In this shift from truth to relationality, friendship for Foucault became "a relation without form," a pattern of relations that must be "invented from A to Z."[12]

There's something impossible about the formless forms of friendship, as read by Foucault and others with "continental" reading habits. Penelope Deutscher, in her deconstructive reading of some of Luce Irigaray's writings on friendship, quotes Irigaray's figuring of an ideal "encounter with the other, friend or lover, as s/he who resists my knowledge and appropriation of them."[13] In the conventional understanding of friendship, you recognize something of yourself; it's a recognition based on identification, however partial. Irigaray predicates friendship on difference:

> I am listening to you, as to another who transcends me, requires a transition to a new dimension. I am listening to you: I perceive what you are saying, I am attentive to it, I am attempting to understand and hear your intention. Which does not mean: I comprehend you, I know you, so I do not need to listen to you and I can even plan a future for you. No, I am listening to you as someone and something I do not know yet . . . *with* you but not *as* you.[14]

Irigaray voices the traditional, important, and productive methodological imperative of cultural anthropology: to listen attentively, trying to grasp the language, intentions, and understandings of an other so Other as to merit the word "transcendent," in some other realm or dimension—maybe the eighth? It's this organized reach for but never grasp of transcendent and complete comprehension that cultural anthropology shares with genomics—impossible grounds for friendship, I would say.

It's a stilted exercise, but imagine the "I"s and "you"s in the quote above as the generic placeholders they are, and make a series of quotidian substitutions: "I" a genomicist speaking of or to a "you" that is a genome, or several thousand different genomes; "I" genomicist speaking to "you" fellow genomicist; "I" ethnogrammatologist speaking to "you" genomicist, to "you" genomics text. Each substitution still makes sense, and I think provides a better sense of how genomics and other sciences actually occur. More aspects of these patterns of an impossible friendship—within science, within anthropology, and between anthropology and science—are drawn out in the rest of the chapter.

But from wherever or whomever this reach to comprehend is coming, this attempt to solicit the whole and wholly other, it doesn't . . . really . . . quite . . . happen. *And yet*—attentiveness remains, and solicitude's play (work) returns. Because "as you" is impossible (difference won't resolve into identity and identification) the ethnogrammatologist or genomicist persists in solicitude "with you"—you *toll*-like receptor, you data, you genome-wide association study, you *yous*. Where comprehension is at its most precarious, the "transition to a new dimension"—call it the care of friendship—can be sensed. And comprehension is *always* at its most precarious; genomicists are only ever on the verge, always one step away. It's this abiding, persistent exposure and openness toward the other, and the refusal to anticipate her, his, or its future and instead be positioned to be surprised, that we can call friendship.

My friends are a real pain sometimes, and I'll venture that everyone is inclined to agree—especially my friends. So in advising that we analysts of science exemplify and embody more of a mode of caring friendship with the scientists and sciences we analyze, listen to, and attend to, I am not saying that this is straightforward, nor that this entails relating to sciences and scientists as if they were cuddly puppies. Revisit the discussion in chapter 3 of care as a non-cuddly bundle of affect that includes anxiety, worry, and lamentation. Friendship is a volatile conglomeration and an opaquely patterned set of precarious habits and practices that nevertheless still comports us toward openness and attentiveness. Friendship is trying, always.

I'll supplement these necessarily poetic phrases with some equally necessary prosaic explication of a science studies text, also from this same circa 2000 era when the affective pushes of Big Data in genomics and on genomic scientists were first being sensed. My friend Adam Hedgecoe is a sociologist of science who was analyzing the writings of psychiatrists and other biomedical researchers working on the genetics of schizophrenia around the same time I was first becoming acquainted with Patrick Sullivan and his work in psychiatric genetics discussed in chapter 5.[15] Our analyses were undertaken at different times, from different perspectives, and without knowledge of each other's work (I'd written nothing on any of these matters in 2001, and only read Hedgecoe's article long after I had already begun figuring out my analysis of Sullivan's work in terms of care). For that reason, you could regard one of the main arguments each of us has made as an independently validated finding of social scientific work: this time when the Human Genome Project was on the verge of "completion" was marked by an increasing frequency and intensity of statements of caution, carefulness, and responsibility starting to percolate through the discourse of the genomics of complex conditions such as schizophrenia. From there our readings differ.

Starting again by stressing: Hedgecoe's is a well-supported, valid, robust, productive, important, and simply *good* reading of this scientific literature and its discourse, a careful reading that is good to have as part of our discourse in studies of scientific practice, theory, and culture. I make sure my students read and learn its truths. And our differing readings of psychiatric genomics share a motivation and intent: they each respond to the critical discourse of "geneticization" that had come to dominate science studies in the 1980s and 1990s. Geneticization was the social scientist's term for what geneticists did to complex conditions like schizophrenia, turning what to the social scientists was a fundamentally social phenomena or problem into a biological one, or at least reducing a complex whole to a simple single part. Geneticization, reductionism, and determinism tend to become synonymous in this analysis, or at least get packaged together. And the reasons for packaging them together were cogent ones based on a long, ugly history of naturalizing and pathologizing human differences—the history of eugenics, in a word, and scientific racism in another. Geneticization was stupid-spirited science and mean-spirited politics, and geneticists, and then genomicists, seemed to be doing a lot of it, at least to the many historians, philosophers, and social scientists who were writing about it.

In chapter 4, I gave a brief potted history of research "geneticizing" asthma, and how that crazily complex condition was cast, in both the lab and the media, as essentially reducible to a few soon-to-be discovered genes. Things did not turn out to be nearly so simple, as we saw, and more and more, even supposedly "simple" one-gene Mendelian conditions like cystic fibrosis were being understood to be knotted with complex genomic, biochemical, physiological, and sociocultural forces. As a result, Hedgecoe noted, it had become "very hard, in these post-colonial times, to say anything good about" a geneticizing and therefore essentializing genomics in "its currently ethically loaded context." Hedgecoe's conceptual-methodological strategy for saying something good about genomics was to reframe geneticization "within the more neutral, historical context of 'molecularization,'" casting contemporary genomics as "simply the latest in a long line of attempts to analyze the body in terms of molecules, rather than just an opportunistic tactic employed by doctors to gain power over patients."[16]

With this book I've also wanted to say something "good" about genomics research, in a multidisciplinary culture that remains ethically overdetermined; I try to portray scientists and physicians as something other than opportunists bent on power, or as unwitting enforcers of a powerful "geneticizing" ideology. Where Hedgecoe opted for something "more neutral," however, I am exploring and trying to understand the differences that explicitly adopting "friendship" might make.

One difference: focusing on review articles on the genetics of schizophrenia, Hedgecoe interprets the kinds of cautionary language I read through the analytic of scrupulousness in Patrick Sullivan's writings in psychiatric genetics (chapter 5) as geneticists' strategy to adopt a "narrative of enlightened geneticization." Geneticists demonstrated caution, for example, by "showing how others have gone wrong in the past," or "spend[ing] over one page explaining the specifics of the level of proof required to suggest linkage" between a gene and a diagnosis of schizophrenia.[17]

Prolix narrations that I regard as a fitting necessity and epistemic virtue sound more to Hedgecoe like "the scientist doth protest too much, methinks." Hedgecoe reads such statements of caution, care, or what I name scrupulousness primarily in terms of their strategic value: "A strategy of caution deflects criticism that researchers are over-enthusiastic and guilty of genetic hype, and allows readers to see authors as responsible and objective in their assessments."[18] This helps them construct a "narrative of enlightened geneticization": a "discourse surrounding the genetics of schizophrenia" that is "constructed to prioritize genetic explanations, and subtly to undermine non-genetic factors, while at the same time accepting that they have a role in its aetiology." Or in slightly different terms: a "narrative that subtly privileges genetic explanations without succumbing to hard-line genetic determinism."[19]

I want to state again: this is a good and important critical reading, which I advance in my teaching. *And yet*, I am left with the subtle sense that a geneticist still can't win for losing. Why does "enlightened" sound like something to be suspicious of, an affected air put on for shrewdly calculating reasons? Hedgecoe asks at one point, "How enlightened is the narrative of enlightened geneticization?" His answer is that "enlightened geneticization still has a lot in common with Abby Lippman's original definition in terms of a core determinism. Hopefully the term *enlightened* geneticization highlights the sophisticated way in which this narrative is constructed, the apparent acceptance of environmental causation, and the careful avoidance of 'hype.'"[20]

So the "core determinism" of "enlightened geneticization" still shares "a lot" of the same qualities as the "hard-core determinism" found in less-enlightened geneticization. Where a James Watson might have been criticized in the early days of the Human Genome Project (and long before and long after, as well) for not just succumbing to, but positively *reveling* in hard-line genetic determinist statements, the genomicist of 2001 who engaged in elaborate articulations of caution about their own findings would still be brought before the social scientific bar, although having to answer to the relatively more minor charges of "prioritizing" and "privileging." Guilty, all too guilty—albeit "subtly" so.

Indeed, it is all about subtlety:

> The narrative of enlightened geneticization presents a complex, multifactorial vision of schizophrenia with a role for environmental influence. It also subtly: ensures that schizophrenia can only be considered in terms of *some* necessary genetic aetiology (even if it is not sufficient); suggests that a variety of different possible environmental factors should be seen against a "genetic baseline" which is the only single necessary condition for causation; and finally that non-genetic factors are, in contrast to genetic ones, non-specific, unpredictable and thus less researchable.[21]

It's true that, in the dominant cultural framing of science and scientists, "privileging" and "prioritizing" are epistemic vices and not virtues, since even a subtle departure from strict neutrality/objectivity is still a straying from the straight and narrow path of the upright. But what if "prioritizing" and "privileging" are inescapable, welcome, and productive component of scientific *creativity*, an effect of an affective interest-enjoyment that careful genomicists (and careful sociologists, for that matter) get from and bring to their methodological and conceptual work? What if these enthusiasms—subtle ones, to be sure—are an expression of the tacit, *careful commitments essential* to the scientific sense-making work of creativity, intuition, play, thick reasoning, and experimentation, rather than evidence of some kind of departure from an *impossibly* neutral scientificity? What if we read their cautionary statements of care and responsibility less as a rhetorical strategy of persuasion, and more as a *truly* deep appreciation for the *real* nonlinear complexities of a complex condition in a complex diverse population living in a complex diverse set of changing environments—an appreciation won through years of hard, collective, careful work among genomicists? Genomicists whose professional training and job description, one might very well argue, are pretty much defined by prioritizing genes as explanatory elements?[22]

So, alongside the conceptual questions this book asks about care in the sciences, it also asks this methodological one, with motivations patterned similarly to Joan Silk's methodological proposition for primatology: What happens when we ethnogrammatologists take a step beyond moderation, neutrality, and fairness, and in our reading strategies *befriend* genomic sciences and scientists? What becomes of *our* priorities, which have been normed predominantly around critique, in neutral shades? To whom and to what will our privileges attach, bound subtly or otherwise?

The Genomics Community of Those Who Have Nothing in Common

Ethnographers and sociocultural analysts of all sorts have their own patterns of privileging. They tend to assume, for example, that cultural and/or environmental causalities are more subtle, complex, "nonspecific," and "unpredictable" than are biological or genetic causes, which by comparison, at least, are presumed to be deterministic, linear, and unequivocal. This is an epistemic pattern that deserves to be made more complicated—easy enough to do since, in reality, it is more complicated. As we'll see in the NHGRI workshop discussions below, genomicists repeatedly and inexorably came up against limits of specificity, predictability, and their own evolving experimental methods. They had come together for the better part of two days to speak together as a community—an apparently well-gender-balanced community, albeit one pretty skewed toward haplogroup H[23]—about the impossibilities of and thus the necessity of interminable care in speaking the truths of genetic etiology.

Writing as a friend, I have learned to have great affection and esteem for their successful failures in these encounters with the impossibilities I've gestured toward in previous chapters—for being carefully ever on the verge. It's in anxious encounters like these that science's double bindings to care become most apparent, when genomicists contend together as a community (let's leave this stalwart but elusive term in place for now) to solicit the whole of genomics, the whole collective program of analyzing whole genomes in large populations; to undertake scrupulous analytic engagements with every essential part of that whole genomic apparatus; and to reiterate in public the burdens and pleasures of the data curation that infrastructures all these other quotidian genomic infrastructures: the generous NIH funding, the varied -omic technologies distributed across hundreds of labs, and the thoughtfulX-careful genomicists themselves—not the highly public genomic celebrities written about in the *New Yorker* or the ones writing their own books, but these quotidian genomicists gathered here in the name of a wider "community," the ones whose names mostly go unnoted and who are even in many ways hardly known to each other, yet who make the whole genomic-truth-assemblage go.

The question driving it all, the question they are anxiously (worriedly, eagerly) taking up: How do you change a community's pattern of practice, thought, and similar habits when those patterned habits—especially the bad ones—exhibit a pattern of evading control? In more straightforward terms: How do you change a scientific culture? How does a scientific community access what's beyond them *and* fundamental to them, bootstrapping a more careful, better-minded genomics? It will all hinge on conversation about words and statements and their regulation, a conversation meant to be carried forward not by being translated into regulation or policy, but as continued community conversation itself. Because it's not scientific change they're after so much as cultural change, which will turn out to be the same difference.

Some further context to this workshop: the genomics under discussion here is not the genomics of the genome-wide association studies (GWAS) of the previous chapters—although as we'll soon see, it's the GWAS era of the latter 2000s and early 2010s that they are directly, anxiously referencing in their efforts to reinvent care and to mind genomics yet again. As a catachresis, we have to reestablish a meaning for "genomics" in every new instance. Whereas common and complex conditions like asthma were the object of genomics in the GWASs of the later 2000s and on, when genomicists searched for patterns of sameness in large populations of difference, the genomics being discussed at this workshop in 2014 is about efforts to write truths about the different mutations of "the same" gene in one population—all of the women whose test results tell them their genome includes a BRCA1 gene, for example, but who, when that gene is completely sequenced, are shown to have different point mutations that might have significance, a finer-grained truth about more subtle effects that the BRCA1 test itself could not provide. But the study of these kinds of variants in individuals was enabled and driven, as GWAS were, by "next-generation sequencing technologies" that made the requisite in-depth sequencing doable and affordable, another data avalanche on top of, or within, or after genomics' already Big Avalanching Data.

So the recent GWAS wave served as a cautionary tale, a pebbly scrupule in the genomicists' shoes to remind them of the failures in their past successes that needed worrying. In the workshop's opening presentation, Mark Daly related the analytic story of a gene named dysbindin that had been—and still was, to some researchers—a promising candidate for explaining the etiology of schizophrenia. But in that case, a bevy of genomicists being ever on the verge had turned . . . well, it had not turned into anything. So the dysbindin pursuit had to be counted, *in retrospect,* as a failed venture:

> But the main reason, you know, to bring this example back to the floor, is that it's not the only example. In fact, there are myriad choices I could have chosen. I just happened to have slides on this one and was most familiar with it as an example. But there are huge numbers of genes that still get a tremendous amount of attention. And, in fact, still a tremendous amount

of publications in 2011 and 2012 focused on models of this disease, and in which abnormal cell types they must be exerting their action, and all this stuff, all motivated by that original association report in schizophrenia. We fail as a field when we accept this as the standard. And this was the standard for a very long period of time.

And not to pick on this example in particular, but this is a collective community failure. . . . You can imagine with this many papers how many years of graduate students and postdocs have been dedicated to this; how much funding has been dedicated to efforts to study the relationship between dysbindin and schizophrenia. We can't accept that, and happily, we didn't accept it, and we moved on, which is back to common associations. . . . There was one and only one way of getting past our history of failure, and that was to embrace extraordinarily strict thresholds for declaring association.[24]

Daly's presentation enacted the solicitous care discussed in chapter 6: out here beyond the limits of formal genomic protocols, a "collective community" had to take the floor and assess the entire genomic research apparatus, and themselves in it, to figure out what things had gone awry so that it was judged, not as success, not even as leading to another promising verge, but judged as failure. That collective judgment, however, was made and, it seems to me, could be made only in retrospect. There were still at present "huge numbers of genes" getting "tremendous" attention and reward, "myriad" similar genomics studies for which the jury, so to speak, was still out. They remained on the verge for now, but maybe not forever. But having solicited this judgment on dysbindin research, it might suggest new ideas for how to better mind the genomic study of sequence variants in particular genes.

Later in his presentation, Daly reflected further on how the community would have to become more careful, again stressing the collective quality of the effort required:

One of the reasons I am so motivated by, you know, working as a group on this particular topic area is because I, like you, read a lot of these papers, and . . . we are just, you know, repeating all of the things that got us into trouble when we were doing candidate gene association studies 10 years ago, and just assuming that everything would come out nicely at the end, because it sort of made sense, and . . . we know it simply won't.

I mean, the most aggravating thing is that so many papers that don't come right out and say it but . . . basically write the paper in a way that you're supposed to think that every one of these de novo mutations that has been found must be important to the disease or must be important to something. And that just is not the case.[25]

There was something going on here with the genomics community that, in my reading, is beyond a dispassionate acting out of falsifiability or reproducibility-testing. Some members of the community diagnosed a kind of repetition compulsion in the broader genomics community that, as so many unconscious forces do, got them into "trouble": driven to know, a genomicist assumes a sense of things that makes sense (sort of) *and yet* also knows that it simply won't. It's aggravating to see the community you're part of—your friends, in the expansive sense I am trying to develop here—be caught in this pattern of finding something surprising, something interesting that must be important, scrupulously (you tell yourself) working out its sense that you have also sort of already assumed, trying to convince not only yourself but others what you know to be the case, going to the verge and... doing it all over again. Some members of the community are at least somewhat aware of all this, at some level, and yet the community writ large keeps repeating it.

And they keep attending to it, minding genomics at the next level up the impossible staircase. There's a reason why some analysts think the work of analysis is interminable.

Heidi Rehm, whose epistemic virtues include staunch commitment to open sharing of genomic (meta)data, especially data on these rare genomic variants that may have clinical relevance, presented next to the collective.[26] That commitment takes the form of her long-standing involvement in founding and maintaining two related databases in the National Center for Biotechnology Information, ClinGen and ClinVar; her presentation that day revolved around this work. Rehm prefaced her remarks by noting that the people gathered around the table would be hearing "common themes across certain talks here." What was held in common, in her view? What, besides terabytes of data, was being shared? "We all hear the same pain and feel the same pain." Even though her perspective came from "running a clinical lab for the last, about, 10 years," the pain she was speaking of did not seem to relate directly to the lab's "genetic testing for various diseases, rare and somatic cancer, various things," although surely that situation abounds with painful stories (these are mostly rare conditions, the most unexpected of events manifesting mostly in children). The pain of which Rehm spoke, sounding like she was speaking to a circle of friends, was related more to the researcher's anxious worrying about having been careful enough, even if they had a "fairly robust process to evaluate variants that we have to write into clinical reports." Rehm described some of the scienceXcare work that went into the database, beginning with the curation work shared among members of her lab in a Charybdis-like swirl of activity:

> Before we got into full swing of next-gen sequencing, we were doing about 300 novel variant assessments a month, and to date have curated 25,000 variants that have been put into clinical reports. So we do a lot of this, and I have to hire a lot of staff to do it. In fact, if you come within a three-foot radius of my lab, you sort of get sucked into the vortex of novel variant assessment in my lab, and all the rotating residents and fellows, and everybody does this. And we even time, you know, how long it takes them to evaluate a variant, because we have to staff for this in a very robust way, and hit our turnaround times for clinical testing. And so we time them, and it takes about 22 minutes to evaluate a variant that's never been reported or seen before. And it's just a matter of searching every database we do, and doing in silico assessment, and then if there's population data, it takes 25 [minutes], and if there's publications on the variant, it takes on average two hours, depending obviously on how many papers. And that's the first-line process that mainly our PhD molecular fellows go through. And then our genetic counselors draft a report with the variant information in it and put it into a clinical context for the patient. Geneticists review the data, sign out the case, and for our somatic cancer, pathologists add annotations as well. So that's our process.[27]

Rehm was describing the curatorial care effort that keeps this vast corner of vaster genomic infrastructure going; there's a lot of it and it takes a lot of people that we are unlikely to ever get word of. We also see the layer upon layer of interpretive work: assessments, evaluations, annotations that put the meta- in the (meta)data to make it "data."

The timing, and therefore placement, of this interpretive curation-care would be the object of later questioning and discussion. Workshop cochair Teri Manolio asked a "dumb epidemiologist's" question, since friends aren't afraid of revealing their ignorance among other friends (or at least not afraid to play at being dumb): Wouldn't you want to interpretively annotate as much as possible? Here we learn that connected to the logical limits of (meta) data are more quotidian limits, such as money:

> TERI MANOLIO: Let me just ask, as a dumb epidemiologist: I hear a lot about HGMD [Human Genome Mutation Database][28] and the challenges and the problems with it, and I guess in my own mind I'm thinking, this

is probably a group that wants to capture all the information that's out there, and have other people do some judgments on it. But is there not some way to have it capture some of the judgments as well? Has anybody approached them about that?

REHM: Yeah, we have, and they've said it has to be published in the literature for them to put that reference in there. They will not override what's been published. So it's been very frustrating.

MANOLIO: One wonders if there isn't some way to work with that?

UNIDENTIFIED: They just say—they're very honest. They don't have the resources to do that. They would have to operate a 24/7 call center to be taking info from all of us who are doing this stuff, about what we think about their variants.

REHM: Yeah, as it is it's about $27,000 a year to use that database for clinical use. For clinical service.[29]

There is also the matter of different databases hosting differently curated data on these genomic variants. Even the experts convened at this workshop, those closest to these matters, were unfamiliar with the full landscape:

DANIEL MACARTHUR: I'm sort of intrigued by the sheer number of in-house, or curated or fixed, versions of HGMD databases that exist in these clinical labs for clinical populations, but I have no idea how many that is. Is there communication between clinical labs that can be used to help merge those efforts?

HEIDI REHM: Yeah, there's been—you know, that's something that we're trying to work on. I've actually had contact with HGMD and said—because some of the U1 groups that have been funded, we're all sharing these stories, and we all have these in-house databases, and we've gone through all these efforts, and we each have our list of HGMD variants. HGMD is actually planning a scientific advisory board, so they've asked me to be on it. And I said, I'd be happy to organize some efforts to bring everyone's data that has conflicted with HGMD's classifications, and try to bring that to you, so that it can be put in there. And they haven't agreed to do that yet, but I think at least agreeing to form a scientific advisory board—I think there's some hope that we can get some traction there.

UNIDENTIFIED: It seems like the interpretation of very rare stuff is very much a numbers game, right? One of the things I liked about one of the slides we had up there—it almost looks like a medical note, where you're, like, writing a note about a variant, right? Maybe HGMD isn't the appropriate nexus for organizing an antagonistic community about the problem. What you want is something where everyone can contribute annotations from wherever they are, about variants they observe in their patients, in some addendum-type way, to get those numbers, to be able to add interpretation over time.[30]

We're a long way now from Walter Gilbert's "DNA sequencing is boring and yet everyone is doing it" of 1986 (chapter 2)—and not. Care's pattern has shifted but remains recognizable: *Curating avalanches of sequence data is* toll!*—interesting not interesting—yet everyone is doing it.* Sequencing itself is fully automated and out of view, it's the interpretations that have begun to avalanche: "everyone" is binding more and more (meta) to the always already (meta)data, infrastructuring the avalanche repeatedly over time, caring for it so that sense might be made of it. And in this case, at least, it's a woman whose work is never done: infra/structuring "all these efforts," establishing governance structures and forming smaller advisory communities within the larger "antagonistic" community to manage or at least solicit those differences.

Maybe "friendship" is a not-bad word for what turns an antagonistic community of differences into one in which everyone "contributes."

Continuing this discussion, Rehm began to weave together and overlay the patterns of care into an interrupted pyramidal structure under the sign of curation (see figure 7.2). As she talked about how to care for these multiple antagonistic interpretations accumulating over time, her language became full of qualifications, doubling backs, and a series of "but"s, "and"s, and "and yet"s:

REHM: Yeah, absolutely. And maybe Donna [Maglott] can speak in terms of the NCBI ClinVar viewpoint, in terms of creating a truly publicly accessible environment to be able to do that, and we've been trying to work with them and make some inroads in that area.[31] . . . If you can put up slide #14 on the screen? So in working with Donna, and thinking about how to annotate the curation level, and who has curated that data, this is the scheme we came up with, where within ClinVar, there may be data at the bottom that is uncurated like data in dbSNP and ESP cohorts, but then the clinical laboratories we're working with would come in with assertions, but it would be marked as single-source curation. So somebody said something about it, and you see who said it, so you can perhaps pass some judgment

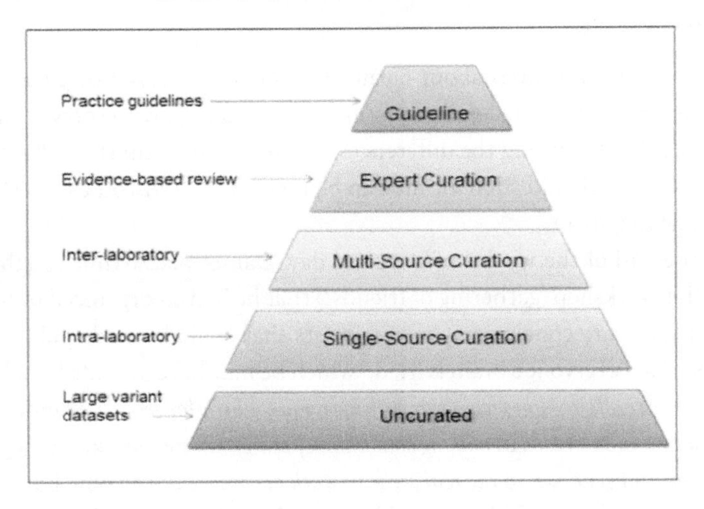

FIGURE 7.2 · Heidi Rehm's Pyramid of Curation. Slide from National Human Genome Research Institute, "Implicating Sequence Variants," at 11:53.

on who that was. And then there's this inter-laboratory multi-source, where you're actually collecting multiple different curations that actually may be the same or may be different, but there's some ability to find consensus, because there's multiple ones. But they're still done independently. And then you get to the expert curation, where you truly bring a group of experts together, who in a consensus evidence-based manner, collectively decides a level of evidence on a variant and classifies it. And then there's some variants that now exist today within clinical practice guidelines, that can be marked and referenced to those guidelines. But that was our attempt to really set up a scheme so that you could have a place that has all data, everything from only the population frequency on up to the most clinical grade data, and not have to go to eight different sources that are different grades, and yet still have some way to show a difference in the amount of curation that's happened on that data. . . .

EUAN ASHLEY: Some of it is simply crowdsourcing. We have several people doing just what you described, Heidi, every day in our lab. . . . Some of it is just going to be creating standards in a forum like here, and then bringing together all the people who are doing that, to speak to the same standard and just pull that data together. Because it's happening, I think we just need to harness what's already happening and bring it together into one thing.[32]

Genomic friendship is about harnessing the differences, bringing them "together into one thing" and yet "still hav[ing] some way to show a difference." It's about marking the differences, a crowdsourced effort, so that who said what at such and such a time can be registered "and you can perhaps pass some judgment."

At the end of the workshop's second day, Daniel MacArthur sought to bring this workshop (gathering of friends?) that he had co-organized to some kind of summary conclusions or statements that could be made public. He guessed that "everyone's brain is fried," which he may have intended as a kind of *I know we're all tired but please stay with me for just a little longer . . .*, or more like *Now that we've succeeded in making smoothies of our brains we can access the eighth dimension and get down to the real creative work of care . . .*, or maybe both. He projected an outline of what would become the published workshop report up on the screen. The final exchanges among these friends would be a hashing out of the new languages that would structure the language of genomic publications concerning DNA sequence variants with clinical significance, a language that would be sensible to journal editors and journal readers alike, not to provide answers or definitive solutions but so that this community could enter into and continue these interpretive loops of care, and "perhaps pass some judgment."

The very first "general point" presented by MacArthur went by with nary a blink: "Adopting an explicitly statistical approach is critical in all cases—this can be done even for rare disease (by considering patient sequence in the context of very large sets of reference sequence data)." No one doubted that statistical analysis was critical. But the next point building on this one—"Experimental or computational support for a variant's impact on biological function should not be regarded as a replacement for compelling statistical evidence for variant association with disease"—drew immediate questioning.

"Maybe there is still some controversy about that point," MacArthur allowed, "but, you know, I felt that seemed to be the general consensus of the group." Russ Altman was the first to suggest a departure from the general consensus. Something was "bugging" him, he began, some tiny scruple in his mind-body's shoe that . . . *eh, maybe it's nothing, I can't tell*:

RUSS ALTMAN: The only thing that's bugging me, and this is really a question, is whether the asymmetry of the compelling statistical evidence versus experimental—because I can imagine compelling experimental data that should be regarded as a replacement for the lack of compelling statistics. And in stuff like penetrance, right, where you have a variable penetrance, and so you have, in the penetrant case, you have a really great story. And then the statistics don't bear it out because, actually, you know, it depends on other things. But—and I can't, I really can't decide if this asymmetry that statistics should trump experimental. I mean, even saying it, it seems like anathema.[33]

That last bit was spoken with a huge grin, face lit up despite (or perhaps because of) any brain-frying that might have occurred (see figure 7.3). Daniel MacArthur immediately responded by noting his own predilections: "I mean, that's my bias, and so I think it's—I felt that there were probably more people leaning in that direction than otherwise. But I'd obviously like to discuss it with you guys." If we translate the already informal "you guys" as "my friends," we can perhaps hear intimations of the kind of community necessary for an open, searching discussion of the complexities of care: who else would you open up to about a "bias" that is not supposed to be part of the adjudicating equations but which is impossible to exclude? The subtleties and contradictions of care in/as science need the careful ears, tolerant of halts and backtracks and holes and opacities, that friends provide.

FIGURE 7.3 · Russ Altman entertaining anathema. National Human Genome Research Institute, "Outline of White Paper," at 7:54.

The discussion that followed involved the solicitous weighing of evidence and interpretations, with an eye toward developing language appropriate in different contexts:

MARK DALY: ... I just come back to—I resonate with the point that Joel made, is that there's a distinct difference between two types of papers. One is, we started with a patient with a severe phenotype or we started with a disease that we're focused on, we screened the genome, and we're trying to prove that this is the gene for that disease. That's one type of paper that needs to rely on statistics. There's no question about it. There is another type of paper that, you know, groups that have a keen interest in a certain piece of functional biology or a certain gene from a functional standpoint, sometimes are wrapping in some human data in an effort to make, you know, their functional story or their genes seem more exciting. This is where a lot of people get into trouble and where it's a little harder to say there could be a lot of stories and a lot of functional data and models that are actually very much worth publishing because they enlighten us on the biology of certain genes and so forth. And the human genetics may not be the leading part of that, but we can't also allow it to sort of just be used as a throwaway just to, sort of, elevate the profile. And I think that's where the difficulty comes in.[34]

The statistical analysis that is part of the human genetics analysis is a way to offset the "trouble" that the "excitement" of "keen interest" in the functional biology; data in this context is the "wrapping" around a "story" that elevates its profile. But even though, in this view, the story may prove to be not only unexciting but an empty one, it could also be the "other way around":

SUZANNE LEAL: Okay, and I think with the statistical evidence you have to be a little bit careful because you could get some very magnificent p-value. But that could also just be because of all kinds of bias you introduced into your analysis. So it maybe is, even if you have a very significant p-value that is probably not—that's definitely not enough on its own, you would like to see replication, and also some other evidence would be nice. So I think that replication is also quite key. Although, of course, in some cases that's difficult—I realize that. But we have to realize that just a very small p-value by itself doesn't necessarily mean that really is the p-value.

EUAN ASHLEY: And I think it's kind of a similar point, but I've been in rooms with basic science colleagues having the same discussion where they have this the other way around. They say that an association, no matter how good the p-value, is never causation. I mean, that's—you know, so it doesn't matter how close the association is. And until you have mechanistic data from several different places you cannot decide causality. So there we'd have this actually the other way around.[35]

The double bind has kicked in again, and the one way also goes the other way around. What's essential is supplemental, and the supplement becomes essence. The discussion continued apace, turning first to the meaning of statistical evidence and then to how that meaning would translate outside this group of like-minded friends—at least some of whom, remember, have reached not only the limits of language but the limit where fresh brains have become fried, and not in the good eighth-dimensional kind of way:

DAVID DIMMOCK: I'm just going to say I think we all know what we mean by statistical evidence, sitting around this table. But I actually think it's very hard to understand the term, particularly for someone who doesn't live in a genetics world. So I think we need to actually think about our expressions. I don't want to split hairs, but I just—you know, because we have statistical evidence, that the functional studies prove something. I think we need to be just a little bit more clear and actually think about what we—what is a better term that will make sense to wider audience about a sequence-based statistical genetic approach or something like that. My brain's a little fried, but I think we need a better term for that.[36]

There were other times when the limits of language provoke, not hairsplitting or linguistic invention, but the laughter that can ripple through a group of friends:

GONCALO ABECASIS: One thing I did like is David Goldstein's suggestion, having some statement or some language where you say, "How compelling is the statistical evidence?" And you don't have to say in every paper, "The statistical evidence definitely implicates this variant or this gene." You could say, "The statistical evidence makes this gene interesting," or "It says this gene is where you need further consideration." You could have different levels, and then you can build on that and do other things. But it is nice to be able to separate; and that actually let's people do whatever they want. You can still publish a paper where the statistical evidence is completely not compelling. You just have some phrasing to describe what it is. You know, I'm sure within the creative set of people we have here, we could have a phrase where, say, "It's not compelling at all," actually sounds not too bad, and someone wouldn't feel too offended to write it in their own paper.

[Laughter.]

DANIEL MACARTHUR: We can devise our secret code for labeling variants. [More laughter.] That sounds reasonable.[37]

I'm confident MacArthur would have appended "</sarcasm>" to the last comment had he tweeted rather than spoken it, but readers can watch the video and judge for themselves.

Things quickly turned serious again, only to run up again against the impossibility of "actually" limiting language:

MARK GERSTEIN: I would just urge . . . that it not be a secret code. If people want to actually create a bar for the word "causal," or create a bar for a particular word, that they actually define it. Having an actual, practical definition is really useful. . . . I think that's actually where the problems all lie: where one person thinks "suggestive" means this, another person thinks "associative" means this. I think if things can be defined, then a lot of problems disappear.[38]

Except that, at the end of this extended enactment of solicitude among friends, the problems didn't disappear. The demand of or for care returned again and again in this effort to draft new guidelines for new epistemic virtues. The conversation came back repeatedly to the problem of catachresis: naming things that don't have a proper name, and stipulating in a "mini-glossary" what the names should mean:

GREG COOPER: . . . one very useful thing that might come out of this paper might be some sort of, like, mini-glossary of, you know: causal comes with certain consequences in terms of burden of proof, and it needs to be clear where that proof is coming from, in your mind. And "damaging" means this and "deleterious" means that—so you can imagine just trying to lay out—*obviously it can't be a comprehensive sort of listing* [emphasis added], but some sort of concise but useful list of what we should expect a certain term to imply.[39]

The comprehensiveness that genomics and genomicists *must* have, one name for everything and everything with a name—well, *obviously* that can't happen. Everyone knows that's impossible; all the friends here seem to have been in on that joke. And so you do the best you can, substituting a useful concision for the full prolix presence, and prepare to solicit it all again, and again, in this semiotic territory where a word like "strong" is "too wimpy" *and yet* also risks being too strong:

DAVID DIMMOCK: I almost wonder if we can do with an intermediate term. I don't know whether, like, "implicate" would be a good word, or something else; we can probably argue about that again. But it might be nice to have a way in which you can say, "Well, we think this leads to this disease, but we don't meet the standards of causality." So, there's something—there's a punchy single word that you can put in a title that says, "We've got some degree of proof, but it's not all the way there." That is, an accepted word that everyone knows means, "We've done a lot of work and this is very exciting, but it's not at a point where it's a slam-dunk."

MALE SPEAKER: Maybe "likely candidate" or something like that? "Strong candidate?"

DAVID DIMMOCK: It's not—it's too wimpy. It's not going to—

MALE SPEAKER: You don't want it to be too strong, though, either.[40]

And again, the funniness of the limited semiotic capacity of ex-apes is not lost—

JEFFREY BARRETT: Yeah, I mean, the problem with those things is that, you know, someone who reads the paper is not going to necessarily think about the extreme care you chose in the exact word. There's just like, "Aha, this is the gene for blah blah blah." [laughter] So, I think you need to be very careful about that.[41]

And the only fitting thing to say is, *Be very careful*. . . . Indeed, perhaps the only thing funnier than carefree language in constant need of care is thinking that numbers and statistics will solve or dissolve these double binds:

DANIEL MACARTHUR: Jeff, are you arguing for a numerical score that we should abide by . . . [inaudible, lost in outburst of laughter][42]

What we watch in this and other GenomeTV videos is the work of caring for the sociolinguistic infrastructure of an infrastructural genomics, where friends—who can speak openly and honestly, risking uncertainty and bumping into limits and occasionally joking because they recognize what a funny place they're in—together try to invent a new lexicon and language for constantly evolving analytic meanings and practices.

The published guidelines from the workshop, collectively authored by most of the participants in these conversations, embodied all the hallmarks of double-binding care: the need for interminable curation work on multiple levels; the guidelines for scrupulous attention to detail that can never be quite scrupulous enough; the whole solicitous effort of thinking the entire system that enables and limits the research that produces it. It's all appropriately, wonderfully, carefully open-ended and deserves to be read in full. I will quote only one especially indicative passage, further emphasizing two particular phrases in it:

> Unambiguous assignment of disease causality for sequence variants is *often impossible*, particularly for the very low-frequency variants underlying many cases of rare, severe diseases. Consequently, we refer in this manuscript to the concept of implicating a gene or sequence variant: that is, *the process of integrating and assessing the evidence* supporting a role for that gene or variant in pathogenesis. We emphasize the primacy of strong genetic support for causation for any new gene, which may then be supplemented and extended with ancillary support from functional and informatic studies [emphases added].[43]

The lengthy discussion opened up by MacArthur's proposed statement that experimental (functional) evidence should not replace statistical (genetics) evidence has been rerendered in terms of the "primacy" of the latter while the former is recognized as supplement. This is the only appropriate tack for the genomicist and the deconstructionist alike, given the recognized impossibility of the clarity/ambiguity double bind. But I also want to emphasize the equating of "the concept" of a sequence variant implicated in disease with "the process of integrating and assessing" evidence. This implication of noun and verb, concept and practice, is the double bind of science and care, the impossible unambiguous and the interminable soliciting of scrupulous curatorial care that makes something like genomics possible.

Moreover, we should extend this elevation of process over arrival, to pertain as well to the friendly conversational workshop and the professional published guidelines. The latter were (and are) undoubtedly important and productive for the further enacting of the care essential to genomics (the article has been accessed over 85,000 times, according to *Nature*, placing it "in the 99th percentile [ranked 757th] of the 198,433 tracked articles of a similar age in all journals and the 90th percentile [ranked 98th] of the 1,006 tracked articles of a similar age in *Nature*").[44] But it was the process of producing those published guidelines that was the more important and formative, I would contend, for the group that articulated them. It's that scene of friendship that was the whole point of the exercise, not their arrival at another set of guidelines; the latter only matter to the extent that they provoke a new set of deliberations and exchanges among other, wider circles of genomics friends who cannot do genomics by themselves. The process of drawing up guidelines to be published will have to be repeated because the difficulties they are meant to resolve are impossible. Does a network or a consortium really have the wherewithal to endure that repetition? Marshaling the spirit of commitment, openness, and play that care requires, that *is* care, seems to need something like the force of friendship.

Imposition

"We should not, by means of artifice, pretend to carry on a dialogue."

I was much better friends with the writings of Maurice Blanchot when I was younger; after many years of being apart these words of his, from his very short récit "Friendship," found me again. It's a piece of writing that seems to say more about distance and death, and even (so Blanchot!) oblivion, than it does about friendship. But his writing works on me and teaches me without seeming to say or mean anything (also so Blanchot).

So, another reminder about methodology and writing: this ethnogrammatologist has met none of those whose words figure here in this chapter, has never corresponded or spoken with them teletechnologically, has no relationship whatsoever to any of them. Text is artifice. It's a marked departure from the genre conventions of ethnography, not to mention its methods, which so foreground its I-was-thereness and thus ground it in a presence, even if a now-absent one. *And yet*, in what is surely an imposition, I've written of them under the sign of friendship, use their archived words to pretend to have a conversation with them and to query the meaning of friendship in, with, to the sciences.

And yet: What clearer sign of friendship than taking the risk of imposition? *I just happened to be in the neighborhood and saw the lights on . . .*

It's no accident that there are no recordings of NHGRI grant review panels on YouTube. I can't really imagine what this "community" looks and sounds and feels like during those events, but probably something like friendship would be harder to read. A review panel like that would heighten the more agonistic aspects of science, and the more negative affects in play. But I never promised a strong theory of all of genomics or science, and I happened upon this scene in my efforts to understand what Stefan Helmreich calls the "informalisms" doing vital work alongside or within the formalized procedures of the sciences.[45] In contrast to what would probably happen during a grant review panel, it's easier to read for friendship in this kind of event where hesitancies and vulnerabilities—even if these are of the methodological sort—are, so to speak, the heart of the matter.

Why use friendship rather than "love" to understand these affective-cognitive assemblages in a science like genomics? It seems to me that "love" is more predicated on or gravitates toward identification, possession, and the collapsing of distance and difference, whereas friendship is composed differently. Prone to idealizations of its own, friendship is at least more resistant to them, built up from a more quotidian and less transcendental set of affects and relations better suited to the minding of genomics.

And, in turn, to my minding of genomicists, to my attempt to come to better terms with the relationship between the analyst and his scientist subject. Love feels overblown and overdetermined; maybe I'm just not ready for that kind of commitment to genomicists. Antagonistic critique? Not tonight, it gives me a headache. Maybe later. A neutral middle ground? Affectless and thus not only boring, but wrong. Unscientific and untruthful.

A group of scholars in the United Kingdom who also work in the fields of the life sciences offer the trope of neighborliness to denote a nonidentifying relationship very much like what I am pursuing with friendship. They consider it essential to cultivate neighborliness in and with the interdisciplinary collaborations that are now essential to genomics and other life sciences, a relationship that, like friendship, keeps its distance and keeps difference in play:

> To be neighbourly, then, would mean to recognise our differences and to respect them, whilst seeking to welcome each other without losing our sense of ourselves and our own commitments, responsibilities and proclivities. It is fundamentally an ethical disposition, which does not mean shying away from conflict, but rather making conflicts and their causes part of how we collaborate. In this regard, to be neighbourly to each other in an interdisciplinary collaboration would involve working together to identify our differences, to explore how we are differently vulnerable and how there might be different relations of power involved in our collaborative work. By doing so, we can make this relevant to the decisions that we make not only about how our collaborations are organised but also about the research and innovation itself.[46]

Friendship, then, is a style both of doing (impossible) science and of writing about the doing of science in a way that opens a relationship that can be called "authentic" because of its respect for the difference and distance that it engages, with care. Through those enduring engagements we might also begin to appreciate how friendship may, to reiterate part of the quote from Jacques Derrida at the head of this chapter, move us toward a rethinking of thinking altogether, beyond the Cartesian cogito so foundational to our scientific inheritance, a rethinking glimpsed in the conversations of the scientific collectives presented above: "I think, therefore I think the other; I think, therefore I need the other (to think); I think, therefore the possibility of friendship lodges itself in the movement of my thought insofar as it requires, calls, desires the other, the necessity of the other, the cause of the other at the heart of the *cogito*." Minding genomics is an act of science, an act of care, an act of friendship.

POEM-LIKE *TOLLS* 3 ·

an appendix

MRNA Is an Endogenous Ligand for
Toll-Like Receptor 3

KATALIN KARIKÓ, HOUPING NI, JOHN CAPODICI,
MARC LAMPHIER, AND DREW WEISSMAN

TLRs play
an important role in innate
immunity by recognizing
microbial infection and
orchestrating
an appropriate immune response. Mammalian TLRs respond not only
to pathogen-associated structures but also
to host-derived molecules that are released from injured
tissues and cells. The growing list of endogenous
ligands that activate TLR include
heat shock proteins
surfactant protein A
fibrinogen
hyaluronan
heparan sulfate

defensin and
chromatin-IgG complexes

We identify
mRNA as a new endogenous ligand by
demonstrating that RNA, associated or released
from necrotic cells, can modulate
DC activation and that RNA activation of DCs
requires TLR3 signaling.

What could be
the physiological relevance for TLR3
activation by cellular RNA? Examples certainly exist
for nucleic acids being
involved in induction
of immune responses

Mammalian chromatin can
induce the secretion of anti-DNA
immunoglobulin by B cells through dual
interaction with TLR9

Systemic lupus erythematosus is marked
by autoantibodies against DNA and RNA. It therefore comes
as no surprise
that deficiencies
in the removal of chromatin by DNase-I has been linked to
systemic lupus erythematosus. Despite the ubiquitous
distribution of RNases, additional pathophysiological conditions
are associated with
the appearance of extracellular RNA. An example of this is the
demonstration
that RNA is a constituent of the plaques found in patients with
Alzheimer's, a disease
that is increasingly being viewed as a chronic inflammatory
process.

Given these precedents,
it is reasonable to suppose

that RNA might have a role
similar to that already known
to be played by chromatin.

Such RNA could come
in contact with the DCs or
other immune system cells through release by necrotic cells or
the phagocytosis of necrotic cells or
their components. Such liberated cellular
RNA could conceivably stimulate TLR3 signaling pathways

The potential of RNA to induce immune reactions also
suggests
that physiologic mechanisms for the regulation
of RNA-induced immune activation
must exist.

Once antigen is acquired, DCs
have the peculiar capacity
to shunt the exogenously
captured material into the major histocompatibility
complex class I pathway via a mechanism that is dependent
on cytosolic antigen processing machinery.

Multiple mechanisms
have been suggested.
Considering that TLR3
and TLR9 are unique among the TLRs in their capacity
to respond to nucleic acids with antigen encoding potentials, and
considering the data
presented here and previously, these results
suggest
that mRNA could be
a novel source of antigen for presentation and cross-presentation by
professional antigen-presenting cells.

mRNA therefore joins
the list of endogenous ligands for TLRs and
is the first endogenous ligand described for TLR3.

Pathogens continuously mutate
and evolve to escape
immune surveillance, including
detection by TLRs. Whether the same
cellular mRNA that activates
DCs through TLR3
can also serve as a source of encoded antigen is currently
under study.[1]

Suppression of RNA Recognition by Toll-Like Receptors: The Impact of Nucleoside Modification and the Evolutionary Origin of RNA

KATALIN KARIKÓ, MICHAEL BUCKSTEIN, HOUPING NI, AND DREW WEISSMAN

It has been known
for decades
that selected DNA and RNA molecules have
the unique property
to activate the immune system.

It was discovered
only recently
that secretion of interferon in response to DNA is mediated
by unmethylated CpG motifs acting upon
TLR9 present on immune cells

For years,
bacterial and mammalian DNA were portrayed
as having the same
chemical structure, which hampered
the understanding of why only bacterial, but not mammalian, DNA
is immunogenic.

Recently, however,
the sequence and structural

microheterogeneity
of DNA has come to be
appreciated.
For example, methylated cytidine in CpG motifs
of DNA has proven to be the structural basis of
recognition
for the innate immune system.

Given
that multiple TLRs respond to RNA, a question
emerges as to whether the
immunogenicity
of RNA is under the control of similar types
of modification.
This possibility is not unreasonable

given
that RNA undergoes nearly one hundred different nucleoside modifications.
Importantly, the extent and quality
of RNA modifications depend
on the RNA subtype and correlate
directly with the evolutionary level of
the organism from which the RNA is isolated.
Ribosomal RNA, the major constituent (~80%) of cellular RNA, contains
significantly
more nucleoside modifications when obtained from
mammalian cells versus
bacteria.

A substantial number of nucleoside
modifications are uniquely present in either
bacterial or mammalian RNA, thus providing an
additional molecular feature
for immune cells to discriminate
between microbial and host RNA. Considering that cells
usually contain five to ten times more
RNA than DNA, presence of such distinctive
characteristics on RNA could make them a

rich molecular source for sampling by the immune system, a notion
becoming evident by the identification of
multiple TLRs signaling
in response to RNA.

We sought to
determine whether
naturally occurring nucleoside
modifications modulate the
immunostimulatory
potential of RNA and the role TLRs
might play in this process.

We demonstrate that
selected natural RNA isolated
from mammalian and bacterial cells and RNA transcribed
in vitro or synthesized
chemically activate human dendritic cells and
stably transformed cells expressing human
TLR3, TLR7, or TLR8.
Such activation was reduced or
completely eliminated
with RNA containing naturally occurring modified nucleosides, such as
m5C, m6A, m5U,
pseudouridine,
or 2′-O-methyl-U. Insights gained
from this study could advance
our understanding of autoimmune diseases where
nucleic acids play
a prominent role in the pathogenesis, determine
a role for nucleoside modifications in viral RNA, and
give future
directions into the design
of therapeutic RNAs.[2]

POSTSCRIPT

Stupefaction and astonishment open the scene of any possible knowing. . . . I want to say simply that science truly amazes us. It fascinates. Which is to say, also, it blinds us and may itself be blinded to its own trajectories, axiomatic presuppositions, procedures and premises; not to speak of the unmarked status of scientific desire, whether in crisis or, on the contrary—but this is not a contradiction—self-assured and well-funded, state supported. So much blindness compels an inward turn, or at least a staggering movement to something like a philosophical outside. · AVITAL RONELL, *THE TEST DRIVE*

Science is the drive to *go beyond*. · SHOSHANA FELMAN, *JACQUES LACAN AND THE ADVENTURE OF INSIGHT*

And Yet Another *Toll!* Tale

The surprisingly rapid development, in a little less than a year, of not just *a* successful vaccine but several vaccines for SARS-CoV-2 in 2020 prompted some science writers to dig into the back story of the messenger-RNA-based vaccines from Moderna and BioNTech.[1] A key figure in these stories was Katalin Karikó, an expert and unwavering believer in mRNA technology who migrated from Hungary to the United States in 1985, when *toll* had only just been characterized as a gene crucial to a fruit fly's larval development, and "genomics" didn't have its not-bad name. Sometimes troped toward a narrative of "she never doubted it would work," sometimes as "dismissed idea becomes leading technology," they all stressed the professional, personal, and technical difficulties Karikó endured in her work on mRNA through the

1990s and 2000s. I'll skip the part about Karikó meeting immunologist colleague Drew Weissman at the copy machine. I'll omit the stories about the start-up biotech company they would eventually found (and, presumably, made bank on). And there's no time or space to narrate the years of excited and tedious tinkerings in between, trying to get this still young mRNA biotechnology to work in vaccines. I'll highlight only the part of the story that grabbed my attention. But it's a part that was key to the eventual success of their work, the success of their business venture, and the success of the COVID-19 vaccine utilizing their science that is also a technology.

Karikó, Weissman, and their collaborators figured out that viral RNA is recognized by some human *toll*-like receptors, specifically TLR3, TLR7, and TLR8. Recognized, the TLR initiates a complex set of molecular signaling down multiple pathways, stimulating immune responses that cut up the RNA (good when it's from a virus) but that can also initiate cytokine storms, widespread cellular inflammation, and even death. The TLRs were rendering mRNA-based vaccines ineffective (the mRNA would be cut up before it could be transcribed to shape immune response) and risky. Eventually they figured out that modifying the nucleosides in the mRNA—methylating them, say, or using an isomer like pseudouridine instead of uridine—prevented recognition by the TLRs, prevented the inflammatory response, and allowed the mRNA to be translated into an effective and safe immune response—the response that helped keep untold millions of people, including readers of this book and its writer, alive or at least mostly well through 2021 and beyond.[2]

I know that this and my other *toll!* tales read like another extolling of the "basic," "pure," and/or "curiosity-driven" research that has been the dominant style (or at least the dominant cover story) in US science for the past seventy years. I've traced the gene *toll* in some of its becomings as an epistemic object—from a gene sending development signals to a gene in immunological pathways, from flies to humans, from one gene to a disseminated system of multiple, one-off *toll*-like receptor genes, and the even more multiple variants of them found distributed throughout populations—as a metonymic sign for a larger narrative of numerous scientists working and playing the excesses produced in and by a genomics infrastructure whose eventual ends and uses remained at least partly undefined, an open space of surprise and excitement.

And I've dotted this text with "*toll!*" the expression, playing with it as a minimal, ambivalent, multiply messaging sign just this side of nonsense that, despite or because of its multiplicity, nevertheless signals a cognitive-affective event that humans not only welcome but *want* and maybe even *need*. These *toll!*-like responses are at the molecular end of the complex semiotic pathways connecting to the vast molar structures of that basic, infrastructure-dependent, money-dependent, curiosity-dependent research system of the United States during the Cold War. So, to say by way of conclusion that I have written to elicit a reader's response of *genomics is toll!* would also be to say that, notwithstanding all the complexities and the contradictions and the qualifications, I have written *for* genomics.

Maybe these affirmations of *toll!*-like genomics and by extension its value as a social good leaves you a little anxious. It does me. *Toll!*-like sometimes feels *Candide*-like, pollyanna-ish and naïve. True to care's pattern and affective effect, having now written *for* genomics, I'm left with a persistent worried anxiety that I also feel responsible for rubbing in. The anxiety signals, as care always does, a double bind in need of minding: What's the use value of uselessness?

Science, the Endless Impossible Staircase

While I was in Albany, New York, for that 2017 March for Science that opened this book, a busload of some of the most mindy and elite scientists in the United States from the Institute for Advanced Study (IAS) left Princeton, New Jersey, to attend the main march in Washington, DC. The *New Yorker* magazine sent a writer to accompany them on the bus and at the march, because of course it would.[3] Nerds on a bus? Like snakes on a plane, it's *New Yorker* narrative gold, the comedy of contradiction.

The disjuncture was doubled by the thin book carried by some of these scientists, Abraham Flexner's *The Usefulness of Useless Knowledge*, the 1939 report that the IAS (founded by Flexner) had just then recently reissued. The IAS director at the time, string theorist Robbert Dijkgraaf, had written an introduction to the new edition and seemed well aware of the ironies Trump's election had brought into the timeline: "I now feel like I have to write a prequel: 'The Usefulness of Useful Knowledge,'" he told the *New Yorker* writer, who acknowledged that, indeed, "the concept of useless knowledge had begun to seem a little quaint." Not to mention self-serving. I worriedly imagined I had entered some variant timeline and now was just another nerd on some bus in 1950s Jim Crow America taking my marching orders from Flexner and traveling to DC to help Vannevar Bush carry out his plan to install "basic research" as a key node in the conceptual-ideological-cultural fabric legitimating the federal funding of the mindy elite scientist's free curiosity-driven inquiry, through his much-hyped, much-critiqued, and highly influential report, *Science, the Endless Frontier* . . .[4]

Maybe that era of US science hasn't ended so much as returned again and again with a difference, forcing us to anxiously renegotiate the nonnegotiable, impossible double bind of basicXapplied, usefulXless science.

In a 2020 article, science policy analyst Roger Pielke Jr. acknowledged Bush's "rhetorical brilliance" powering this short 1946 report, and it's enduring "role in shifting how we think and speak about science, technology, and government." Pielke also recognized "Bush's semiotic wizardry" that installed a "contradiction at the heart of *Endless Frontier* between no-strings-attached science and promises of social benefit." Fundamental to Bush's wizardry was "the adoption of the frontier metaphor" that

> served to rationalize research exclusively on the basis of the motivation of the individual scientists, "dictated," in Bush's words, "by their curiosity for exploration of the unknown." In so doing, the metaphor has been an intellectual soporific, absolving the scientific community from the need to consider its broader responsibilities. . . . For policy-makers the emphasis of the phrase [basic research] is on research being *basic* to the achievement of political and societal objectives such as national security, jobs, economic growth, and health. For scientists, the emphasis is on *research* that "is performed without thought of practical ends"—what prior to Bush was typically called "pure research."[5]

Pielke was writing in response to the Endless Frontier Act, introduced into the US Senate toward the end of the Trump administration in 2020, eponymizing Bush's seventy-year-old report and promising a $100 billion increase in the National Science Foundation budget. In Pielke's analysis, the problem with continuing to think and enact policy in the terms established by Bush's report was that "the language it introduced serves contradictory purposes." Pielke's own rhetoric is replete with expressions of "uncomfortableness" with the "fundamental dissonance" and "jarring disconnect" of Bush's (now our) language. The result is that "the themes that *Endless Frontier* discusses never seem to get resolved," and this was "not in fact a paradox. It is a design flaw."[6]

Pielke is far from alone in calling for a redesign of the contemporary scientific order, one that would emphasize a "socially responsible" team approach, not science done by individuals "without thought of practical ends." Continuing to operate under the spell of Bush's semiotic wizardry by valorizing basic-fundamental-pure-individual-curiosity-driven research, it's implied, means favoring irresponsible science and a continuation of a kind of settler-colonialist relationship to epistemological and pragmatic frontiers.

I'm uncertain where US science culture is going, but I am certain it will not involve a decision between two different orders. As chairwoman of the House of Representatives Committee on Science, Space, and Technology and one of the sponsors of the Endless Frontier Act, Representative Eddie Bernice Johnson (D-TX) would contend in a later issue in the same journal, "The pursuit of knowledge and curiosity are not mutually exclusive with the quest to solve grand challenges or to translate research results into use."[7]

And yet, even if curiosity and utility are not mutually exclusive, they are also not easily reconcilable. Dissonance, disconnect, and discomfort are not the product of design flaws in science ideology; they are, in fact, the affective signs of science's enduring limits, as infrastructured by paradox and impossibility. These infrastructured patterns of US science policy and research since the 1950s never seem to get resolved because they are a double bind, and the language of basicXapplied curiousXresponsible individualXcollective science is self-contradictory because it can't be otherwise.

Believing that we scientists can't ever be made more comfortable with our enduring discomforts, that these are the result of infra/structural paradox and not design flaw, this book has worked on the premise that we can only keep learning to better mind their contradictions. Maybe I place too much stock in rhetoric and semiosis (and isn't all semiosis a wizardly conjuring of forceful meaning, a non-quantum-physics version of "spooky action at distance"?), but I think the way forward will not be through a redesign of science policy that will somehow harmonize the conceptual and political dissonances and bring comfort to all, scientists and legislators and citizens alike, but through learning how to live with these dissonances, to think and act with these contradictory signs in ever more careful ways. I hope the readings of genomics double binds that I've given here provide some sense of how that might make sense, and how some scientists have indeed taken a good deal of responsibility for caring for and minding their science. I don't pretend that *Science, the Endless Impossible Staircase* could ever have the same appeal and zing that *Science, the Endless Frontier* had, but I do think it better suggests the contradictions that inhere in all experimental sciences like genomics, their ever-stepping into a beyond that is not beyond, and why care's patterns recur within science's ever-returning (infra)structural limits.[8]

I didn't join a March for Science in the spring of 2021 because there weren't any. The political crisis that simplified matters of science, its truths, its culture, and its governance just four years earlier, and made a decisive cut through these knotted binds necessary, had passed (at least as far as science is concerned), and we are fully engaged once again with these difficulties and impossibilities. To which I can only say: *toll!*

Infra-Infrastructure

To stay with the discomfort for one more turn, I'll admit that an image of endless marches up the impossible staircases of multiple sciences, no matter how truthful, doesn't easily translate into knowing or saying which particular sciences to march *for*. A more pointed question would be *What is genomics good for?* What value can be assigned to its disjunctured useful uselessness?

In the spring of 2010, around the tenth anniversary of the ritualized if not realized "completion" of the Human Genome Project, the *New York Times* featured an article titled "A Decade Later, Genetic Map Yields Few New Cures," by its lead science writer Nicholas Wade. (Wade was almost always an uncritical fan of anything genomic, especially when he read it as upholding the reality of biological race and validating its utility, even when the majority of genomicists had moved on—as an unusually large number of them would make clear in very public critical responses to his later 2014 book, *A Troublesome Inheritance*.)[9] Wade recognized at least one value of genomics in the preceding decade: "For biologists," he wrote, "the genome has yielded one insightful surprise after another." He reiterated this assessment later in his article, allowing that "research on the genome has transformed biology, producing a steady string of surprises."[10]

But the dominant message was already announced by the headline: precious few of the biomedical benefits that HGP advocates had promised had actually been realized. As a "translational science," knowledge that moves from "bench to bedside," genomics' promises may not have been broken, but they still required a lot more keeping. But Harold Varmus, the Nobel laureate soon to become head of the National Cancer Institute, provided an alternative narrative: "Genomics is a way to do science, not medicine."[11]

Wade's article conveyed a collective affective loss of interest-excitement, a sense of disappointment that readers were meant to share. To the genomic elite, and to only slightly less elite readers of genomics in the academy, surprises and exciting biological insights might be worth the hundreds of millions of dollars they cost, but to readers of the *New York Times* they were a dime a dozen, last year's news, and need not figure in the calculations of worth that were being anxiously assessed. Many historians, feminist philosophers, sociologists, bioethicists, and more than a few genomicists themselves had been making similarly deflating, cautionary remarks about the expected medical benefits of the HGP for as long as there had been an HGP. They—*we* were not surprised when the relationship between genomes, illness, clinical practice, and public health initiatives turned out to be far more complicated than hoped for or predicted by the more hardcore genome-as-Holy-Grail mythmongers—whose semiotic wizardry, it must be said, conjured up all that money to begin with. Shouldn't a $3 billion investment (to use the most convenient, repeated, and not too inaccurate price tag attached to the HGP) translate into something more than a better "way to do science"? Shouldn't there be something of real use value beyond a surplus *jouissance* of anxious surprise and interest-excitement in genomicists and their ethnogrammatologist friends?

First, "cures" are only one way to measure outcomes, in the same way that a cure is not care's only meaning or goal. The HGP is a poor metonym for a catachrestic "genomics" that was always at least as much about building infrastructure as it was about arrival at a destination like personalized asthma treatments. And that infrastructure of genomics technologies, data and databases, diverse practices of truth-making, and the growing ranks of careful expert genomicists has had a transformative effect, at both "basic" and "applied" levels, on fields too numerous to detail, from virology and geomicrobiology to forestry and conservation biology. And I've drawn attention to genomics' infrastructural dimensions as a constant reminder of the importance and even the primacy of quotidian care in, of, and for sciences—care as the infrastructure of genomics infrastructures, the underlying work that overlies it all, the just-barely-sensible infra- that permeates the conceptual, technological, and social structures of science. In this sense, this book has been an infra-narrative of the infrastructures of mindful care that make genomic sciences possible, assembled from reading between the lines of writings in genomics, and in its margins, for its paradoxes, impossibilities, and double binds.

It's not that the pure, basic, curiosity-driven sciences that genomics has metonymized here lack value, in comparison to more valuable, more applied sciences like drug development; it's more that they are situated beyond the applied. Asking where genomics as "basic science" ends and genomics as "biomedical applications" begins is like asking where an infrastructure becomes a structure: a contrived accounting, by number and narrative, that has to be supplemented with a more solicitous attention to something approaching, and exceeding, a whole.

Second, and I think more importantly, it may be that surprises themselves are quite precious, resisting valuation not because they lack value but because they exceed it. Alongside this book's infrastructural refrain has been a periodic sounding of "*toll!*" as a sign of a scientist's minimum interest-excitement response to something unexpected. Not *eureka!* and not *aha!*, *toll!* is not an expression of wonder or curiosity but an a-grammatical exclamation of an interest-surprise yet to acquire a syntax and narrative—and value. As we rethink and tinker with the Cold War science ideology that privileged "basic science," its recognition that scientific infrastructure *should* support the production of surprise, and surprised scientists who at least occasionally say "*toll!*," is something somehow worth keeping, in keeping with its promissorial quality.

We need more diverse empirical materials from the sciences to advance more careful hypotheses of a scientist-subject deeply invested in and cathected to objectivities, confounded by impossible contradictions. For that rethinking, we'll need to better mind our theories of who scientists are and can be, how they learn to do what they do and want what they want. This impossibly whole affectiveXcognitive thing we multispecies multicultural multiversal humans are—how does it *run*? How does it accelerate and *avalanche*? How might something like an epistemophilic drive—tinkered (that other sign of catachrestic care) together from the likes of *toll*-like receptors and spookily powerful signs in the fanning biocultural evolutionary trajectories of our species—not just make us curious, in the usual mindy sense of that term, but make us *feel* curious? Make us want to know, to relate to objects through object relations that have already somehow (and this is the really *toll!* part) been tinkered in to us?

Care in/of/for Sciences: Affects, Ethics, and Epistemophilia

Theorizing an epistemophilic drive and the relations of care composed with it through genomicists, as I've begun to do here, shouldn't be taken to imply that scientists are uniquely or at least unusually curious, *toll!*-seeking, truth-making subjects. They are no more and no less *toll!*-seeking and knowledge-making subjects than are, say, historians or elementary school teachers—or their students, for that matter. But scientists may be the knowledge-makers most resistant to suggestions that there is anything other than epistemology happening in their minds and labs, that something like care is woven throughout their work and worlds, or that their cognitive achievements are also patterned by matters of affect. So in addition to the fact that I am not an ethnogrammatologist of historians or elementary school teachers but of genomicists, I take some *toll!*-seeking satisfaction of my own by reading for those traces and signs of affect that are not supposed to be at work and in play in the sciences, and drawing attention to their infrastructural forces.

At the same time, we've also seen how some scientists are quite sensitive readers of both the nested systems in which they work and their nested selves within and beyond them. Bruno Lemaitre, the geneticist-turning-immunologist whose work leading to the elucidation of the pattern-recognizing *toll*-like receptors and the innate immunity systems of flies and humans I breezed over in chapter 1, recognizes how the forces of an unconscious must affect scientists and sciences. I learned this only long after I had read Lemaitre's scientific publications when, checking my notes and references, I visited his lab's website and found his other writings.

Like many scientists (and me), Lemaitre finds physical chemist Michael Polanyi's writings on the tacit dimensions of science (which Polanyi renamed "personal knowledge") to be true to his own experience. Tacit, personal knowledges are essential not only to learning laboratory skills, but also to how a scientific community arrives at its truths. "It is impossible," Lemaitre writes, "to evaluate a scientific statement in a purely objective manner without being influenced by many tacit factors, one of them being the convincing power of the author." Narcissism and "innate traits associated with dominance," in Lemaitre's view, drive some of that power.[12]

Lemaitre diagnoses "a rise in narcissism" in the contemporary life sciences, a limited and formulaic diagnosis, perhaps, but an interesting one nonetheless. My somewhat idealized characterization of the "careful genomicist" doesn't attend carefully enough to the problems presented by the narcissist, of which genomics has a few. Maybe this is another of friendship's double binds: narcissism is intolerable; narcissistic friends deserve some slack, if only because they tolerate my own narcissism. In the name of friendship with genomics and genomicists, this text has mostly turned a blind eye toward genomic narcissism and neuroses, and toward much else in genomics that is less than admirable (a word John Dewey valued). But in a text trying to do the work of caring for the more careful, admirable side of genomics that I for one had long been blind to, I couldn't find a place for all of that, or didn't want to. I have instead privileged the counterpart to the narcissist scientist whom Lemaitre names the "meticulous scientist" or the "classic scientist," pointing to the slightly nostalgic and mournful feel of this figuration—and of mine, too. We share these infra-structurings and their limits. Still, the difficulties these kinds of figures signal are real:

> The remaining, meticulous scientists might feel in many respects disadvantaged compared to these new rock star [narcissist] scientists. . . . This type enjoys science for the sake of experimenting, testing hypotheses and making discoveries. Financial gain and institutional power are all rather secondary to him; his true reward is the daily intellectual challenge. This scientist is keen to collaborate scientifically, but is not particularly interested in networking and attending all those artificial and boring strategic meetings. . . . He has too great a sense of community and feels guilty when he uses too many resources. He cannot compete because he does not have the competitive character.[13]

I hope I've written a slightly more complicated careful genomicist, who can find interest and even enjoyment in the boring, or at least recognize their amalgamation. The institutional resources and symbolic capital that a careful genomicist enjoys are certainly the envy of many other life scientists; I would understand if they felt "guilty" about this, an emotion Silvan Tomkins bound to the affect of shame. And that clean division between collaboration and competition—well, our friends tend to be messier, don't they? Still, Lemaitre identifies the affective dimensions of his world and helps us recognize them, too. And many younger researchers are in fact exiting the fields of life sciences research, often citing the dearth of pleasure and excitement they thought "basic research" was supposed to provide, and an increasingly competitive system that makes advancement difficult and unpleasant.

The truth is we need more empirical data to better recognize and explore the dense tangle of affects, emotions, and tacit and formal knowledges that constitute the contemporary scientist-subject. Such a project to "re-affect science" can be part of "a strategy to pursue diversity in science and knowledge politics," writes anthropologist Sandra Calkins, a project relevant not only in the privileged worlds of US genomics but in the sciences of nations of a Global South like Uganda, where Calkins worked with plant biologists. Ethnographic work like Calkins's recognizes the importance of "attending not only to material equipment and geopolitical location but also to the moods or affective atmospheres that infuse these places, such as atmospheres of boredom, seriousness, excitement, or playfulness." These affective atmospheres matter to the sciences and scientists of the Global South as much as anywhere else, and are crucial to the scientific creativity and the relationalities of care that humans seem to need and thrive on, as does an anthropocenic planet in its diverse, quotidian multiplicity of places.[14]

And to supply some necessary meta- for reading that necessary data, we could use some psychoanalytic theories that have moved beyond such classic Freudian concepts as narcissism. There are encouraging developments on this score that there's neither space nor time to incorporate here. The scholarly literature on care that I have built on makes little use of psychoanalytic theory, if any. My analysis partly resembles that of Lorraine Daston and Peter Galison, who describe the "epistemic virtues" of scientist-subjects able to declare or depict truths according to changing historical configurations of objectivity. But although they show us why every objectivity posits a subjectivity capable of achieving it, the scientist-subject they posited is essentially that of German idealism. Esha Shah has critiqued this postulated subject for its "neo-Kantian" idealizations, and its lack of any kind of unconscious or body (same difference).[15] Shah theorizes her more complex scientist-subject (one prone to pathologies as well as virtues) largely through Jacques Lacan, whose ideas about science and scientists have shaped many subsequent analyses.[16] My future work may draw more fully on Lacan's insights about impossibility, anxiety, and the Real, but for now I'll close this present work with a brief discussion of epistemophilia, or what Freud called *Wisstrieb*, the knowledge drive.

Epistemophilia: such a mindy word. Why not *curiosity*, for which there is a much larger literature to explore, and which most readers would find more familiar, more recognizable? But recognition is the first thing to dodge: already knowing (we think) what curiosity is, it doesn't invite much further probing into its causes or nature, only its effects and objects. There's not much need to be very curious about curiosity itself, only about its troubles (dead cats) and rewards (Eureka!). Curiosity is in fact even mindier: all mental, an identifiable psychological phenomenon or cognitive faculty like imagination. Epistemophilia's jarring, *toll!*-like weirdness might interrupt those familiar circuits: a knowledge instinct (that other translation of Freud's *Trieb*)? Sounds sociobiological . . .

There's no mention of epistemophilia in Evelyn Fox Keller's *Gender and Science* (1985), one of the first and still one of the few efforts by scholars of science (including scientists, like Keller) to theorize a scientific self from a psychoanalytic perspective. Keller drew on the work of Ernest Schachtel to analyze the "cognitive-emotional styles" of the masculine scientist-subject who was, historically and currently, the most common and dominant figuration of the scientist. Keller wanted to open a space to think, not a feminine or feminist but a gender-free scientist-subject. Simultaneously, she was writing about the life and work of maize geneticist Barbara McClintock, whom she thought exemplified this gender-free style of what Keller called "dynamic objectivity."[17] Dynamic objectivity is "not unlike empathy," in that both are "premised on a continuity" *and yet* "recognize the difference between self and others." Keller used that differentiation (I would say double bind) "as an opportunity for a deeper and more articulated kinship" with—indeed, a "love" of—objects of the natural world. Dynamic objectivity thus makes "use of subjective experience . . . in the interests of a more effective objectivity."[18]

In its diffracted patterns care, too, is "not unlike empathy," and with the binding of careXscience I've put forward here I, too, am trying to understand (as Silvan Tomkins was) this amalgam of "cognitive-emotional," their difference-in-continuity, that happens at the limit of the biological. Through Schachtel, Keller explored only a limited range of cognitive-emotional styles that might compose a (masculine) scientist—the paranoid style, the obsessive-compulsive style, and the sadistic style—to produce a particular style of objectivity. While many of the specifics here resonate with my own analysis ("the paranoid resembles the quintessentially meticulous scientist"), I want only to emphasize the broader point Keller was making, that "the connections among these three components of scientific ideology lie not in their intellectual cohesion but in the cohesion of the emotional needs to which they appeal."[19] If the affective does not trump the epistemic, it is at least its binding medium.

Anxiety—that other name for name-exceeding care—did not figure in Keller's analysis, who after all was writing long before "the affective" wave in academia had begun to build. Keller's analysis of scientific thinking and knowledge proceeded on the "higher-order" level of complex emotions and its clinical presentation in formations of sadism and obsession-compulsion.[20] But anxiety had figured centrally in the psychoanalytic practice and writings of Melanie Klein, who worked and thought more in the pre-symbolic order, and with very young children exhibiting forms of autism rather than full-blown paranoia or sadism. It's at this more affective level, more aligned with the body than with the clinic, that Klein may help us think a scientist-subject in terms of epistemophilia.

In Freud's scattered remarks first positing an epistemophilic drive, it emerges as "an imperfect negotiation of anxiety that is simultaneously induced and relieved by knowledge."[21] Klein developed Freud's few thoughts into a more organized theory in which this *Trieb* for knowledge assumed a position equal in fundamental importance to the life and death drives. In Klein's theories, the early world of a young infant is a fairly terrifying place, full of aggression and violence, and thus a constant cause of anxiety as a largely inchoate infant becomes a more organized child through its relations with an ever-growing population of "objects" both material and fantasized—hence, the object-relations theory that Klein inaugurated and which for the most part succeeded classical Freudian theory.[22]

What Klein called the paranoid-schizoid position is the primary mode of relating to objects, a defense against anxiety in which to know is to control ambivalent objects through splitting them into (e.g.) good or bad, or (later in development) objective and subjective, and any and every other doubly bound real entity. For Klein and Kleinians this is a normal and inescapable developmental event, and to position it as something "bad," ascribing it to something only other people do—scientists, say—is to adopt the same splitting style. (In other words, only a splitter could describe someone else's epistemophilic drive as focused exclusively on "splitting," while thinking themselves innocent and free of such vulgarity and violence.) The paranoid-schizoid subject stuck in this position alone "must be alert to every possible clue. Nothing—no detail, however minor—eludes scrutiny. Everything *must* fit." "Normally," Keller points out, "scientists recognize that their interpretation cannot account for every detail," that there must be a "trade-off between logic and realism," but the tendency of this position is "to leave no room for an alternative explanation; the pieces are locked into place by the closeness of their fit."[23]

The congealing of these patterns and (infra)structures over time produces the scientific epistemologies characterized by Keller as paranoid, obsessive-compulsive, and so on. But it's not only the sciences in which this mode or position becomes dominant. Eve Kosofsky Sedgwick used Klein's work to show how this cognitive-emotional style dominates in the critical, cultural, and literary theory in which so many of us—and I include myself, as Sedgwick did herself—read and write, that is, "theorize." From the paranoid-schizoid position a scientist, a humanist, a social scientist, or a paranoid becomes relentlessly anticipatory, trying to avoid any surprises, trying to think as far ahead and as widely as possible for every possible unknown cause and hidden connection. Genomicists—some of them, anyway—in their paranoid-schizoid position split genes from environment and privilege the former as the true cause and only plausible interpretation; gaps in the interpretation can be due only to the failure to have pursued genetic logics with sufficient zeal and at sufficient length. This is an epistemophilic drive that tends toward the absolute: only total zeal, exerted unendingly, can be paranoid (and scrupulous) enough. And for their part, historians, philosophers, anthropologists, or any of us in between these disciplines have a tendency, in this position, to already know that genomicists are determinists-essentialists who can only think reductively and geneticize. Determined to find all the conceptual and social causes of this kind of behavior, we too can never be paranoid (scrupulous) enough. Genomics and genomicists, in the mostly paranoid-schizoid worlds of science studies, need their every social, cultural, or rhetorical determinant to be tracked down, scrupulously analyzed, and publicly exposed. (Exposure is the overarching logic and final goal of paranoia, writes Sedgwick, that acts "as though its work would be accomplished if only it could finally, this time, somehow get its true story known."[24])

But genomicists, I hope I have shown, are indeed unexpectedly surprised, despite their most admirable paranoid tendencies. They design experimental systems that generate those surprises, interrupting and displacing their paranoid-schizoid emplacement. An ambivalent *toll!* signs for an anxiety doubled as care. What happens then, according to Klein's analysis, is that an infant (ethnogrammatologist, genomicist), "rather than continue to pursue the self-reinforcing because self-defeating strategies for *forestalling pain* offered by the paranoid-schizoid position," might achieve the "often risky positional shift" and "move toward a sustained seeking of pleasure (through the reparative strategies of the depressive position)." The depressive-reparative position tries to bind together what had previously been split into partial-objects. A kind of solicitude, this position is oriented to the whole, and "inaugurates ethical possibility" in the form of "a guilty, empathetic view of the other as at once good, damaged, integral, and requiring and eliciting love and care."[25]

Remember that to the epistemophilic drive "the other" here is any one and any *thing* that is not the self/subject. For Klein and many psychoanalysts who came after, the first and most important "others" are a breast, its milk, and then the mother to which these are attached. For a genomicist, the most important "other" epistemophilic drive-object is avalanching data—an alien, excessive data-deluge that genomicists themselves initiated, that is clearly "good" and "integral"—but is also just as clearly "damaged" (precarious) and in need of care (curation). Anxiety's *toll!* is ambivalent and doubled, worrying and exciting, needing to be known and controlled *and yet* needing *more avalanching*...

For the ethnogrammatologist or ethnographer, "the other" is a genomicist and her genomic science, both of which have also avalanched in number and kind, part of our own empirical deluge, a sublime anxiety of our own. Genomics has become a key part of the research infrastructure for subjects and objects far beyond this book's focus on human: genomics infrastructure has helped transform virology, geomicrobiology, ecology, forestry, conservation biology, evolutionary theory, fisheries management, synthetic biology, and other sciences. It's a rich and diverse territory for anthropologists to more fully make the "risky positional shift" away from the paranoid-schizoid position, which casts genomic(ist)s as a partial object that is *bad*, an object deserving only our skepticism, criticism, dismissal, or other expressions and modes of negative affect, and to place ourselves more in Klein and Sedgwick's depressive-reparative position. Anthropologist Amber Benezra's work has been exemplary in this regard, undertaken in deep collaboration with genomicists of the microbiome and other transdisciplinary friends, to further a more interesting, more exciting, and more socially just genomics.[26]

The depressive-reparative position kicks the epistemophilic drive into another cognitive-emotional style, one in which "thinking is poignant and qualified by a certain sadness," as Deborah Britzman writes, "and where the ambivalence made from both loving and hating the same object devastates neither the ego nor its other."[27] As scientists and humanists, our knowing is "never fully able to escape the shadow" of anxiety, even in its earliest forms experienced as "terror" (in Klein's view) by the infant scientist-humanist; anxiety "is the prerequisite for thinking to even emerge." But with the addition of the depressive-reparative position, that affect-bound thinking "can come to tolerate, in very creative ways, the anxiety that sets it to work. This is what Klein means by learning, and perhaps what Sedgwick means by reparative reading."[28]

And it's what I mean by minding genomics: genomicists anxiously caring for the way they analyze genomes and make better genomics truths of them, and we analysts of genomics learning how to solicit our own narrative and the excesses they anxiously cover, to see if we can get something to shake out that surprises us.

Too often in the science studies literature, "science is read as 'bad boy' to the humanities' denigrated yet superior sensitivities," Vicki Kirby points out, "a bad boy whose penetrative and instrumental logic must be distinguished from more poetic, creative, and generous curiosities."[29] I've tried to read differently, analyzing genomics with care to trace where they become indistinguishable. My minding of genomics didn't aim for a comprehensive history of genome-wide association studies, or a thick ethnographic account of research into gene-environment interactions, or even one focused only on the actions of *toll*-like receptors in the bodies of exposed asthmatics. I did try to be careful to account for a "whole" genomics, one that can't simply be split or severed into the familiar analytic categories: reductionist, determinist, eugenicist, neoliberalist, elitist. And when so split, to have at least marked the lines where the sutures might be sewn. My account has had its more critical passages, but it is most serious about eliciting a sense of surprise and excitement, or at least interest, toward genomics. It has its paranoid tendencies—wasn't I supposed to be scrupulous?—but at least I tried to put these into oscillation with the reparative tendencies of care. For as Sedgwick insists, "It is sometimes the most paranoid-tending people who are able to, and need to, develop and disseminate the richest reparative practices."[30] They are doubly bound.

I've analyzed the multiple double binds of the scienceXcare of genomics, how within these impossible limits of contradictory demands, scientists have created new, rich practices of care in/of/for sciences: multiple patterns of care, (infra)structural relational forces that energize and enliven the sciences in multiple ways and on multiple levels. Care *in* the sciences: to worriedly care for (meta)data, databases, and *toll*-like receptors, pulling them together and keeping them together, cleaning and ordering and reordering, anxiously troubled and excited by their present state, attending to their incessant needs. Care *of* the sciences: to discipline and create oneself as a scientist-subject, cultivating a self capable of the scrupulous care necessary to sustain fragile objects, rapidly evolving but always imperfect methods and technologies, and the precarious truths they offer—in a profligate economy of ease in which carelessness has many rewards. Care *for* the sciences: to establish and grow collective resources and infra/structures, to build the consortia and communities capable of honest analysis and self-analysis, to do the hard work together of minding an impossibly whole science, exceeding any one person's care.

Both knowledge *in* the genomic sciences (the kind of knowledge generated by genomic scientists) and knowledge *of* the genomic sciences (the kind of knowledge generated by genomic scientists themselves, and their ethnogrammatologist friends) are each generated by an epistemophilic subject who, driven to know, engages the world from both a paranoid-schizoid position and a depressive-reparative position, both product and producer of anxious care. The latter subject position has too often been overlooked, underappreciated, marginalized, trivialized, disavowed, or simply gone uncared for. We should care enough to compensate for this systemic privileging, while attending to the limitations and blind spots that are produced by it alongside its insights.

"No less acute than a paranoid position, no less realistic, no less attached to a project of survival, and neither less nor more delusional or fantasmatic," Eve Sedgwick wrote, "the reparative . . . position undertakes a different range of affects, ambitions, and risks."[31] I've tried to craft a text with care that might spark that different range of affects toward genomic truths and the quotidian (infra)structures that sustain them, and interest in the genomic scientists who sustain those infrastructures in turn. We are not so caught by the paranoid-schizoid position that we can't learn about the reparative practices of care in genomics from our friends, the only partly paranoid-schizoid scientists who invented them, and that will in turn, I hope, enrich and strengthen all of our capacities for thinking and doing genomics with care.

NOTES

Chapter One. Fors

1 Fortun and Fortun, "Scientific Imaginaries and Ethical Plateaus."
2 Guarino, "March for Science Began." The originating Reddit thread is archived at https://www.reddit.com/r/politics/comments/5p5civ/all_references_to_climate_change_have_been/dcoi17w/. Like any mass movement, this one had organizational challenges, controversies over diversity and representation, and questions of long-term sustainability beyond the scope of this introduction, and book. See the Wikipedia entry at https://en.wikipedia.org/wiki/March_for_Science, and the podcast created by the Science History Institute, https://www.sciencehistory.org/distillations/podcast/political-science. Wolfe, *Freedom's Laboratory*, reads the March for Science in the context of (post-)Cold War US history.
3 Fortun, *Promising Genomics*.
4 Robert Proctor, quoted in Mooney, "Historians Say the March."
5 I'm not *against* "postgenomics" as a term, and chapter 2 is based on an essay previously published in an edited book by that title (Richardson and Stevens, *Postgenomics*). I've come to dislike "post-truth" as a designator of a current era supposedly changed from a time when "truth" was simple and determined straightforwardly. Just as it's true that for as long as there has been truth we have also always already had to deal with being post-truth—that supposedly stable ground was always a much more volatile lavaXland (Fortun, *Promising Genomics*)—this book, being *for* genomics, proceeds on the friendly assumption that genomics has always been at least somewhat in tune with the complexity, multiplicity, and indeterminacy that postgenomics is meant to signal more directly.

6 Neil deGrasse Tyson (@neiltyson), "The good thing about Science is that it's true whether or not you believe in it," Twitter, June 14, 2013, 10:41 a.m., https://twitter.com/neiltyson/status/345551599382446081.

7 I should add that I use "ethnographic" and "anthropologist" loosely here, as friendlier or at least more familiar names to those for whom this book may be a stretch of their reading habits. Later I identify myself as an ethnogrammatologist, a neologism that mashes up the ethnologist with the deconstructionist, but it's too early in the book to risk alienating readers.

8 See Keller, "Drosophila Embryos as Transitional Objects."

9 Stere, "The *Fushi Tarazu*."

10 Hansson and Edfeldt, "Toll to Be Paid," 1085.

11 Weissmann, "Pattern Recognition and Gestalt Psychology."

12 *Toll* is not one of the genes announced in the Nusslein-Volhard and Wieschaus 1980 *Nature* article, "Mutations Affecting Segment Number and Polarity." The first mention I could find, and the earliest reference publication for the *toll* gene catalogued in FlyBase, is Anderson et al., "Establishment of Dorsal-Ventral Polarity," in 1985. If that first encounter with *toll* was so exclamation-mark-worthy, why wasn't it in the first write-up? The encounter must have occurred down the developmental pathway of discovery, only after the initial results had been published.

13 Weissman, "Pattern Recognition and Gestalt Psychology," 2138.

14 Quoted in Weissmann, "Pattern Recognition and Gestalt Psychology," 2137.

15 In other words, I try to make the signaling pathways of the book parallel or mime the way biological events happen, semiotics to biosemiotics. The cross talk between the life sciences and the sciences of language is long and complex; see Derrida, *Life Death*; Doyle, *On Beyond Living*; Keller, *Refiguring Life* and *Making Sense of Life*; Rheinberger, "Experimental Systems"; Kay, *Who Wrote the Book of Life?*

16 Stanley Corngold's *Franz Kafka* (esp. pp. 90–104) is a particularly good explication of catachresis. Corngold reads Kafka's writing as thoroughly catachrestic, a pushing of metaphors to their extreme—"metamorphosed metaphor"—before reversing their movement in a chiasm that ruptures familiar ontologies and their reassuring stabilities.

17 Mol et al., "Care: Putting Practice into Theory," 8 (emphasis original).

18 On model organisms in genomics, see Ankeny and Leonelli, "What's So Special about Model Organisms?"; and Leonelli and Ankeny, "Re-thinking Organisms."

19 Kuska, "Beer, Bethesda, and Biology," 93.

20 Mitchell, *Relationality*, 134.

21 I'm grateful to a Duke University Press reader who reminded me of historian of science Lily Kay's work, *Who Wrote the Book of Life?*, and how it, after seeming to opt for an analysis in which "information" is understood as a catachresis (p. 2), quickly reverts to the master trope of metaphor, setting geneticists

up for missing the mark of the real and instead settling for the second-best, borrowed term of "information" metaphorically standing in for a truth that genetics forever lacks. By contrast, Richard Doyle's *On Beyond Living* analyzes many of the same events narrated by Kay through a more complex rhetoric that includes catachresis. A truly deft reading of this history is Jacques Derrida's *Life Death*, a transcription of his 1975–76 seminar in which he (affirmatively) deconstructs "life" as both an effect and instance of writing, in large part through a reading of Francois Jacob's 1970 *La logique du vivant: une histoire de l'hérédité*. Many thanks to Elizabeth Wilson for pointing me to Derrida's text.

22 Fortun, *Promising Genomics*.

23 What I and others call care closely resembles the "attention" of Bernard Stiegler, *Taking Care of Youth* (but see also Stiegler's "Care").

24 Another confounding of the poetic and the prosaic, I'll add here, is in the "Poem-Like *Tolls*" (a play on "Toll-like receptors") that have been inserted before, in the middle, and at the end of the text proper. Friedrich Nietzsche's *The Gay Science* has been a formative influence, so even though my book is far from aphoristic, I wanted it to have something in verse form, and I am grateful to Duke University Press's readers for encouraging this. I have reset portions of scientific articles on the *toll* gene and on *toll*-like receptors in the form of poems, as an experiment in shaking our usual habits of reading genomics, and to help readers read science otherwise by drawing out the inherently poetic quality of even the most poetics-scrubbed prose of the scientific article. I have omitted some terms and reordered some passages, but these poems are otherwise constructed verbatim from the articles. Sometimes I went for rhyme, sometimes for cadence, sometimes simply to accentuate the wording. They are also a way to get readers more interested in "the scientific literature," and to read long excerpts from articles that might otherwise, if simply quoted in the text, remain inert and easily skimmed or skipped. I hope that the simple reforming of sentences with thoughtfully chosen line breaks forces a slower, more careful reading, more attentive to the lively combinations of everyday and expert idioms. This experiment to bring forward the poetic in scientific prose was also inspired by Muriel Rukeyser's *Highway 1*, with its transduction of quotidian documents into poetry; and, in a different vein altogether, James Merrill's *Changing Light at Sandover*, large portions of which were clearly (as the book puts it) "poems of science." That book, like *The Gay Science*, affected me deeply, and I am forever grateful to Rich Doyle for introducing it to me and me to it, at the start of our friendship.

25 Hilgartner, *Reordering Life*.

26 Pickering, *Mangle of Practice*.

27 https://rationalwiki.org/wiki/I_Fucking_Love_Science, accessed 4/15/2021. The site now lives at the more innocuous URL http://ifls.com. For some further information about the IFLS phenomenon, including pointed remarks about

gender and science from former NSF director Rita Colwell, see Dell'Amore, "Why Is a Woman."

28 Subramaniam and Wiley, "Introduction to Science."

29 Tyler, "Post-Modern Ethnography," 140, 134.

30 McNamer, "Origins of the Meditationes vitae Christi."

Chapter Two. Labyrinth Life

This chapter is based on an essay previously published as "What *Toll* Pursuit: Affective Assemblages in Genomics and Postgenomics," in Richardson and Stevens, *Postgenomics*, 32–55.

1 See Fortun, "Making and Mapping Genes"; Fortun, "Projecting Speed Genomics"; Fortun, "Human Genome Project."

2 Latour and Woolgar, *Laboratory Life*, 247.

3 Latour and Woolgar, *Laboratory Life*, 280.

4 Jacob, *Statue Within*, 8

5 Gilbert, "Towards a Paradigm Shift," 99.

6 It is also a philosopher's question, or at least those philosophers similarly attempting to make some sense of what has not yet arrived "and yet" is already here: "What Derrida refers to as the 'to come' and Foucault as the 'actual,' Deleuze calls absolute deterritorialization, becoming, or the untimely. It is the pure 'event-ness' that is expressed in every event and, for that reason, immanent in history. It follows that every event raises with greater or lesser urgency the hermeneutic question, 'what happened?'" Patton, "Events, Becoming, and History," 42.

7 Gilbert, "Towards a Paradigm Shift," 99.

8 Gilbert, "Towards a Paradigm Shift," 99.

9 Watson, *Avoid Boring People.* For the most careful treatment of Franklin and her work, see Maddox, *Rosalind Franklin.* This is as good excuse as any to tell a James Watson story that doesn't get told nearly enough. At Harvard in the 1960s, Watson continued his pattern of "girl chasing" and the various misogynies he had honed in 1950s UK Cambridge. To make that long story short, he began dating and then became engaged to Elizabeth Lewis, a Radcliffe undergraduate who was also his lab assistant. They wed in La Jolla, California, in 1968 and, while honeymooning at the La Valencia hotel, the forty-year-old Watson penned a brief message on a hotel postcard to his friend Paul Doty, who had helped recruit Watson to Harvard: "Nineteen year old now mine." See McElheny, *Watson and DNA*, 140–41. Watson was quite racist to boot, although unlike his misogyny this has become widely known only relatively recently; Gabbatiss, "James Watson," provides a brief summary; for a sharp overview from a Black scientist, see Ogbunu, "James Watson."

10 Watson, "Remarks."

11 Watson, "Remarks."
12 Meselson quoted from Watson, "Remarks."
13 See Bretislav et al., *One Hundred Years*.
14 Author's interview with Robert Moyzis, October 22, 1991. See also Fortun, "Human Genome Project," for a rereading of Moyzis's comments regarding his "surprise" at the readiness of many scientists in 1985 to entertain the idea of completely sequencing a human genome.
15 See Fortun, "Human Genome Project."
16 US Senate, *Human Genome Project*, 130.
17 US Senate, *Human Genome Project*, 91–92.
18 Cook-Deegan, "Origins," 98.
19 Interview with the author, Whitehead Institute, Cambridge, Massachusetts, 1992. This is an opportunity to mark the limits of friendship, the particular pattern of care detailed in chapter 7, putting (my) friendship in tension with (my) scrupulousness. That two-hour interview thirty years ago is the sum total of time I have spent in Lander's presence. He was genial and generous to me, and that encounter, combined with reading his scientific publications over the years (some discussed later in the book), gave me an enduring high regard for him, as a careful genomicist open to a modicum of friendship from an ethnographer. Over the years, I watched Lander ascend the genomics ranks, becoming a scientific and organizational leader in the Human Genome Project. In the later section of this chapter dealing with the surprises of the HGP, I quoted at length remarks of his about how personally rewarding it was for him to manage a large genome sequencing center (see the interview at http://library.cshl.edu/oralhistory/interview/genome-research/surprises-hgp/lander-surprises-hgp/; recorded June 2, 2003).

But his scientific and public persona would become more complicated: by 2016 Lander had "morphed from science god to punching bag" (Begley, "Why Eric Lander Morphed") after his article in *Cell*, "Heroes of CRISPR," lionized the work of his male colleague at the Broad Institute while writing Jennifer Doudna and Emmanuelle Charpentier (who would be awarded the Nobel Prize in 2021) out of CRISPR history. That morphing was chronicled in real time on Twitter, of course, under the hashtag #landergate. In 2018, in another Twitter "explosion," Lander was roundly criticized by fellow scientists (especially women) for his kind toast of racist and misogynist James Watson on his ninetieth birthday (Begley, "As Twitter Explodes"; see also Begley and Joseph, "Scientists Threw James Watson a Party"). My second thoughts about including more of Lander's voice in this manuscript turned to thirds, especially when his organizational style and work was the point, rather than genomics as infrastructure. Then he was tapped to become President Biden's director of the White House Office of Science and Technology Policy (OSTP), a position Biden elevated to cabinet level. The advocacy organization 500 Women Scientists spoke out early against his nomination, not only on Twitter but in the pages of *Scientific American* (500 Women Scientists, "Eric Lander Is Not the Ideal Choice"). Subsequent events validated their criticisms: less than a year later Lander would first be forced to publicly apologize for having "bullied and demeaned" women OSTP staff (Alfaro and Pager, "Top White House Scientist Apologizes"), and then, amid calls for his firing, he resigned (Rogers, "Biden's Top Science Adviser Resigns"). In the late stages of editing, I deleted that passage on his HGP experience that I had in the later part of this chapter, no longer trusting his account, and use this note to document some of these matters.

20 Tilghman, "Keynote Address."

21 Tilghman, "Keynote Address."

22 Roberts, "Plan for Genome Centers," 205.

23 Knorr Cetina, "Objectual Practice," 181. I'll return to the dynamics of Knorr Cetina's "object of knowledge" and its affective associations later in the chapter, where she interprets these beautifully articulated dynamics of indefinite unfolding and open extension through a seemingly contradictory logic of lack.

24 See Mayer, "Discovery of Modular Binding Domains."

25 Lemaitre, "Road to Toll," 523.

26 Lemaitre et al., "Dorsoventral Regulatory Gene Cassette." Jules Hoffman was awarded the Nobel Prize in 2011 for his work showing "that the so-called Toll-gene is active in the development of receptors which are crucial for the immune system of the fly" (https://www.nobelprize.org/prizes/medicine/2011/hoffmann/facts/). Lemaitre (the lead author) believed the award represented an injustice. Janeway (see chapter text) was dead and was thus not eligible, but Lemaitre—a postdoctoral fellow in Hoffman's lab at the time—claims to have done the actual genetics work. In Lemaitre's own version of these events, for which he created a website with supporting materials, he states that Hoffman "was not very supportive of the genetics approach I had undertaken. Subsequently, he has never been able to fully recognize my contribution, yet somehow it is he who is now collecting the honours for my work" (Lemaitre, "Mixed Feelings"). This is just one of the episodes in science that Lemaitre, drawing on the work of Michael Polanyi (among others), analyzes to argue that science is driven by narcissism; see Lemaitre, "Science, Narcissism." A longer version of that essay was published as Lemaitre, *An Essay on Narcissism and Science*, an essay I return to in the Postscript.

27 Lemaitre noted the element of serendipity involved in these experiments: "I now realize that our success in identifying the function of *Toll* in the Drosophila immune response was partly because we routinely used a mixture of Gram-negative and Gram-positive bacteria to infect flies, whereas other groups only used Gram-negative bacteria." Unknown to them at the time, the *toll* mechanism evolved to recognize only gram-positive bacteria, which "strongly activated the *Toll* pathway and enabled us to discern the role of *Toll*." Lemaitre, "Road to Toll," 524–25.

28 Medzhitov et al., "Human Homologue of Drosophila Toll."
29 Medzhitov, "Approaching the Asymptote."
30 Medzhitov, "Approaching the Asymptote."
31 Weissmann, "Pattern Recognition," 2140.
32 Again, "pattern recognition molecule" needs to be read not as metaphor but as catachresis: molecules don't "recognize" "patterns," actually or metaphorically (what would they be metaphors *for*?). See chapter 3 for a conversation Gayatri Spivak had with a cell biologist who recognized that "recognition" was a catachresis, at least when it came to cellular . . . uh, signaling. No, communication . . . er, messaging. See chapter 3.
33 The story also illustrates why I decided to use genomics throughout this book, rather than postgenomics. If we were to characterize the knowledge about *toll* developed in 1996 as "postgenomic," then postgenomics began well before the Human Genome Project was nominally completed in 2001, that is, in the time of genomics. My revisionism can be considered a small act of friendship toward genomics, characterizing it as already attuned to nonlinearity and complexity, even if less than perfectly, rather than awaiting its "new and improved!" post- version.
34 Nelson and White, "Metagenomics," 172.
35 See Fortun, "Celera Genomics."
36 Fleischmann et al., "Whole-Genome Random Sequencing."
37 Zimmer, "Yet-Another-Genome Syndrome."
38 Two different (verging on Manichean) insider accounts of those events can be found in Venter, *A Life Decoded*, and Sulston, *The Common Thread*.
39 "GSK Completes Acquisition of Human Genome Sciences," GlaxoSmithKline website, August 3, 2012, https://www.gsk.com/en-gb/media/press-releases/gsk-completes-acquisition-of-human-genome-sciences/.
40 Nowak, "Bacterial Genome Sequence Bagged," 469.
41 Nowak, "Bacterial Genome Sequence Bagged," 469.
42 Zimmer, "Yet-Another-Genome Syndrome."
43 Sternberg, "On the Roles," 155.
44 See Ganser et al., "Distinct Phenotypes in Zebrafish"; Rice et al., "Developmental Lead Exposure." The startle reflex in humans is prone to a variety of pathologies; see Dreissen et al., "The Startle Syndromes."
45 See Wilson and Frank, *Silvan Tomkins Handbook*; Frank and Wilson, "Like-Minded"; Sedgwick and Frank, *Shame and Its Sisters*.
46 Tomkins, *Affect Imagery Consciousness*, 985.
47 Tomkins, *Affect Imagery Consciousness*, 620, 652.
48 Tomkins, *Affect Imagery Consciousness*, 273.
49 Keller, *Refiguring Life*, 22.
50 Tomkins, *Affect Imagery Consciousness*, 652.
51 Tomkins, *Affect Imagery Consciousness*, 188.
52 Tomkins, *Affect Imagery Consciousness*, 189.

53 Tomkins, *Affect Imagery Consciousness*, 992.
54 Tomkins, *Affect Imagery Consciousness*, 992.
55 Tomkins, *Affect Imagery Consciousness*, 996.
56 Frank and Wilson, "Like-Minded," 876. Wilson and Frank, *A Silvan Tomkins Handbook*, is a welcome and welcoming overview of Tomkins's thought, and a stunningly smart work of analysis in its own right.
57 Bio-IT World, "Making a Genesweep"; Pennisi, "Low Number Wins."
58 Interviews found at http://library.cshl.edu/oralhistory/interview/genome-research/surprises-hgp/roe-surprises-hgp/. Long after writing this passage, I went back to this collection of interviews and found another clip from Bruce Roe's interview, in the category "Science and Spirituality." That clip ends, I'll just teasingly note, with Roe uttering an everyday, delightful, somewhat paradoxical, "You know, I don't know." Check it out yourself: http://library.cshl.edu/oralhistory/interview/cshl/research/roe-science-spirituality/.
59 Collins, *Language of God*.
60 Interview found at http://library.cshl.edu/oralhistory/interview/genome-research/surprises-hgp/surprises-hgp/; recorded June 1, 2003.
61 Shubin, *Your Inner Fish*. Anthropologist Jonathan Marks (*What It Means*) cares less for this literature than I do, and for his usual good reasons.
62 Interview found at http://library.cshl.edu/oralhistory/interview/genome-research/surprises-hgp/olson-surprises-hgp/; recorded June 1, 2003.
63 Rheinberger, "Experimental Systems," 291. *Tatônnement* is Francois Jacob's term, who is also the source for Rheinberger's "play of possibilities" (he notes that the English translation of Jacob's book *The Possible and the Actual* (Seattle: University of Washington Press, 1982) does not convey this connotation from the French title, *Le Jeu Possibles*). The Derrida quotes are from *Of Grammatology*.
64 Lemaitre, "Mixed Feelings."
65 Tomkins, *Affect Imagery Consciousness*, 188.
66 Tomkins, *Affect Imagery Consciousness*, 659.
67 Stiegler, "Care," 104–5. See also Stiegler, "Relational Ecology."

Chapter Three. Double Binds of Science

1 Smith, "Last-Gen Nostalgia," 146.
2 Author's transcript of 1986 Cold Spring Harbor meeting, quoted in Fortun, "Making and Mapping Genes."
3 On Amgen's acquisition of deCODE, see Greely, "Amgen Buys deCODE." Fortun, "Zombie Corporate Vikings," is a four-part series of blog posts that tells some of the story of deCODE's bankruptcy and also traces some of the connections among deCODE's first CFO Hannes Smárason, the Russian nogoodnik Felix Sater, the real estate schemes of Donald Trump, and the heists

of/by Iceland's banks and the resulting financial meltdown of the nation (and world economy).

4 Smith, "Last-Gen Nostalgia," 146. Such language would be out of place in the genre of the scientific journal article, of course, but lately scientists have more recourse to other genre options—"perspectives," editorials, or opinion pieces in scientific journals, as well as popular books, blogs, and various social media.

5 All this is taken up further in the Postscript, but see in particular Keller, *Reflections*, and "Drosophila Embryos."

6 Or, more carefully: that it can elicit such affective responses in a scientist who has been encultured into those practices and has learned to read and want and respond to these signs, and has developed the unnatural physiology of an embodied scientist-subject.

7 See, for example, any of the fine essays in Gitelman, *Raw Data Is an Oxymoron*.

8 These culinary metaphors are developed more fully by Geoffrey Bowker in *Memory Practices*.

9 The text has arrived in a crucial but complicated territory that I am entering slowly here, so if readers have been surprised enough by the appearance of these psychoanlytic terms to have flipped the pages to this note, this is all to the good. When I first began writing this book I was thinking and writing almost exclusively in terms of a knowledge "drive" (Freud's term was *Wisstrieb,* which for better and worse James Strachey translated as "epistemophilia" rather than "knowledge drive"), due to the formative influence of Avital Ronell's *The Test Drive*. Stephen Mitchell, *Relationality*, is an excellent guide to the long, slow, and definitive if still unsettled shift in psychoanlaytic theory from internalized drives to externalized object relations, beginning with Klein. I still use the former sometimes in the text but hew mostly toward the latter. I'll point the reader again to the Postscript for further treatment.

10 Smith, "Last-Gen Nostalgia," 146.

11 Smith, "Last-Gen Nostalgia," 146.

12 Since I am trying to break my Heidegger habits (I've been pretty clean lately), I will mention only in passing here his analysis of addiction as a kind of degraded or inauthentic form of care; see Ronell, *Crack Wars*, 34–42. For an analysis of the habit of naming almost anything an addiction, see Sedgwick, "Epidemics of the Will."

13 The few excerpts and glosses in this section do not do justice to the extensive care literature, especially as it pertains to sciences. The authors here write and think within different genealogies of care, analyzing it within different contexts, in relation to different objects. I would point readers to Almklov, "Standardized Data"; Atkinson-Graham et al., "Care in Context"; Clark and Bettini, "'Floods' of Migrants"; Danby, "Lupita's Dress"; Davies, "Mobilizing Experimental Life"; Denis and Pontille, "Material Ordering"; Friese, "Realizing Potential"; Gabrys, "Citizen Sensing"; Gill, "Caring for Clean Streets";

Goodman et al., "Ten Simple Rules"; Hartigan, "Plant Publics"; Kenner, *Breathtaking*; Kittay and Feder, *Subject of Care*; Lappé, "Taking Care"; Lavau and Bingham, "Practices of Attention"; Lavoie et al., "Nature of Care"; Martin et al., "Politics of Care"; Mattern, "Maintenance and Care"; Mol et al., *Care in Practice*, "Care"; Murphy, "Unsettling Care"; Nadim, "Data Labours"; Olarte-Sierra and Pérez-Bustos, "Careful Speculations"; Pinel et al., "Caring for Data"; Pols, "Accounting and Washing Good Care"; Puig de la Bellacasa, "Matters of Care," "'Nothing Comes without Its World,'" "Making Time for Soil," *Matters of Care*; Schillmeier, "Cosmopolitics"; Schrader, "Abyssal Intimacies"; Stiegler, "Care"; Subramaniam, "Cartographies"; Thompson, "Repairing Worlds"; Tronto, "Beyond Gender Difference," *Moral Boundaries*; Trundle, "Tinkering Care"; Vinsel and Russell, *Innovation Delusion*; Viseu, "Caring for Nanotechnology?"; Watson, "Listening"; Zegura et al., "Care and the Practice of Data Science."

14 Tronto, *Moral Boundaries*, quoted in Engster, "Rethinking Care Theory," 50.
15 Engster, "Rethinking Care Theory," 55 (emphasis original).
16 Engster, "Rethinking Care Theory," 56.
17 Haraway, *When Species Meet*, 82.
18 Friese, "Realizing Potential."
19 Haraway, *When Species Meet*, 287.
20 Haraway, *When Species Meet*, 76.
21 Puig de la Bellacasa, "'Nothing Comes without Its World,'" 197–98.
22 For a particularly compelling and cogent overview of the importance of care in regard to everything infrastructural, see Vinsel and Russell, *Innovation Delusion*.
23 Puig de la Bellacasa, "'Nothing Comes without Its World,'" 197.
24 Puig de la Bellacasa, "'Nothing Comes without Its World,'" 198.
25 For more on paleonymy as a general feature, one might even say law, of any writing system, see Derrida, *Dissemination*, 3–4, 30–31.
26 Saint-Amour, "Weak Theory."
27 Quoted in Wilson and Frank, *Silvan Tomkins Handbook*, chapter 8, "Theory, Weak and Strong," 93–94.
28 Fleck, *Genesis and Development*, 49.
29 I use Jonathan Marks's invention "ex-apes" as a way to name our evolutionary status as a divergence in a lineage, difference crossed with sameness (Marks, *Tales of the Ex-Apes*). I also read Sarah Blaffer Hrdy's primatological research and writing as an argument that the emergence of care was a crucial event in the evolution of primates; see Hrdy *Mother Nature*, and *Mothers and Others*; Hrdy and Burkart, "Emergence of Emotionally Modern Humans."
30 Although the older, *cura* etymological strand of care is similarly doubled. Extending back to the Roman Empire, *cura* differed from itself by connoting worry and anxiety, as well as the more "positive connotation of care as attentive conscientiousness or devotion." Bioethicist Warren Reich tracks this

rhetorical "struggle" within care through Virgil and Seneca, originating in a mostly forgotten Greco-Roman mythical figure, "Care" (Cura), until Martin Heidegger revived it in discussing Goethe's *Faust* (Reich, "History of the Notion of 'Care,'" referencing Burdach's "Faust und die Sorge" as Heidegger's key source). In the myth, Care (Cura) makes a human being out of mud and asks Jupiter to give it the "spirit of life." Care, Jupiter, and Terra argue about whose name the new being should take. Saturn decides that Jupiter would get the spirit and Terra the body after death but, "Since Care first fashioned the human being, let her have and hold it as long as it lives." After reading this in Burdach's 1923 article, Heidegger stopped using *Bekümmering* (sorrow, grief) as a translation for *cura* and began using, as Herder and Goethe had, the more ambivalent *Sorge* (care), which became *the* most basic mode of being of humans, those entities renamed *Dasein* for the 1927 *Sein und Zeit*. It's this doubled meaning of care, in John Hamilton's view, that underwrites Heidegger as he "almost effortlessly moves from the 'ontic generalizations' to a rigorous 'existential-ontological interpretation.' . . . With Heidegger, [George] Steiner reformulates the founding Cartesian premise: 'I care, therefore I am.'" It is exactly this transcendentalization, or reification, or ontologization, or, more accurately, onto-theo-logization that I am trying to interrupt as quickly and forcefully as possible, even as it underwrites my own writing here. I hope that, reading care as a catachresis for these "ontic," quotidian acts of relatedness that indeed define us and our world completely, we might resist the onto-theo-logization that I think haunts so many treatments of care.

31 Tomkins, *Affect Imagery Consciousness*, 537.

32 Reich, "History of the Notion of 'Care.'"

33 Spivak, "Translation as Culture," 13–14.

34 I also learned about catachresis as an essential feature of language when I began reading more Icelandic literature. In *Promising Genomics*, I channeled the great Icelandic writer Halldor Laxness at length, in particular the wise fool Pastor Jón of *Under the Glacier*, who tells the emissary of the bishop, or Embi, who has come like an anthropologist to investigate strange goings-on up at Snaeffellsjokull:

> It's a pity we don't whistle at one another, like birds. Words are misleading. I am always trying to forget words. That is why I contemplate the lily-flowers of the meadow, but in particular the glacier. If one looks at the glacier for long enough, words cease to have any meaning on God's earth.
>
> EMBI: Doesn't the dazzle cause paralysis of the parasympathetic nervous system?
>
> PASTOR JÓN: . . . Sometimes I feel it's too early to use words until the world has been created.
>
> EMBI: Hasn't the world been created, then?

PASTOR JÓN: I thought the creation was still going on. Have you heard that it's been completed?

EMBI: Whether the world has been created or is still in the process of being created, must we not, since we are here, whistle at one another in that strange dissonance called human speech? Or should we be silent?
PASTOR JÓN: [No, it's just] that words, words, words and the creation of the world are two different things; two incompatible things.

35 This paragraph draws from historian Tani Barlow's commentary on Spivak's commitment to catachresis:

> Catachreses are ubiquitous key words or terms which appear repetitively in primary source evidence and seem to attract attention precisely because they denote a stable meaning that subsequently proves resistant, even aspecific. . . .
>
> Spivak argues that catachreses are a special kind of misuse of terms, where concepts are unstable because there is no one-to-one correspondence between concept and referent. . . . It is an important and sophisticated commonsense point. The catachresis or concept-metaphor without adequate referent is what she at one time called "a master word" to highlight general historical terms that have no literal, or particular, referent, because, as she put it very colloquially, there are "no 'true' examples of the 'true worker,' the 'true woman,' the 'true proletarian' who would actually stand for the ideals in terms of which you've mobilized" them. Calling attention to how catachresis works is a good way to illustrate how universals and particulars operate in ordinary language and therefore how much historical and theoretical meaning simple terms carry (Barlow, *Question of Woman*, 32–33).

36 Or N-binds, since even Gregory Bateson, in his early articulations of a double bind theory, cautioned that counting binds was a difficult task, and likely to extend well beyond counting to two; see Koopmans, "From Double Bind to N-Bind."

37 Bateson, "Toward a Theory of Schizophrenia," 208.

38 My analysis here draws on Gibney, "Double-Bind Theory," and Visser, "Gregory Bateson on Deutero-Learning."

39 Gitelman, *Raw Data Is an Oxymoron*. I don't think the essays there, or elsewhere as far as I know, ask, what kind of a thing is data such that it always has to be cooked and can never be raw? My answer: it's a thing, like all things, that differs from itself, a thing that defers or temporizes itself, that always already requires supplementing, a "given," which means a Gift which means a poison, a pharmakon.

40 Gibney, "Double-Bind Theory," 50.

41 Deleuze and Guattari, *A Thousand Plateaus*, 24.

42 On science as a practice of "muddling through," see Fortun and Bernstein, *Muddling Through*.

43 For a spectacular analytic history of diffraction as both phenomenon and analytic trope, and its careful development by feminist scholars such as Trinh Minh-ha, Gloria Anzaldua, and Donna Haraway at the University of California, Santa Cruz, see Barad, "Diffracting Diffraction."

44 Thus far I have avoided using this identifier, opting instead for ethnographer, anthropologist, or historian. Ethnogrammatology acknowledges the deep debt this work owes to the playful works of Jacques Derrida, how much it has inherited from him conceptually and methodologically. Severe allergies to, misunderstandings, and misrepresentations of Derrida and deconstruction (now *there's* a catachresis!) abound in scientific communities (but not only scientific communities), and I knew that every mention of his name risked losing some readers, readers who, admittedly, may have existed only in my fantasies. But if they are here and they've made it this far, I think they are in it for the long haul and ready for my preferred name for what I think I am doing. On the other hand, many anthropologists and ethnographers are only too eager to coin a new expression as a sort of branding strategy, and so I wanted at least to defer this neologism to a point in the text where its immodesty might be less on-the-surface. Ethnogrammatology puts the grammatology in ethnology, the science of marking/inscribing/writing into the science of human difference. Analytically, I privilege scientists' writing over their saying or doing, an inversion of what most ethnographers do. I read deconstructively (a practice, like any science, always needing to be reinvented) what scientists write, to solicit the limits (see chapter 6) of these text-like objects, to analyze the productivity of their inevitable margins and their necessary contradictions, and to uncover what they also block or blind us to.

Poem-Like *Tolls* 2

1 *Annual Review of Immunology* 29 (2011): 447–91.

2 *Annual Review of Microbiology* 67 (2013): 499–518.

Chapter Four. Curation

1 US Department of Energy, "Report."

2 Larkin, "Politics and Poetics of Infrastructure," 328. Infra/structure is to structure as infrared is to red: part of red, but not red—beyond red. Whatever and wherever the difference is, the difference between a structure and its infra-, the difference quivering within durable infrastructure, it is, as AbdouMaliq Simone says, a "fluid and pragmatic" one (Simone, "Infrastructure"). And it's infra/structure all the way down, to employ an old anthropological and

philosophical saw. The truths of genomics require a genomics infra/structure to produce them, to support them, to connect them and to move them; in order for there to be genomics knowledge we need, not only a "Route 1," but an entire interconnected highway system, the industrial capacity to produce its diverse materials, and the work crews to make it all happen.

3 For a detailed study of model organism databases and the practices that sustain them, see Leonelli and Ankeny, "Re-thinking Organisms."

4 Author/participants: Helen Berman, Elbert Branscomb, Peter Cartwright, Michael Cinkosky, Dan Davison, Kenneth Fasman, Chris Fields, Paul Gilna, David Kingsbury, Thomas Marr, Robert J. Robbins, Thomas Slezak, F. Randall Smith, Sylvia Spengler, Ed Uberbacher, Michael Waterman. Agency observers: David Benton, John Wooley, David Smith, Jay Snoddy.

5 Zorich, "Data Management," 432.

6 Zorich, "Data Management," 432.

7 A few further remarks on my love for the mindy word "vicissitude." The title of this section is a play on Sigmund Freud's 1915 essay "Instincts and Their Vicissitudes," a classic of psychoanalytic theory. A lot of energy has been expended (yay!) on the translation of *Trieb* as "instinct" or "drive." Much as I side with those who prefer drive—and putting "data" in the place of drive sets up the right tendencies, in my view—I am more interested in the multiple translational possibilities of the second compound term in Freud's title, "Trieb und Triebschicksale." "Vicissitudes" alone, as nice as that word is, loses the compounding effect that is so crucial to the theory: drive-vicissitudes. Bruno Bettelheim (*Freud and Man's Soul*, 105) considered both terms "grievous mistakes," and opened up additional possibilities. Vicissitudes for him was a "bookish" term (Spivak would probably say "mindy"), without the "emotional reaction" that more vernacular terms like "fate" or "destiny" would have evoked. But this is precisely why I like vicissitudes: it connotes, as Bettelheim pointed out, the "nonhuman occurrences" that are more appropriate to "stuff" and other things like data or genomes or a Human Genome Project. Bettelheim would have preferred something more along the lines of "change" or (better still, I think) "mutability." I'd like "Data and Its Vicissitudes" to be read as "data and its mutabilities," to suggest the inhuman forces inherent in data and other stuff that keep them constantly (open to) turning, changing, mutating, deteriorating, and/or disorganizing were they not subject to constant care and attention. Like other infra/structures, data are always subject to and even subjects of that other favorite Freudian word *Zufall*: chance, coincidence, accident, the happenstance that befalls something or someone. See Rottenberg, *For the Love of Psychoanalysis*.

8 "The Maintainers" is the wonderfully heroic-sounding name given to the people who do this kind of care work, documented in a blog/network of that name maintained by Andy Russell, Jessica Meyerson, Lee Vinsel, and Lauren Dapena Fraiz (https://themaintainers.org). For a thorough valorization of

care work of all sorts but particularly that in engineering, see Vinsel and Russell, *Innovation Delusion*.

9 Vukmirovic and Tilghman, "Exploring Genome Space," 820. See also Fortun, "Genes in Our Knot."

10 Vukmirovic and Tilghman, "Exploring Genome Space," 820. Tilghman's "Keynote Address" to the National Human Genome Research Institute almost ten years later (quoted in chapter 2) conveyed these same thoughts and sentiments almost exactly. Those conversations with physicists and engineers she reporting having back in the late 1980s clearly left an enduring impression.

11 Vukmirovic and Tilghman, "Exploring Genome Space," 820.

12 The hypothesis, I admit, is in need of theoretical-empirical development toward a richer psychoanalytic theory of the scientific subject. The Postscript takes this a little further. Here I'll note that part of the broad shift from drive-based theories to theories of object relations in psychoanalysis drew on studies of the first months of infant life when, before symbolic capacities are acquired and even before object relations are stabilized, the world is experienced as disorienting excess. Jessica Benjamin and Galit Atlas explore some connections between early childhood experience of the excessive "too muchness" of sexuality, to the affect of excitement and to different attachment styles; see Benjamin and Atlas, "'Too Muchness' of Excitement."

13 Leonelli and Ankeny, "Re-Thinking Organisms."

14 On the simultaneous importance and undervaluation of scientific infrastructure see Edwards, "Meteorology," and "Infrastructure and Modernity"; Plantin et al., "Infrastructure Studies Meet Platform Studies"; Larkin, "Politics and Poetics of Infrastructure"; and Star, "Ethnography of Infrastructure."

15 Ashburner et al., "Gene Ontology," 2.

16 On Robert Boyle's "literary technology" as a mechanism of conflict avoidance in seventeenth-century England and its sciences, see Shapin, "Pump and Circumstance." For a complicating treatment of the "modest witnessing" that was another component in the new social machinery of experimental sciences, hinging on complications of genderings, see Haraway, *Modest_Witness*, 26–32.

17 This is as good a place as any to note in passing that "accurate" and "accuracy" are also related to care, stemming from the Latin *accūrātus*, "performed with care, exact"; the adjective was formed from the "past participle of *accūrāre,* to give attention to, to perform with care" (OED). To be accurate is to be caring; accuracy requires curation. Only in the seventeenth century does the term begin to pertain to instruments and methods.

18 Salimi and Vita, "The Biocurator," e125.

19 Bourne and McEntyre, "Biocurators," e142.

20 Burkhardt et al., "Biocurator Perspective," e99.

21 Burkhardt et al., "Biocurator Perspective," e99.

22 Brooksbank and Quackenbush, "Data Standards," 94.

23 Johanna Drucker's argument, in "Humanities Approaches to Graphical Display," that data (what is given) should be replaced by the Latin *capta* (what is taken), has been influential in the digital humanities. But while this reversal better recognizes the complexity and dynamics of data and their inevitable entanglement with human affairs, it does so while leaving the semiotic system intact: *capta* are social constructions that are human, all too human, and the resistant alterity that "data" signifies is left unapprehended. My analysis of data, and my writing it as (meta)data, is more in tune with Matthew Lavin's conceptual intervention of "situated data" (Lavin, "Why Digital Humanists Should Emphasize Situated Data over Capta"). Through a much longer and more complex genealogy of data, including its pre-Latin Greek form, Lavin theorizes data that is both given and taken, so to speak, opening "more opportunities for conversation with other disciplines," like the sciences, that *capta* on its own would shut down.

24 Brooksbank and Quackenbush, "Data Standards," 95. On microarrays, microarray data, and the demands of biocuration that accompany them, see Keating and Cambrosio, "Too Many Numbers."

25 Leonelli, "Documenting the Emergence of Bio-Ontologies," 112 (emphasis added).

26 Leonelli, "Documenting the Emergence of Bio-Ontologies," 114.

27 Leonelli, "Documenting the Emergence of Bio-Ontologies," 115.

28 Brooksbank and Quackenbush, "Data Standards," 94.

29 Stevens, *Life Out of Sequence*, 81.

30 Stevens, *Life Out of Sequence*, 81.

31 Stevens, *Life Out of Sequence*, 81.

32 Stevens, *Life Out of Sequence*, 99.

33 Stevens, *Life Out of Sequence*, 90.

34 Stevens, *Life Out of Sequence*, 89.

35 Stevens, *Life Out of Sequence*, 90.

36 Interview with Lori Hoepner, March 2013. My friend, colleague, and then just recently former student Alison Kenner helped me create this data, doing the preparatory research and arranging the interview, one among several that we conducted that day. The others were with "higher profile" researchers, first or last authors on publications who are usually credited with the discoveries or results presented in them. Those other interviews did not particularly excite me, and I have yet to quote them in any publication; I was much more interested in Lori's thinking and doings, and much more excited about presenting this work of care here.

37 ISAAC was "the largest worldwide collaborative research project ever undertaken" (http://isaac.auckland.ac.nz/), a twenty-year study of global asthma patterns that enrolled more than two million children in more than a hundred countries. For a richer account of ISAAC, see Asher et al., "International Study of Asthma and Allergies in Childhood (ISAAC)," and Williams et al., "Worldwide Variations."

38 Interview with Lori Hoepner, March 2013.

39 That seemed so *toll!* to me that, nerd that I am, I had to look it up later. This particular data curation task was developed in the context of occupational safety and health—an appropriately infrastructural context of hard-to-see scientific care: "Analysis of urine samples for compounds such as lead, phenol, or 4,4'-methylene bis (2-chloroaniline) (MBOCA) are often adjusted for specific gravity following a recommendation in a NIOSH Benzene Criteria Document. The adjustment normalizes all results to a specific gravity of 1.024 by multiplying the analytical result in microgram/liter by 24/G where G is the last 2 digits of the specific gravity of the urine sample." Graul and Stanley, "Specific Gravity Adjustment."

40 I imagined precarious must be etymologically related to care; it is not. "Latin, precarious: obtained by begging or prayer, depending on request or on the will of another. . . . Derived originally from the Latin precari, it first signified 'granted to entreaty,' and hence, 'wholly dependent on the will of another.' Thus it came to express the highest species of uncertainty." *Websters 3rd.*

41 Dewey, *Experience and Nature*, 59–60, 70–71.

42 Data as a flow and as movement is nicely demonstrated by Antonia Walford, in "Data Moves," using the case of climate data in Brazil and beyond.

43 Fortun and Bernstein, *Muddling Through*.

44 Gitelman and Jackson, "Introduction," in *Raw Data Is an Oxymoron*, 2; emphasis added. The paranoid style as a driver of critical theory (or the kind of object relations critical theory embodies) is discussed further in the Postscript, but for a reading of how paranoid critical theory emphasizes vision and exposure, see Sedgwick, *Touching Feeling*.

45 It's no solution, either, to summon up Levi-Strauss's structure in its originating triangular entirety of raw-cooked-rotted, as Boellstorff has proposed ("Making Big Data"). Or rather, it's a solution (although Boellstorff insists it is not), but that's exactly the problem. That schema is certainly more faithful to Levi-Strauss, and even accommodates something like the precarity of data that I discuss here (although Boellstorff only discusses this in terms of "bit rot"). It *feels satisfying* to anthropologists, especially when a Geertzian notion of interpretation is also thrown in, to have their theories validated as the forgotten and ignored but thicker and superior truth. Shifting the terms of the debate to "the interpretive" allows anthropologists to play on their home field, comforted by the salve of theory quite some distance from the troubled anxieties of the data scientist working in the midst of the thickness rather than outside it.

46 I work collaboratively along other pathways to produce, curate, and openly share ethnographic data; see Fortun et al., "Civic Community Archiving"; Fortun et al., "What's So Funny?"; Khandekar et al., "Moving Ethnography."

47 "About," Dr. Lori Hoepner, accessed October 15, 2022, https://www.drlorihoepner.com/about.

Chapter Five. Scrupulousness

Epigraph: Huxley, *T. H. Huxley on Education*, 53. Huxley goes on to an anodyne (i.e., careless) definition of the scientific method: science starts with observation, proceeding to comparison and classification, then on to deduction, and finally to verification. How can the same habit have a careless and a scrupulous expression or instantiation? What kinds of discipline might it take to move from the former to the latter? These are questions he does not ask, perhaps sensing the double bind: to be a scientist one must more scrupulously cultivate the habit of more scrupulous application of methods that one has been less than scrupulous about to begin with. The Victorian scientist-subject was not as sensitive to or interested in recursive, paradoxical structures of thought as today's may be, but I still take it as an early insight that scrupulousness is another name for the paradoxes of care.

1 An excellent overview and social analysis of these busy and dangerous intersections is Nelson, *The Social Life of DNA*.

2 Fausto-Sterling, "Refashioning Race," 4–5.

3 The social science literature on the various intersection of genomics with "race," race/ethnicity, and the use of racialized categories or variables has grown to be quite extensive and diverse; see, for example, Benezra, "Race in the Microbiome"; Bolnick et al., "Science and Business of Genetic Ancestry Testing"; Braun et al., "Racial Categories in Medical Practice"; Fujimura et al., "Introduction: Race, Genetics, and Disease"; Fujimura and Rajagopalan, "Different Differences"; Fullwiley, "Race and Genetics" and "The Biologistical Construction of Race"; Goodman et al., *Race: Are We So Different?*; Kahn, *Race in a Bottle*; Koenig et al., *Revisiting Race*; M'charek et al., "Trouble with Race"; Montoya, "Bioethnic Conscription"; Nelson, "Bio Science" and *Social Life of DNA*; Smart et al., "Standardization of Race and Ethnicity." On race as a persistent "ghost variable," see Karkazis and Jordan-Young, "Sensing Race."

4 Shields et al., "Use of Race Variables."

5 Foucault, *Uses of Pleasure*, 27.

6 Karin Knorr Cetina's "Care of the Self" also used the "care of the self" to think about scientific practice in high-energy particle physics and molecular biology, in ways similar to what I am calling scrupulousness. She seems more committed to a clearer dividing line between self and world, with the former being the rather exclusive focus of care, than is my account here, and her own object of concern never strays from the "epistemic."

7 To be sure, scrupulousness has been a scientific virtue for a long time (or at least since T. H. Huxley) in virtually all areas of scientific activity. An earlier ethnographic essay of mine (Fortun, "Fluctuating about Zero") tried to attend to the scrupulousness (although I did not use the word at the time) by which experimental physicist Steve Lamoreaux measured the previously

unmeasurable "Casimir force" exerted by or through the quantum vacuum, known colloquially as "nothing." Measuring the force of nothing is, as you might well guess, a delicate operation in which many things can go wrong, and numerous experimental assumptions and conditions demand attention, including whether or not the occupant of the lab below yours was playing music on the radio while you were fine-tuning your apparatus and taking your measurements. So scrupulousness only names what any careful scientist and some of the people who study scientists already know: doing science involves microattention to a world of details and their vicissitudes if it is to be done well, carried out carefully.

8 That performative usage dates to 1604; Shakespeare *Hamlet* iv. iv. 9 + 31; quoted in OED.

9 Annemarie Mol, Ingunn Moser, and Jeannette Pols equate care and tinkering ("Care: Putting Practice into Theory"). In his historical study of the problem of deciding when a physics experiment has ended (i.e., has identified a phenomenon correctly and eliminated any artifact of the experimental apparatus), Peter Galison (*How Experiments End*, 74) characterizes the dilemma of scrupulousness through the mytheme of Scylla and Charybdis; see the discussion later in this chapter on why this popular Greek mythological reference is not the same, when we read it scrupulously, as a double bind. On Nietzche's "eternal return" as repetition, not of the same, but as repetition of difference, see Deleuze, *Nietzsche and Philosophy*.

10 All quotes from Sullivan et al., "Genetic Case-Control Association Studies," 1015.

11 Sullivan et al., "Genetic Case-Control Association Studies," 1016–17.

12 Sullivan et al., "Genetic Case-Control Association Studies," 1018.

13 See Derrida, *Of Grammatology*, esp. 141–64.

14 Tsilidis et al., "Evaluation of Excess Significance Bias."

15 But see Traweek, "Generating High-Energy Physics in Japan," and other (more historically inclined) essays in Kaiser, *Pedagogy and the Practice of Science*.

16 See Tomšič, "Better Failures."

17 For Millennium in China, see Xu et al., "Genomewide Search"; for Sequana in Tristan da Cunha see Zamel et al., "Asthma on Tristan Da Cunha." Sequana tried to establish a research operation in Iceland under a similar set of assumptions about isolation and founder effects, but Sequana CEO Kevin Kinsella was snubbed by Kari Stefansson, founder and CEO of deCODE Genetics, who used Sequana's research proposal to persuade the Icelandic parliament to back his own efforts (see Fortun, *Promising Genomics*). The Millennium project in China had its own complex scientific, ethical, political, and social challenges; see Sleeboom, "Harvard Case of Xu Xiping."

18 Zamel et al., "Asthma on Tristan Da Cunha," 1902.

19 Zamel et al., "Asthma on Tristan Da Cunha," S132.

20 Vogel, "Scientific Result without the Science."

21 Xu et al., "Genomewide Search," 1274.
22 Xu et al., "Genomewide Search," 1276.
23 Collaborative Study on the Genetics of Asthma, "Genome-Wide Search," 392.
24 Collaborative Study on the Genetics of Asthma, "Genome-Wide Search," 392. On the spirometer and the ATS "correction," see Braun, *Breathing Race into the Machine*, esp. 199–204.
25 Ober et al., "Genome-Wide Search," 1394.
26 Lander and Kruglyak, "Genetic Dissection of Complex Traits," 241.
27 For Silvan Tomkins, "glazed-eyes" would indicate the absence of the interest-excitement affect, visible on the face as an intense or riveted gaze.
28 Lander and Kruglyak, "Genetic Dissection of Complex Traits," 241.
29 Lander and Kruglyak, "Genetic Dissection of Complex Traits," 241.
30 Quoting from the wonderfully inventive translation of Emily Wilson, the first woman to translate Homer, *The Odyssey* (W.W. Norton and Co., 2018), 309. I follow Wilson's translational choice, in which she replaces the adjectives "wiley" or "clever" that are usually chosen to describe Odysseus with "complicated." It's also worth noting that there is no middle path available on the return trip, either, and Odysseus has to steer straight into Charybdis.
31 Lander and Kruglyak, "Genetic Dissection of Complex Traits," 241.
32 Lander and Kruglyak, "Genetic Dissection of Complex Traits," 241.
33 Lander and Kruglyak, "Genetic Dissection of Complex Traits," 246.
34 Lander and Kruglyak, "Genetic Dissection of Complex Traits," 245, 246.
35 Lander and Kruglyak, "Genetic Dissection of Complex Traits," 244.
36 Hoffjan et al., "Association Studies for Asthma."
37 Hoffjan et al., "Association Studies for Asthma."
38 Ioannidis, "Why Most Published Research Findings Are False," and "Why Most Discovered True Associations Are Inflated."
39 There are now good social scientific studies of the phenomenon; among the more creative for its "mixed methods" approach to this event is Nelson, "Mapping the Discursive Dimensions." An analysis that closely parallels the one I would imagine doing, were I to actually try to work it out, with a similar message, comes from a transdisciplinary team including a neuroscientist and a few philosophers is Redish, "Opinion: Reproducibility Failures Are Essential."
40 *Guardian* columnist Ben Goldacre wrote about these events ("Quackbuster Causes Too Much Flak for University"), and on his relocated blog, Colquhoun documented some of the responses to that article, including a letter from UCL's provost defending his actions (Colquhoun, "What a Paean"). I'll add that although I much admire Colquhon's writings on statistics, science, and even "quackery," like many scientists he has fallen for Alan Sokal's profitable brand of quackery, namely his hack analyses of science studies, feminist theory, deconstruction, poststructuralist theory, and other work in those neighborhood (see Colquhoun, "The Diary: June 2008–May 2009"; the jeremiad

against postmodernism can be found in the entry dated January 13, 2009, discussing UCL's awarding an honorary degree to the "notorious postmodernist writer" Luce Irigaray). Sokal's and Colquhon's (and Richard Dawkins's, and Steven Pinker's, and too many other scientists') near-complete incomprehension of what they lump together under the label "postmodernism" doesn't keep them from making the most poorly reasoned, wrongheaded statements about scholars whom they have barely read and, if read at all, read carelessly and through the thickest of ideological lenses. It's writings by scientists like these that pose the greatest test to the spirit of friendship sketched in chapter 7.

41 Colquhoun, "Investigation of the False Discovery Rate," 1, 2.

42 Colquhoun, "Investigation of the False Discovery Rate," 3.

43 Colquhoun, "Investigation of the False Discovery Rate," 11.

44 I've stuck with the colloqualism of "drilling down" to the level of data, but these metaphorics are part of the "data as solid foundational ground" epistemic structures that are in the process of being revised and reframed. In the sciences-to-come, the cat's-cradling relationalities of care will have superseded the more rigidly structural ones through which we currently understand "science." Ludwik Fleck provided a prescient start to such a reframing not long after he was freed from Buchenwald, in his 1946 dialogic sketch of "the problem of the science of science"—the reflexive paradox of a positivist science unable to provide its own account of itself. In this rich and extended dialogue, Simplicius (name-checking a Galilean persona) worries about how an "ultra-criticism" will lead to a "barren skepticism"; science in this conventional framing requires a "certain solid and stable foundation," he says, otherwise it would become "top-heavy." Fleck's other persona, Sympatius, argues that science is not "a terrestrial building which stands on a foundation, with an attic at the top." Science is better metaphorized as "rather like a round fruit, with a juicy pulp, and a thick, indigestible skin. It may be turned at will, the base can be the top, or the top the base, depending on our desire, but they are both equally tough and indigestible. Only the center of science is useful. . . . In order for this miraculous fruit to grow, it must be taken between two fires: the hot, though dark, fire of romanticism, and the cold, but bright, fire of skepticism. . . . The aim of my inference is not to belittle the value of science but, on the contrary, to raise it." Fleck, "Problem of the Science of Science," 117.

45 Ioannidis et al., "Commentary," 205–9.

46 Ioannidis et al., "Commentary," 209.

47 Ioannidis et al., "Commentary," 209.

48 Khoury, "Case for a Global Human Genome."

49 Ioannidis et al., "Road Map," 4, 5.

50 Spivak, in Sipiora et al., "Rhetoric and Cultural Explanation," 300.

51 Bossé and Hudson, "Toward a Comprehensive Set."

52 Vercelli, "Discovering Susceptibility Genes," 176.
53 See Fortun, "Genes in Our Knot."
54 Martinez, "Gene-Environment Interactions," 27.
55 Martinez, "Gene-Environment Interactions," 30.
56 See "GABRIEL: A GABRIEL Consortium Large-Scale Genome-Wide Association Study of Asthma," accessed October 20, 2022, https://www.cnrgh.fr/gabriel/index.html.
57 Single nucleotide polymorphisms or SNPs are the smallest unit of difference, meaningless on their own but meaningful in context and in comparison to other SNPs. On any 1000 base pair stretch of DNA in a person's genome, one of those As, Ts, Cs, and Gs will be different from the same stretch of DNA in their neighbor's genome, which gets especially interesting when their neighbor is geographically and genealogically quite distant. Those several million differences in a person's genome aren't responsible for anything, and only matter because they can be used to map those segments of DNA in different populations, and to group them according to sameness or difference—haplotypes and haplogroups. The Wikipedia entry at https://en.wikipedia.org/wiki/International_HapMap_Project "needs improvement," but is not a bad introduction. Our friendship for the HapMap would have to undergo tests too severe and time consuming to undertake here, but a charitable beginning might be found in Reddy, "Good Gifts," and "Caught in Collaboration." For more critical readings of the HapMap project, see Hamilton, "Revitalizing Difference in the HapMap," and Reardon, *Race to the Finish*.
58 Moffatt et al., "Genetic Variants Regulating ORMDL3 Expression."
59 Vercelli, "Discovering Susceptibility Genes," 179–80.
60 On translational science and medicine, and translationality in general, see Sunder Rajan and Leonelli, "Introduction."
61 I learned why "asthma" is a catachresis from the scientists who care for it in their labs and the clinicians who care for it in their patients. A 2008 editorial in *The Lancet* stated, "With every new piece of the puzzle, the notion of asthma as one unifying disease concept is disappearing further into the realm of historical oversimplification. . . . 2 years ago, we made a plea to abandon asthma as a disease concept. This plea is now more justified than ever. Asthma is at best a syndrome with different risk factors, different prognoses, and different responses to treatment" (*Lancet*, "Asthma," 1009). This prompted several researchers to write back in partial agreement: "Nominalists," among whom they counted themselves, "recognise that asthma has no existence apart from the patients who have it, and that the causal implications of a diagnosis of asthma are widely varied. . . . Hence, the term asthma remains useful, provided we have an understanding of how we use the word. It needs to be maintained, but physicians and researchers need to appreciate that it describes an abnormality of function and not a specific phenotype defined as

a set of observable characteristics of an individual or group" (Hargreave et al., "Asthma as a Disease Concept," 1415–16).

62 Vercelli, "Discovering Susceptibility Genes," 178–79.

63 Visscher et al., "Five Years of GWAS Discovery," 8.

64 Visscher et al., "10 Years of GWAS Discovery."

Chapter Six. Solicitude

1 Ober and Vercelli, "Gene-Environment Interactions."

2 Shostak, *Exposed Science*; see also Fortun and Fortun, "Scientific Imaginaries and Ethical Plateaus."

3 To understand why holism's impossibility is necessary to anthropological practice and theory, and therefore generative of productive experiments in thought and writing, see Marcus, "Holism."

4 Green, "Stress in Biomedical Research," 178.

5 This is a horseshoe-shaped area of thought wherein some scientists are in close proximity to some theorists in the humanities. In a thoughtful series of columns for *Genome Biology*, Itai Yanai and Martin Lercher sketch out the dimensions of what they call "night science," a concept they inherit from Francois Jacob that closely resembles what I've put under the names of care. They characterize night science as the "unstructured" but "most exciting part of science. . . . The most creative and arguably most significant part of our work" (Yanai and Lercher, "Night Science"); it requires "two languages" (Yanai and Lercher, "Two Languages of Science"), arises out of "contradictions" (Yanai and Lercher, "Novel Predictions Arise from Contradictions") and is "improvisational" (Yanai and Lercher, "Improvisational Science"), that is unformalizable. In a different language of abstractions, Rosi Braidotti connects creativity or what I am calling solicitude to a whole, albeit a "virtual totality" (a block of past experiences), while also pointing to its time-out-of-jointness, ever on the vergeness, and too lateness—and also to a commitment to critical thinking and affirmative critique: "Critical thinking is about the creation of new concepts, or navigational tools to help us through the complexities of the present, with special focus on the project of actualizing the virtual. This signals an intensive, qualitative shift in becoming that I connect to affirmative ethics. Creativity—the imagination—constantly reconnects to the virtual totality of a block of past experiences and affects, which get recomposed as action in the present, thereby realizing their unfulfilled potential. This mode of affirmative critique is an exercise in temporary and contingent synchronization, which sustains, in the present, the activity of actualizing the virtual. This virtual intensity is simultaneously after and before us, both past and future, in a flow or process of mutation, differentiation or becoming, which is the vital material core of thought. We know by now that there is no Greenwich Mean Time

in knowledge production in the posthuman era" (Braidotti, "Theoretical Framework," 37).

6 Quoted in Yanai and Lercher, "Night Science," 179.

7 Green, "Stress in Biomedical Research," 176.

8 See, inter alia, Derrida, *Aporias*.

9 Green, "Stress in Biomedical Research," 178.

10 Green, "Stress in Biomedical Research," 177.

11 Penrose and Penrose, "Impossible Objects."

12 The artist Rhonda Roland Shearer (who was also the wife of Stephen Jay Gould) has shown that Lionel and Roger may have taken inspiration from their close relative Roland, the first British collector of Marcel Duchamp's art; his collection included Duchamp's 1916 *Apolinère Enameled*, which featured an impossible bed. Whether or not this was the case, it is certainly the case that Lionel was extremely interested in quirky reversals, like his reversible musical compositions "Was It a Rat I Saw?" and "Palindrome in F Major." It is also certain that Lionel and Roger shared their figure of an impossible staircase with M. C. Escher, who adapted it for his 1960 work *Ascending and Descending*. See Roland Shearer 1997, available in a different form at http://www.marcelduchamp.org/ImpossibleBed/PartI/page1.html.

13 Penrose and Penrose, "Impossible Objects," 31.

14 Strauss, "Work and the Division of Labor."

15 Fujimura, "Constructing 'Do-Able' Problems."

16 Nick Seaver ("Care and Scale") also "scales up" care to think it outside of its common framing as detailed attention at the more micro- or at least human level.

17 Derrida, "Force and Signification," 5.

18 Spivak, "Translator's Preface," xvi.

19 Derrida's *Life Death* holds that both texts and what Francois Jacob called "living systems" are structured or constituted by the logics and forces of writing. In a somewhat different vein, expressed through the philosophy of Charles Sanders Peirce, theoretical biologists like Marcello Barbieri (*Code Biology*) and Jesper Hoffmeyer (*Biosemiotics*) are key figures in the small community of biosemioticians who have theorized "life" not as an instance or effect of writing, or as an emergence of "vitality" out of inert molecular matter, but fundamentally as a transmission or transduction of meaning within a thoroughly semiological universe.

20 Rheinberger, "Experimental Systems—Graphematic Spaces," 291.

21 Vukmirovic and Tilghman, "Exploring Genome Space," 822.

22 Keller, *Refiguring Life*, 22.

23 Vukmirovic and Tilghman, "Exploring Genome Space," 822.

24 Pritchard et al., "Inference of Population Structure," 945.

25 See, for example, Wilson et al., "Population Genetic Structure."

26 Bolnick, "Individual Ancestry Inference."

27 Pritchard et al., "Documentation for STRUCTURE Software," 3.

28 Pritchard et al., "Documentation for STRUCTURE Software," 5.
29 Pritchard et al., "Inference of Population Structure," 949.
30 Pritchard et al., "Documentation for STRUCTURE Software," 6–17. There are many other programs now that cluster (structure) data in this way. Each of those clustering programs has its own virtues, because each has its own limits. "No single method or software can optimally solve all of these problems," one review article concluded (Liu et al, "Softwares and Methods," 1). Joan Fujimura and Ramya Rajagopalan ("Different Differences") examine another of these programs, Eingenstrat, in far greater detail than I have examined STRUCTURE here.
31 For a brilliant deconstruction of ad infinitum and its ontotheological dependencies, that takes "God out of mathematics and put[s] the body back in," see Rotman, *Ad Infinitum*.
32 Quoted in Hyde, "Erika von Mutius," 1029.
33 Quoted in Hyde, "Erika von Mutius," 1029.
34 Hyde, "Erika von Mutius," 1029.
35 von Mutius et al., "Prevalence of Asthma and Allergic Disorders," 1395.
36 von Mutius et al., "Prevalence of Asthma and Allergic Disorders," 1397.
37 von Mutius et al., "Prevalence of Asthma and Allergic Disorders," 1398.
38 Hyde, "Erika von Mutius," 1029.
39 The multiple-millennia-old concept-strategy of "hypothesis" also suggests a long-standing recognition (quickly repressed or forgotten) of the intimate connection between a thesis and impossibility. Like Peirce's act of abduction, hypothesizing accesses (so to speak) an outside, and not some deep internal truth, essence, or Being. Reading German Jewish philosopher Hermann Cohen's reading of Plato's *hypotheton*, Jacques Derrida states that what occurs "under the name of hypothesis" is a "determination of the idea as an opening to the infinite, an infinite task for 'philosophy as a rigorous science.'" Derrida, "Interpretations at War," 58.
40 Strachan, "Family Size, Infection and Atopy," S2.
41 See Cambrosio and Keating, *Exquisite Specificity*.
42 Interestingly, the Th1 and Th2 cells referred to by Strachan, von Mutius, and others here are themselves a kind of historical vicissitude, a product of particular research trajectories on a particular disease with particular model organisms, all coupled to the current state of technological development. Although the distinction is still useful, contemporary immunologists are less likely to refer to them as two different types of T-lymphocytes and more likely to see them as an environmental effect themselves, a product of lymphocyte plasticity in response to a changing cellular environment. According to immunologist Tirumalai Kamala, "We immunologists identified Th1 and Th2 using a variety of highly artificial *in vitro* cell culture techniques, we observed some apparent evidence of this dichotomy in CD4 helper T cell function *in vivo* with the experimental mouse model of leishmaniasis, the

early promise of this mouse model did not pan out along predicted lines, the mental construct imposed by this dichotomy nevertheless persisted and we began to classify an increasing plethora of CD4 T cell subsets such as Th3, Th9, Th17, Treg and Tfh over the past two decades. In other words, we let our increasing technological proficiency dictate the course for understanding the underlying biology rather than the other way around. Doing so, it took years before we came full circle to appreciating CD4 *helper T cell plasticity* (17), i.e. that CD4 helper T cells do not actually differentiate terminally into non-overlapping or mutually exclusive or even antagonistic secretion of particular groups of cytokines with different effector functions such as Th1, Th2, Th3, Th17, etc but rather they evolve particular overlapping features of functionalities depending on the activating cell(s), site(s) and microbiota, i.e. the CD4 T cell's environment. Coming full circle in this fashion? Now that's what I call irony." "What's the difference between Th1 and Th2 helper T-cell subsets?," Quora, accessed August 5, 2016, https://www.quora.com/Whats-the-difference-between-Th1-and-Th2-helper-T-cell-subsets. See also Bonelli et al., "Helper T Cell Plasticity."

43 Riedler et al., "Exposure to Farming," 1129.
44 Riedler et al., "Exposure to Farming," 1132.
45 Riedler et al., "Exposure to Farming," 1133.
46 Eder et al., "Toll-Like Receptor 2," 482–83.
47 Eder et al., "Toll-Like Receptor 2," 487.
48 Eder et al., "Toll-Like Receptor 2," 487.
49 Ege et al., "Gene-Environment Interaction."
50 Solicitude calls for what Maurice Blanchot called a passivity beyond passivity, a "passivity that is the *pas* ["not"] in the utterly passive, and which has therefore abandoned the level of life where *passive* would simply be the opposite of *active*." Blanchot, *Writing of the Disaster*, 13. We seem to be in koan territory again: solicitude is both passive and active, and neither. It tries to take in the whole as it is, all of it, on nothing but its own terms, and to do so it has to grasp it all, all at once, forcing and squeezing and pulling, until it just threatens to break.
51 Penrose and Penrose, "Impossible Objects," 31; emphasis added.
52 Taking a step toward a more interesting psychoanalytic theory of the scientific subject, geographers Paul Robbins and Sarah Moore advance a similar vision built around Lacan's impossible Real as the "whole" that is an object of desire: "Lacan, in his discussion of scientists in 'Subversion of the Subject,' argues that academics are prone to the kind of melancholy that has at its base a drive to understand what he calls the *objet petit a*. This *objet petit a* is the 'object of anxiety par excellence'—the 'essential object which isn't an object any longer, but this something faced with which all words cease and all categories fail.' While the *objet petit a* is variously defined in Lacan's work, it is generally the object cause of desire. In this case, the desire for knowledge, the admi-

rable core of the scientific urge, can never be completely fulfilled because it centers on the elusive, non-symbolizable *objet petit a*. This has implications for the study of ecology in the Anthropocene where nature itself, and explanations for natural phenomena, are the object cause of desire. As such, nature is an ever-receding object that escapes the scientist's grasp (physical and mental) and generates anxiety through the impossibility of possession." Robbins and Moore, "Ecological Anxiety Disorder," 10.

53 My discussion of the limits of GWAS here may be tiresomely long for some, but in many ways I have soft-pedaled the critiques of GWAS by scientists that circulated in this period and in retrospect. Perhaps the most serious limitation of GWAS was their almost exclusive use of data sets and the technologies depending on them, like the Illumina chips, that consisted almost entirely of genomes of European ancestry, that is, White people's genomes. It is highly unlikely genomicists can "simply extrapolate the results of... hundreds of GWASs to the rest of the world's population" (Roberts, "Deep Genealogy"). "Caution should be exercised in applying any genetic risk prediction model based on [GWAS] outside of the ancestry group in which it was derived" (Carlson et al., "Generalization and Dilution," e1001661). "The lack of ethnic diversity in human genomic studies means that our ability to translate genetic research into clinical practice or public health policy may be dangerously incomplete, or worse, mistaken" (Sirugo et al., "Missing Diversity"); see also Bustamante et al., "Genomics for the World." David Goldstein's early critique ("Common Genetic Variation") established the discourse on what came to be called "the missing heritability problem" in GWAS; see also Manolio et al., "Finding the Missing Heritability"; Gallagher and Chen-Plotkin, "Post-GWAS Era"; Flint and Ideker, "Great Hairball Gambit."

54 NIH, "Analysis."

55 NIH, "Analysis."

56 Dick et al., "Candidate Gene–Environment Interaction Research," 37.

57 Dick et al., "Candidate Gene–Environment Interaction Research," 37.

58 Khoury "Editorial: Emergence of Gene-Environment Interaction Analysis."

59 McAllister et al., "Current Challenges and New Opportunities," 753.

60 I read the "steps" in the Penrose's staircase as experimental-ethical plateaus, as conceptualized by Deleuze and Guattari, building on Gregory Bateson's work, and Michael Fischer; see Fischer, *Emergent Forms of Life*, 30; and Laurie, "Epistemology as Politics."

Chapter Seven. Friendship

1 Genome.gov., "Implicating Sequence Variants in Human Disease," September 12, 2012. https://www.genome.gov/27550063/implicating-sequence-variants-in-human-disease.

2 National Human Genome Research Institute, "Outline of White Paper, Summary of Key Messages and Next Steps—Daniel MacArthur & Teri Manolio," YouTube video, September 12–13, 2012, https://www.youtube.com/watch?v=278IeY7B8wE.
3 Gunter and MacArthur, "Great Sequencing Power—Great Responsibility."
4 Vermeulen et al., "Understanding Life Together," provides a good overview of the growth of collaboration at different times in different life sciences, in both the lab and the field.
5 Terrell, *Talent for Friendship*, 228.
6 Terrell, *Talent for Friendship*, 216.
7 Terrell, *Talent for Friendship*, 122.
8 Terrell, *Talent for Friendship*, 93.
9 Silk, "Using the 'F'-Word," 440. See also Silk, "Cooperation without Counting."
10 Silk, "Using the 'F'-Word," 422.
11 Silk, "Using the 'F'-Word," 422.
12 Foucault, "Friendship," 136, 137.
13 Deutscher, *Politics of Impossible Difference*, 177–78.
14 Quoted in Deutscher, *Politics of Impossible Difference*, 177.
15 Hedgecoe, "Schizophrenia." Calling Adam "my friend" partakes of the performative aspect of friendship and any other naming, like "gender": it's the act, repeated, that makes it. I've never met Adam in person (although we were once on a video call together), and have had only occasional professional exchanges with him via email.
16 Hedgecoe, "Schizophrenia," 877.
17 Hedgecoe, "Schizophrenia," 893, 895.
18 Hedgecoe, "Schizophrenia," 893.
19 Hedgecoe, "Schizophrenia," 876.
20 Hedgecoe, "Schizophrenia," 903.
21 Hedgecoe, "Schizophrenia," 885.
22 Hedgecoe's bond to an ethic and rhetoric of neutrality, in which even minor privileging and prioritizing would be marked as negative, is reenforced in note 58, where he abjures "reflexivity":

> On a reflexive note, it is only fair to acknowledge the rhetorical, discursive nature of this paper and my own construction (through symmetrical analysis) of a position mid-way between genetic "hype" and radical criticism. Just as the authors of the review articles adopt a strategy of caution and responsibility to show how trustworthy and believable their claims are, my moderate position with regard to the extent to which geneticization is happening, and my refusal to assume that geneticization is a negative process, strengthen my empirical claims and position as an "objective" commentator. On a more basic level, this paper is, of course, constructed

to persuade.... That said, the kind of discourse analysis developed by Greg Myers keeps reflexivity at arm's length, and this is a stance which I adopt in this paper. While I could provide a running commentary on the rhetorical nature of my own text, I am not sure how this advances my argument about the geneticization of schizophrenia. Since the main aim of this paper is not a methodological/theoretical one (with regard to discourse analysis), but is an empirical one (with regard to geneticization), reflexive comments will be kept to a minimum. (Hedgecoe, "Schizophrenia," 908–9)

I read for places where the methodological and the theoretical become entangled with the empirical; these are places that in my view need to be foregrounded, not minimized. In what I think of as a deconstructive more than a reflexive analysis, I read Hedgecoe's more neutral, "moderate" and "'objective'" argument about the discourse of geneticization as thoroughly dependent on the rhetoric of his text. It steers that prudent "mid-way" course between Scylla and Charybdis, which is not only a mytheme embedded in a philosopheme, but grounded in a bad reading of myth (see chapter 5). In that sense, the objectivity of my text is an effect, not of an ethic of objectivity that places itself "mid-way between" or "at arm's length," but of an ethic of friendship that fully plays both sides against the middle, as it were, working (and playing!) to heighten contradictions and tensions rather than searching for a middle course. In these mythic terms, then, I may share Odysseus's fate upon returning the opposite way through the contradictions that have no safe middle path, as my text gets sucked down into the swirling Charybdis of excessive admiration and embrace, clinging to a branch above the disaster and looking for one piece of debris to carry me through...

23 It should be said, however, that "White," Western European ancestry is itself more complicated than it is usually construed to be, at least by genomic measures; see the wonderfully, paradoxically titled "Disuniting Uniformity: A Pied Cladistic Canvas of MtDNA Haplogroup H in Eurasia," by Loogväli et al.

24 National Human Genome Research Institute, "The Need for Criteria to Implicate DNA Variants: Real-World Examples—Mark Daly," YouTube video, September 19, 2012, https://www.youtube.com/watch?v=Y_Y_zvo7NQ4, beginning at 5:08.

25 National Human Genome Research Institute, "Need for Criteria," beginning at 28:26.

26 See Hercher, "Interview with Heidi Rehm."

27 National Human Genome Research Institute, "Maintaining Accurate Information in Variant Databases—Heidi Rehm," YouTube video, September 12–13, 2012, https://www.youtube.com/watch?v=Qn6LyTsCDXo, beginning at 00:43.

28 HGMD operates out of the Institute for Medical Genomics at Cardiff. An interesting hybrid in many ways, it consists of a public version that is freely

accessible to registered users, and a "professional" version that is more complete and up-to-date, accessible through their commercial partner Qiagen; see http://www.hgmd.cf.ac.uk/ac/index.php.

29 National Human Genome Research Institute. "Implicating Sequence Variants in Human Diseases: Moderated Discussion," YouTube video, September 12–13, 2012, https://www.youtube.com/watch?v=O1-fDPVtgQk, beginning at 2:34.

30 National Human Genome Research Institute. "Implicating Sequence Variants in Human Diseases," beginning at 6:39.

31 For an explanation of these databases, see https://www.ncbi.nlm.nih.gov/clinvar/intro/.

32 National Human Genome Research Institute. "Implicating Sequence Variants in Human Diseases," beginning at 8:31.

33 National Human Genome Research Institute, "Outline of White Paper," beginning at 07:11. How does anathema come to be experienced affectively and thus register on a human face as enjoyment? I would trace this pleasure (tinged with anxiety) to an experience of anathema's own impossible doubling. According to *Merriam-Webster* it "can be considered a one-word oxymoron"; in its earliest Greek usage, *anathema* was a straightforward "devotional offering," but eventually through the Christianizing intercessions of St. Paul came also to refer to an accursed one.

34 National Human Genome Research Institute, "Outline of White Paper," beginning at 10:48.

35 National Human Genome Research Institute, "Outline of White Paper," beginning at 21:55.

36 National Human Genome Research Institute, "Outline of White Paper," beginning at 24:34.

37 National Human Genome Research Institute, "Outline of White Paper," beginning at 12:50.

38 National Human Genome Research Institute, "Outline of White Paper," beginning at 13:59.

39 National Human Genome Research Institute, "Outline of White Paper," beginning at 14:31.

40 National Human Genome Research Institute, "Outline of White Paper," beginning at 31:20.

41 National Human Genome Research Institute, "Outline of White Paper," beginning at 32:08.

42 National Human Genome Research Institute, "Outline of White Paper," beginning at 32:24.

43 MacArthur et al., "Guidelines for Investigating Causality," 469–70.

44 MacArthur et al., "Guidelines for Investigating Causality," article metrics.

45 Helmreich, "Gravity's Reverb," 468.

46 Balmer et al., "Five Rules of Thumb," 78. Efforts such as these and my own are unique or new in trying to unfold friendship with sciences. I've learned from scientists like Ludwik Fleck and Michael Polanyi that careful (auto) ethnography of scientific thinking and doing can convey its complexities rich with affect and contradictions. Donna Haraway and Evelyn Fox Keller turned their deep, and deeply reflective, experience in the biological and physical sciences into some of the most profound analyses of the sciences and their care. I never understood their writings to be anything other than powerful expressions of, and encouragement of, friendship with sciences and scientists, and am always genuinely shocked to read anyone suggesting otherwise. Closer to anthropology and to genomics, scholars like Stefan Helmreich (whose elucidation of the "informalisms" essential to the formal operations of science is close neighbor to my work-play of care) and Amber Benezra also write genomics and genomicists in the style of friendship. Mike Fischer, too, presents sciences and scientists more as neighbors to befriend than as objects of critique.

Poem-Like *Tolls* 3

1 *Journal of Biological Chemistry* 279, no. 13 (2004): 12542–50, https://doi.org/10.1074/jbc.M310175200.
2 *Immunity* 23, no. 2 (2005): 165–75, https://doi.org/10.1016/j.immuni.2005.06.008.

Postscript

1 See Asmelash and Willingham, "Katalin Karikó's Work in MRNA"; Kollewe, "Covid Vaccine Technology Pioneer"; Garde, "The Story of MRNA."
2 See Karikó et al., "MRNA Is an Endogenous Ligand"; Karikó et al., "Suppression of RNA Recognition"; Karikó et al., "Incorporation of Pseudouridine into MRNA." Excerpts from the first two of these publications are reframed and reprinted in "Poem-Like Tolls 3" in this book. Bibel, "Katalin Karikó," provides an excellent summary of this work.
3 Burdick, "Usefulness of a March for Science."
4 Burdick, "Usefulness of a March for Science."
5 Pielke, "'Sedative' for Science Policy," 44.
6 Pielke, "'Sedative' for Science Policy," 41.
7 Johnson, "Bipartisan Vision." As of March 2022, this legislation has been "ruined" and "wrecked" through watering down in the US Senate, and its future is uncertain; see Hammond, "How Congress Ruined the Endless Frontier Act," and Lopez, "How Congress Wrecked Its Own Science Bill."

8 This passage may help clarify how I read an entire phenomenological and scientific worldview in the image of the Penroses' impossible staircase discussed in chapter 6. My reading comes from reading Maurice Blanchot's *The Step Not Beyond* (*Le pas au-dela*), his fragmented system of truth-speaking and -seeking as steps (*pas*) into a beyond (*au-dela*) *and yet* not (*pas*) beyond, an interminable infinite conversation. See also Derrida's retracing of Blanchot's steps in, for example, "Pas" and *Specters of Marx* (260ff).

9 See Balter, "Geneticists Decry Book."

10 Wade, "Decade Later."

11 Wade, "Decade Later."

12 Lemaitre, *Essay on Narcissism and Science*, 250. See also Lemaitre, "Science, Narcissism."

13 Lemaitre, *Essay on Narcissism and Science*, 139.

14 Calkins, "Between the Lab and the Field," 23. See also Ureta, "Ruination Science," for another promising analysis of a different "affective atmosphere" in Chilean environmental toxicology.

15 Shah, "Scientist-Subject."

16 See, for example, Tomšič, "Mathematical Realism," and "Better Failures."

17 See Keller, "Pot-Holes Everywhere," for her reading of how her work on McClintock has been prone to being misread.

18 Keller, *Reflections on Gender and Science*, 117.

19 Keller, *Reflections on Gender and Science*, 122, 124.

20 Literary theorist Toril Moi has critiqued Keller's analysis as "curiously timid and flawed," and for its reliance on love in opposition to reason as "an unwitting repetition of 'male' Romanticism." (Recall Goethe's romanticizing of feminine "care" that chapter 3 touched on.) I can only note here that my trope of friendship as a pattern of care has tried to straddle that opposition, and defer a more scrupulous analysis of "love" differentiated into *philia, eros, storge*, and *agape* (for starters) to future writings. Obviously, I also agree with Moi's sense that "feminism needs a theory of knowledge which undoes and displaces" such dualisms, one that goes beyond "a mistaken fear of biologism . . . to include the body in thought"; it was "the Freudian theory of epistemophilia, or the drive for knowledge," she proposed, that "provides us with a first outline." Moi, "Patriarchal Thought," 192, 199. Elizabeth Wilson is the contemporary feminist theorist whose writings most forcefully include the body in thought, and move feminist theory beyond a disinterest in or disparagement of the biological; see for example Wilson, "Gut Feminism," and "Another Neurological Scene."

21 Cox, "Epistemophilia," 80. Cox details how epistemophilia has provoked some anxiety among feminist theorists who "have focused on the violent expressions of the drive" when it should be understood as having "both destructive and productive expressions." Epistemophilia in Cox's analysis provides resources both for "psychic defense" and for "psychic strength," and

thus should be central to any pedagogical theory and practice, feminist ones included.

22 The story is a more complex and contested one, needless to say; see Greenberg and Mitchell, *Object Relations*, for an excellent analytic narrative, particularly their extended discussion throughout of the mostly unresolved differences and tensions between drive theories and object relations.

23 Keller, *Reflections on Gender and Science*, 130.

24 Sedgwick, *Touching Feeling*, 138

25 Sedgwick, *Touching Feeling*, 137.

26 For good evidence of the great thinking and doing that friendship with genomicists can produce, see Benezra, "Microbial Kin" and "Race in the Microbiome." For more of Benezra's research where genomics intersects with nutrition, social justice, and getting beyond "race as a ghost variable," see De Wolfe, "Chasing Ghosts"; Ishaq, "Introducing the Microbes"; and (with a lengthy author's list rivaling that of any straight-up genomics collaboration) Greenhough et al., "Setting the Agenda." In a different but also friendly vein, Alexis Walker ("Diversity, Profit, Control") enriches our understanding of the ethical and sociopolitical thinking of scientists working in private genomics companies. Moving out beyond genomics, when I think of anthropologists who model friendship with the sciences, Stefan Helmreich (*Sounding the Limits of Life*) comes first to mind, followed quickly by Sophia Roosth ("Turning to Stone").

27 Britzman, "Theory Kindergarten," 130.

28 Britzman, "Theory Kindergarten," 131.

29 Kirby, *Quantum Anthropologies*, 15.

30 Sedgwick, *Touching Feeling*, 150

31 Sedgwick, *Touching Feeling*, 150.

WORKS CITED

Alfaro, Mariana, and Tyler Pager. "Top White House Scientist Apologizes after Internal Review Finds He Bullied and Demeaned Staff." *Washington Post*, February 7, 2022. https://www.washingtonpost.com/politics/2022/02/07/lander-white-house-apologizes/.

Almklov, Petter G. "Standardized Data and Singular Situations." *Social Studies of Science* 38, no. 6 (2008): 873–97. https://doi.org/10.1177/0306312708098606.

Amanda. "I swear I'm a b*tch." *A Lady Scientist* (blog). March 22, 2008. http://ladyscientist.net/2008/03/i-swear-im-a-btch/.

Anderson, Kathryn V., Liselotte Bokla, and Christiane Nüsslein-Volhard. "Establishment of Dorsal-Ventral Polarity in the Drosophila Embryo: The Induction of Polarity by the Toll Gene Product." *Cell* 42, no. 3 (1985): 791–98. https://doi.org/10.1016/0092-8674(85)90275-2.

Ankeny, Rachel A., and Sabina Leonelli. "What's So Special about Model Organisms?" *Studies in History and Philosophy of Science Part A* 42, no. 2 (2010): 313–23. https://doi.org/10.1016/j.shpsa.2010.11.039.

Ashburner, Michael, Catherine A. Ball, Judith A. Blake, David Botstein, Heather Butler, J. Michael Cherry, Allan P. Davis, et al. "Gene Ontology: Tool for the Unification of Biology." *Nature Genetics* 25, no. 1 (2000): 25–29. doi:10.1038/75556.

Asher, M. I., U. Keil, H. R. Anderson, R. Beasley, J. Crane, F. Martinez, E. A. Mitchell, N. Pearce, B. Sibbald, and A. W. Stewart. "International Study of Asthma and Allergies in Childhood (ISAAC): Rationale and Methods." *European Respiratory Journal* 8, no. 3 (1995): 483–91.

Asmelash, Leah, and A. J. Willingham. "Katalin Kariko's Work in MRNA Is the Basis of the Covid-19 Vaccine." CNN, December 17, 2020. https://edition.cnn.com/2020/12/16/us/katalin-kariko-covid-19-vaccine-scientist-trnd/index.html.

Atkinson-Graham, Melissa, Martha Kenney, Kelly Ladd, Cameron Michael Murray, and Emily Astra-Jean Simmonds. "Care in Context: Becoming an STS Researcher." *Social Studies of Science* 45, no. 5 (2015): 738–48. https://doi.org/10.1177/0306312715600277.

Balmer, Andrew S., Jane Calvert, Claire Marris, Susan Molyneux-Hodgson, Emma Frow, Matthew Kearnes, Kate Bulpin, Pablo Schyfter, Adrian Mackenzie, and Paul Martin. "Five Rules of Thumb for Post-ELSI Interdisciplinary Collaborations." *Journal of Responsible Innovation* 3, no. 1 (2016): 73–80. https://doi.org/10.1080/23299460.2016.1177867.

Balter, Michael. "Geneticists Decry Book on Race and Evolution." *Science | AAAS*, August 8, 2014. https://www.sciencemag.org/news/2014/08/geneticists-decry-book-race-and-evolution.

Barad, Karen. "Diffracting Diffraction: Cutting Together-Apart." *Parallax* 20, no. 3 (2014): 168–87. https://doi.org/10.1080/13534645.2014.927623.

Barbieri, Marcello. *Code Biology: A New Science of Life*. Dordrecht: Springer, 2015.

Barlow, Tani E. *The Question of Woman in Chinese Feminism*. Durham, NC: Duke University Press, 2004.

Bateson, Gregory. "Toward a Theory of Schizophrenia." In *Steps to an Ecology of Mind*, 201–27. New York: Ballantine Books, 1956.

Begley, Sharon. "As Twitter Explodes, Eric Lander Apologizes for Toasting James Watson." *STAT News*, May 14, 2018. https://www.statnews.com/2018/05/14/apology-eric-lander-james-watson/.

Begley, Sharon. "Why Eric Lander Morphed from Science God to Punching Bag." *STAT* (blog). January 25, 2016. https://www.statnews.com/2016/01/25/why-eric-lander-morphed/.

Begley, Sharon, and Andrew Joseph. "Scientists Threw James Watson a Party before Eric Lander's Criticized Toast." *STAT News*, May 17, 2018. https://www.statnews.com/2018/05/17/james-watson-birthday-party-lander-toast/?utm_campaign=trendmd-internal.

Benezra, Amber. "Microbial Kin: Relations of Environment and Time." *Medical Anthropology Quarterly* 35, no. 4 (2021): 511–28. https://doi.org/10.1111/maq.12680.

Benezra, Amber. "Race in the Microbiome." *Science, Technology, and Human Values* 45, no. 5 (2020): 877–902. https://doi.org/10.1177/0162243920911998.

Benjamin, Jessica. *Like Subjects, Love Objects: Essays on Recognition and Sexual Difference*. New Haven, CT: Yale University Press, 1995.

Benjamin, Jessica, and Galit Atlas. "The 'Too Muchness' of Excitement: Sexuality in Light of Excess, Attachment and Affect Regulation." *International Journal of Psychoanalysis* 96, no. 1 (2015): 39–63. https://doi.org/10.1111/1745-8315.12285.

Bettelheim, Bruno. *Freud and Man's Soul*. New York: A. A. Knopf, 1983.

Bhabha, Homi K. "'The Beginning of Their Real Enunciation': Stuart Hall and the Work of Culture." *Critical Inquiry* 42, no. 1 (2015): 1–30. https://doi.org/10.1086/682994.

Bibel, Brianna. "Katalin Karikó—'Solver' and 'Saver' of Synthetic MRNA (like That in Some Vaccines)." *Bumbling Biochemist* (blog). March 19, 2021. https://

thebumblingbiochemist.com/365-days-of-science/katalin-kariko-solver-and-saver-of-synthetic-mrna-like-that-in-some-vaccines/.
Bioephemera. "Pseudonymity: Five Reasons the New Science Blogs/NG Policy Is Misguided." *Scienceblogs* (blog). September 14, 2011. https://scienceblogs.com/bioephemera/2011/09/14/anonymity-among-science-blogge.
Bio-IT World. "Making a Genesweep: It's Official!" 2003. http://www.bio-itworld.com/archive/071503/genesweep.
Blanchot, Maurice. "Friendship." In *Friendship*, translated by Elizabeth Rottenberg, 289–92. Stanford, CA: Stanford University Press, 1997.
Blanchot, Maurice. *The Step Not Beyond*. New York: SUNY Press, 1992.
Blanchot, Maurice. *The Writing of the Disaster*. Translated by Ann Smock. Lincoln: University of Nebraska Press, 2015.
Boellstorff, Tom. "Making Big Data, in Theory." *First Monday* 18, no. 10 (2013). http://firstmonday.org/ojs/index.php/fm/article/view/4869.
Bolnick, Deborah A. "Individual Ancestry Inference and the Reification of Race as a Biological Phenomenon." In *Revisiting Race in a Genomic Age*, edited by Barbara A. Koenig, Sandra Soo-Jin Lee, and Sarah S. Richardson, 70–85. New Brunswick, NJ: Rutgers University Press, 2008.
Bolnick, Deborah A., Duana Fullwiley, Troy Duster, Richard S. Cooper, Joan H. Fujimura, Jonathan Kahn, Jay S. Kaufman, et al. "The Science and Business of Genetic Ancestry Testing." *Science* 318, no. 5849 (2007): 399–400. https://doi.org/10.1126/science.1150098.
Bonelli, Michael, Han-Yu Shih, Kiyoshi Hirahara, Kentner Singelton, Arian Laurence, Amanda Poholek, Tim Hand, et al. "Helper T Cell Plasticity: Impact of Extrinsic and Intrinsic Signals on Transcriptomes and Epigenomes." *Current Topics in Microbiology and Immunology* 381 (2014): 279–326. doi:10.1007/82_2014_371.
Bossé, Yohan, and Thomas J. Hudson. "Toward a Comprehensive Set of Asthma Susceptibility Genes." *Annual Review of Medicine* 58, no. 1 (2007): 171–84. https://doi.org/10.1146/annurev.med.58.071105.111738.
Bourne, Philip E., and Johanna McEntyre. "Biocurators: Contributors to the World of Science." *PLOS Computational Biology* 2, no. 10 (2006): e142. doi:10.1371/journal.pcbi.0020142.
Bowker, Geoffrey C. *Memory Practices in the Sciences*. Cambridge, MA: MIT Press, 2005.
Braidotti, Rosi. "A Theoretical Framework for the Critical Posthumanities." *Theory, Culture and Society* 36, no. 6 (2019): 31–61. https://doi.org/10.1177/0263276418771486.
Braun, Lundy. *Breathing Race into the Machine: The Surprising Career of the Spirometer from the Plantation to Genetics*. Minneapolis: University of Minnesota Press, 2014.
Braun, Lundy, Anne Fausto-Sterling, Duana Fullwiley, Evelynn M. Hammonds, Alondra Nelson, William Quivers, Susan M. Reverby, and Alexandra E. Shields. "Racial Categories in Medical Practice: How Useful Are They?" *PLOS Medicine* 4, no. 9 (2007): e271. https://doi.org/10.1371/journal.pmed.0040271.

Bretislav, Friedrich, Dieter Hoffmann, Jürgen Renn, Florian Schmaltz, and Martin Wolf, eds. *One Hundred Years of Chemical Warfare: Research, Deployment, Consequences.* 1st ed. New York: Springer, 2017.

Britzman, Deborah P. "Theory Kindergarten." In *Regarding Sedgwick: Essays on Queer Culture and Critical Theory*, edited by Stephen M. Barber and David L. Clark, 121–41. London: Routledge, 2002.

Brooksbank, Cath, and John Quackenbush. "Data Standards: A Call to Action." *OMICS: A Journal of Integrative Biology* 10, no. 2 (2006): 94–99. doi:10.1089/omi.2006.10.94.

Burdach, Konrad. "Faust und die Sorge." *Deutsche Vierteljahrsschrift für Literaturwissenschaft und Geistesgeschichte* 1, no. 1 (1923): 1–60.

Burdick, Alan. "The Usefulness of a March for Science." *New Yorker*, April 23, 2017. http://www.newyorker.com/tech/elements/the-usefulness-of-a-march-for-science.

Burkhardt, Kyle, Bohdan Schneider, and Jeramia Ory. "A Biocurator Perspective: Annotation at the Research Collaboratory for Structural Bioinformatics Protein Data Bank." *PLOS Computational Biology* 2 (10): e99. doi:10.1371/journal.pcbi.0020099.

Bustamante, Carlos D., Francisco M. De La Vega, and Esteban G. Burchard. "Genomics for the World." *Nature* 475, no. 7355 (2011): 163–65. https://doi.org/10.1038/475163a.

Calkins, Sandra. "Between the Lab and the Field: Plants and the Affective Atmospheres of Southern Science." *Science, Technology, and Human Values*, December 15, 2021. https://doi.org/10.1177/01622439211055118.

Cambrosio, A., and P. Keating. *Exquisite Specificity*. Oxford: Oxford University Press, 1995.

Carlson, Christopher S., Tara C. Matise, Kari E. North, Christopher A. Haiman, Megan D. Fesinmeyer, Steven Buyske, Fredrick R. Schumacher, et al. "Generalization and Dilution of Association Results from European GWAS in Populations of Non-European Ancestry: The PAGE Study." *PLOS Biology* 11, no. 9 (2013): e1001661. https://doi.org/10.1371/journal.pbio.1001661.

Casanova, Jean-Laurent, Laurent Abel, and Lluis Quintana-Murci. "Human TLRs and IL-1Rs in Host Defense: Natural Insights from Evolutionary, Epidemiological, and Clinical Genetics." *Annual Review of Immunology* 29, no. 1 (2011): 447–91. https://doi.org/10.1146/annurev-immunol-030409-101335.

Clark, Nigel, and Giovanni Bettini. "'Floods' of Migrants, Flows of Care: Between Climate Displacement and Global Care Chains." *Sociological Review* 65 (2 suppl, 2017): 36–54. https://doi.org/10.1177/0081176917711078.

Clarke, Adele E., and Joan H. Fujimura, eds. *The Right Tools for the Job: At Work in Twentieth Century Life Sciences*. Princeton, NJ: Princeton University Press, 2016.

Collaborative Study on the Genetics of Asthma. "A Genome-Wide Search for Asthma Susceptibility Loci in Ethnically Diverse Populations." *Nature Genetics* 15, no. 4 (1997): 389–92. https://doi.org/10.1038/ng0497-389.

Collins, Francis S. *The Language of God: A Scientist Presents Evidence for Belief.* New York: Free Press, 2007.
Colquhoun, David. “An Investigation of the False Discovery Rate and the Misinterpretation of P-Values.” *Royal Society Open Science* 1, no. 3 (2014): 140216. doi:10.1098/rsos.140216.
Colquhoun, David. “The Diary: June 2008–May 2009.” *DC's Improbable Science*, May 2009. http://www.dcscience.net/in-memoriam/the-diary-continued-from-june-2008/.
Colquhoun, David. “What a Paean from Ben Goldacre Can Do.” *DC's Improbable Science* (blog). November 16, 2014. http://www.dcscience.net/2014/11/16/what-a-paean-from-ben-goldacre-can-do/.
Cook-Deegan, Robert. “Origins of the Human Genome Project.” *RISK: Health, Safety and Environment* 5, no. 2 (1994): 97–118. https://scholars.unh.edu/cgi/viewcontent.cgi?article=1182&context=risk.
Corngold, Stanley. *Franz Kafka: The Necessity of Form.* Ithaca, NY: Cornell University Press, 1988.
Cox, Peta. “Epistemophilia.” *Australian Feminist Studies* 25, no. 63 (2010): 79–92. https://doi.org/10.1080/08164640903499745.
Danby, Colin. “Lupita's Dress: Care in Time.” *Hypatia* 19, no. 4 (2004): 23–48.
Davies, Gail. “Mobilizing Experimental Life: Spaces of Becoming with Mutant Mice.” *Theory, Culture and Society* 30, no. 7/8 (2013): 129–53. https://doi.org/10.1177/0263276413496285.
Deleuze, Gilles. *Nietzsche and Philosophy*. London: Athlone Press, 1983.
Deleuze, Gilles, and Felix Guattari. *A Thousand Plateaus: Capitalism and Schizophrenia*. Minneapolis: University of Minnesota Press, 1987.
Dell'Amore, Christine. “Why Is a Woman Who Loves Science So Surprising?” *National Geographic.* March 27, 2013. https://www.nationalgeographic.com/adventure/article/130325-elise-andrew-i-fking-love-science-women.
Denis, Jérôme, and David Pontille. “Material Ordering and the Care of Things.” *Science, Technology and Human Values*, October 2014. https://doi.org/10.1177/0162243914553129.
Derrida, Jacques. *Aporias*. Translated by Thomas Dutoit. Stanford, CA: Stanford University Press, 1993.
Derrida, Jacques. *Dissemination*. Translated by Barbara Johnson. Chicago: University of Chicago Press, 1981.
Derrida, Jacques. “Force and Signification.” In *Writing and Difference*, translated by Alan Bass. London: Routledge, 2001 [1967].
Derrida, Jacques. *Of Grammatology*. Baltimore, MD: Johns Hopkins University Press, 1976.
Derrida, Jacques. With Moshe Ron. “Interpretations at War: Kant, the Jew, the German.” *New Literary History* 22, no. 1 (1991): 39–95. https://doi.org/10.2307/469143.
Derrida, Jacques. *Life Death*. Translated by Pascale-Anne Brault and Michael Naas. Chicago: University of Chicago Press, 2020.

Derrida, Jacques. *The Other Heading: Reflections on Today's Europe*. Translated by Pascale-Anne Brault and Michael B. Naas. Bloomington: Indiana University Press, 1981.

Derrida, Jacques. "Pas." In *Parages*, translated by James Hulbert and Tom Conley. Stanford, CA: Stanford University Press, 2011.

Derrida, Jacques. *Politics of Friendship*. London: Verso, 1997.

Derrida, Jacques. *Specters of Marx: The State of the Debt, the Work of Mourning, and the New International*. Sussex, UK: Psychology Press, 1994.

Deutscher, Penelope. "Mourning the Other: Cultural Cannibalism and the Politics of Friendship (Jacques Derrida and Luce Irigaray)." *differences* 10, no. 3 (1998): 159–84.

Deutscher, Penelope. *A Politics of Impossible Difference*. Ithaca, NY: Cornell University Press, 2018.

Dewey, John. *Experience and Nature*. New York: George Allen and Unwin, 1929.

De Wolfe, Travis J., Mohammed Rafi Arefin, Amber Benezra, and María Rebolleda Gómez. "Chasing Ghosts: Race, Racism, and the Future of Microbiome Research." Edited by Kathryn C. Milligan-Myhre. *MSystems* 6, no. 5 (2021): e00604-21. https://doi.org/10.1128/mSystems.00604-21.

Dick, Danielle M., Arpana Agrawal, Matthew C. Keller, Amy Adkins, Fazil Aliev, Scott Monroe, John K. Hewitt, Kenneth S. Kendler, and Kenneth J. Sher. "Candidate Gene-Environment Interaction Research: Reflections and Recommendations." *Perspectives on Psychological Science* 10, no. 1 (2015): 37–59. doi:10.1177/1745691614556682.

Doyle, Richard. *On Beyond Living: Rhetorical Transformations of the Life Sciences*. Stanford, CA: Stanford University Press, 1997.

Dreissen, Yasmine E. M., and Marina A. J. Tijssen. "The Startle Syndromes: Physiology and Treatment." *Epilepsia* 53 (2012): 3–11. doi:10.1111/j.1528-1167.2012.03709.x.

Drucker, Johanna. "Humanities Approaches to Graphical Display." *Digital Humanities* 5, no. 1 (2011). http://www.digitalhumanities.org/dhq/vol/5/1/000091/000091.html.

Eder, Waltraud, Walt Klimecki, Lizhi Yu, Erika von Mutius, Josef Riedler, Charlotte Braun-Fahrländer, Dennis Nowak, and Fernando D. Martinez. "Toll-Like Receptor 2 as a Major Gene for Asthma in Children of European Farmers." *Journal of Allergy and Clinical Immunology* 113, no. 3 (2004): 482–88. https://doi.org/10.1016/j.jaci.2003.12.374.

Edwards, Paul. "Infrastructure and Modernity: Scales of Force, Time, and Social Organization in the History of Sociotechnical Systems." In *Modernity and Technology*, edited by Thomas J. Misa, Philip Brey, and Andrew Feenberg, 185–225. Cambridge, MA: MIT Press, 2002.

Edwards, Paul N. "Meteorology as Infrastructural Globalism." *Osiris* 21, no. 1 (2006): 229–50. https://doi.org/10.1086/507143.

Ege, Markus J., David P. Strachan, William O. C. M. Cookson, Miriam F. Moffatt, Ivo Gut, Mark Lathrop, Michael Kabesch, et al. "Gene-Environment Interaction for Childhood Asthma and Exposure to Farming in Central

Europe." *Journal of Allergy and Clinical Immunology* 127, no. 1 (2011): 138–44, 144.e1–4. https://doi.org/10.1016/j.jaci.2010.09.041.
Engster, Daniel. "Rethinking Care Theory: The Practice of Caring and the Obligation to Care." *Hypatia* 20, no. 3 (2005): 50–74. doi:10.1111/j.1527-2001.2005.tb00486.x.
Fausto-Sterling, Anne. "Refashioning Race: DNA and the Politics of Health Care." *Differences* 15, no. 3 (2004): 1–37. https://doi.org/10.1215/10407391-15-3-1.
Fischer, Michael M. J. *Emergent Forms of Life and the Anthropological Voice*. Durham, NC: Duke University Press, 2003.
500 Women Scientists. "Eric Lander Is Not the Ideal Choice for Presidential Science Adviser." *Scientific American*, January 21, 2021. https://www.scientificamerican.com/article/eric-lander-is-not-the-ideal-choice-for-presidential-science-adviser/.
Fleck, Ludwik. *Genesis and Development of a Scientific Fact*. Chicago: University of Chicago Press, 1979.
Fleck, Ludwik. "Problems of the Science of Science [1946]." In *Cognition and Fact: Materials on Ludwick Fleck*, edited by Robert S. Cohen and Thomas Schnelle, 113–27. Boston Studies in the Philosophy of Science 87. Dordrecht: Springer Netherlands, 1986. https://doi.org/10.1007/978-94-009-4498-5_6.
Fleischmann, Robert D., Mark D. Adams, Owen White, Rebecca A. Clayton, Ewen F. Kirkness, Anthony R. Kerlavage, Carol J. Bult, et al. "Whole-Genome Random Sequencing and Assembly of Haemophilus Influenzae Rd." *Science* 269, no. 5223 (1995): 496–512. doi:10.2307/2887657.
Flint, Jonathan, and Trey Ideker. "The Great Hairball Gambit." *PLOS Genetics* 15, no. 11 (2019): e1008519. https://doi.org/10.1371/journal.pgen.1008519.
Fortun, Kim, and Mike Fortun. "Scientific Imaginaries and Ethical Plateaus in Contemporary U.S. Toxicology." *American Anthropologist* 107, no. 1 (2005): 43–54. https://doi.org/10.1525/aa.2005.107.1.043.
Fortun, Kim, Mike Fortun, Angela Hitomi Skye Crandall Okune, Tim Schütz, and Shan-Ya Su. "Civic Community Archiving with the Platform for Experimental Collaborative Ethnography: Double Binds and Design Challenges." In *Culture and Computing. Design Thinking and Cultural Computing*, edited by Matthias Rauterberg, 36–55. Cham: Springer, 2021. https://doi.org/10.1007/978-3-030-77431-8_3.
Fortun, Mike. "Celera Genomics: The Race for the Human Genome Sequence." *eLS* (2006). https://doi.org/10.1038/npg.els.0005182.
Fortun, Mike. "Entangled States: Quantum Teleportation and 'The Willies.'" In *Paranoia within Reason: A Casebook on Conspiracy as Explanation*, edited by George Marcus. Late Editions 6. Chicago: University of Chicago Press, 1999.
Fortun, Mike. "Fluctuating about Zero, Taking Nothing's Measure." In *Zeroing In on the Year 2000: The Final Edition*, edited by George Marcus, 121–60. Late Editions 8. Chicago: University of Chicago Press, 2000.
Fortun, Mike. "For an Ethics of Promising, or: A Few Kind Words about James Watson." *New Genetics and Society* 24, no. 2 (2005): 157–74. https://doi.org/10.1080/14636770500184792.

Fortun, Mike. "Genes in Our Knot." In *Handbook of Genetics and Society*, edited by Paul Atkinson, Peter Glasner, and Margaret Lock, 273–86. London: Routledge, 2009. https://doi.org/10.4324/9780203927380-26.

Fortun, Michael A. "The Human Genome Project: Past, Present, and Future Anterior." In *Science, History and Social Concern: Essays in Honor of Everett Mendelsohn*, edited by Garland Allen and Roy MacLeod. Dordrecht: Kluwer Academic, 2002.

Fortun, Michael A. "Making and Mapping Genes and Histories: The Genomics Project in the United States 1980–1990." PhD diss., Harvard University, 1993.

Fortun, Mike. "Projecting Speed Genomics." In *Practices of Human Genetics: International and Interdisciplinary Perspectives*, edited by Michael Fortun and Everett Mendelsohn, 25–48. Sociology of the Sciences Yearbook, vol. 19. Dordrecht: Springer, 1999.

Fortun, Mike. *Promising Genomics: Iceland and deCode Genetics in a World of Speculation*. Berkeley: University of California Press, 2008.

Fortun, Mike. "Zombie Corporate Vikings: First Installment." Blog, December 29, 2015. http://mfortun.org/?p=180 (2015).

Fortun, Mike, and Herbert J. Bernstein. *Muddling Through: Pursuing Science and Truths in the 21st Century*. Berkeley, CA: Counterpoint Press, 1998.

Fortun, Mike, et al. "What's So Funny 'bout PECE, TAF, and Data Sharing?" In *Collaborative Anthropology Today: A Collection of Exceptions*, edited by Dominic Boyer and George E. Marcus. Ithaca, NY: Cornell University Press, 2019.

Foucault, Michel. "Friendship as a Way of Life." In *Michel Foucault: Ethics, Subjectivity and Truth: The Essential Works of Michael Foucault, Vol. 1*, edited by Paul Rabinow, 135–40. New York: The New Press, 1997.

Foucault, Michel. *The Uses of Pleasure*. The History of Sexuality, vol 2. New York: Random House, 1985.

Frank, Adam, and Elizabeth Wilson. "Like-Minded." *Critical Inquiry* 38 (2012): 870–77.

Friedrich, Bretislav, Dieter Hoffmann, Jürgen Renn, Florian Schmaltz, and Martin Wolf, eds. *One Hundred Years of Chemical Warfare: Research, Deployment, Consequences*. New York: Springer, 2017.

Friese, Carrie. "Realizing Potential in Translational Medicine: The Uncanny Emergence of Care as Science." *Current Anthropology* 54, no. 57 (2013): S129–38. https://doi.org/10.1086/670805.

Fujimura, Joan H. "Constructing 'Do-Able' Problems in Cancer Research: Articulating Alignment." *Social Studies of Science* 17, no. 2 (1987): 257–93. https://doi.org/10.1177/030631287017002003.

Fujimura, Joan H., and Ramya Rajagopalan. "Different Differences: The Use of 'Genetic Ancestry' versus Race in Biomedical Human Genetic Research." *Social Studies of Science* 41, no. 1 (2011): 5–30. https://doi.org/10.1177/0306312710379170.

Fujimura, Joan H., Troy Duster, and Ramya Rajagopalan. "Introduction: Race, Genetics, and Disease: Questions of Evidence, Matters of Consequence."

Social Studies of Science 38, no. 5 (2008): 643–56. https://doi.org/10.1177/0306312708091926.
Fullwiley, Duana. "The Biologistical Construction of Race: 'Admixture' Technology and the New Genetic Medicine." *Social Studies of Science* 38, no. 5 (2008): 695–735. https://doi.org/10.1177/0306312708090796.
Fullwiley, Duana. "Race and Genetics: Attempts to Define the Relationship." *BioSocieties* 2, no. 2 (2007): 221–37. https://doi.org/10.1017/S1745855207005625.
Gabbatiss, Josh. "James Watson: The Most Controversial Statements Made by the Father of DNA." *The Independent*, January 13, 2019. https://www.independent.co.uk/news/science/james-watson-racism-sexism-dna-race-intelligence-genetics-double-helix-a8725556.html.
Gabrys, Jennifer. "Citizen Sensing, Air Pollution and Fracking: From 'Caring about Your Air' to Speculative Practices of Evidencing Harm." *Sociological Review* 65 (2 suppl., 2017): 172–92. https://doi.org/10.1177/0081176917710421.
Galison, Peter. *How Experiments End*. Chicago: University of Chicago Press, 1987.
Gallagher, Michael D., and Alice S. Chen-Plotkin. "The Post-GWAS Era: From Association to Function." *American Journal of Human Genetics* 102, no. 5 (2018): 717–30. https://doi.org/10.1016/j.ajhg.2018.04.002.
Ganser, Lisa R., Qing Yan, Victoria M. James, Robert Kozol, Maya Topf, Robert J. Harvey, and Julia E. Dallman. "Distinct Phenotypes in Zebrafish Models of Human Startle Disease." *Neurobiology of Disease* 60, no. 100 (2013): 139–51. doi:10.1016/j.nbd.2013.09.002.
Garde, Damian. "The Story of MRNA: From a Loose Idea to a Tool That May Help Curb Covid." *STAT News*, November 10, 2020. https://www.statnews.com/2020/11/10/the-story-of-mrna-how-a-once-dismissed-idea-became-a-leading-technology-in-the-covid-vaccine-race/.
Gibney, Paul. "The Double-Bind Theory: Still Crazy-Making after All These Years." *Psychotherapy in Australia* 12, no. 3 (2006): 48–55.
Gilbert, Walter. "Towards a Paradigm Shift in Biology." *Nature* 349, no. 6305 (1991): 99. https://doi.org/10.1038/349099a0.
Gill, Natalie. "Caring for Clean Streets: Policies as World-Making Practices." *Sociological Review* 65 (2 suppl., 2017): 71–88. https://doi.org/10.1177/0081176917710422.
Gitelman, Lisa, ed. *Raw Data Is an Oxymoron*. Cambridge, MA: MIT Press, 2013.
Gitelman, Lisa, and Virginia Jackson. "Introduction." In *Raw Data Is an Oxymoron*, edited by Lisa Gitelman, 1–14. Cambridge, MA: MIT Press, 2013.
Goldacre, Ben. "Quackbuster Causes Too Much Flak for University." *The Guardian*, June 9, 2007, sec. Science. https://www.theguardian.com/science/2007/jun/09/badscience.uknews.
Goldstein, David B. "Common Genetic Variation and Human Traits." *New England Journal of Medicine* 360, no. 17 (2009): 1696–98. https://doi.org/10.1056/NEJMp0806284.
Goodman, Alan H., Yolanda T. Moses, and Joseph L. Jones. *Race: Are We So Different?* Hoboken, NJ: John Wiley and Sons, 2019.

Goodman, Alyssa, Alberto Pepe, Alexander W. Blocker, Christine L. Borgman, Kyle Cranmer, Merce Crosas, Rosanne Di Stefano, et al. "Ten Simple Rules for the Care and Feeding of Scientific Data." *PLOS Computational Biology* 10, no. 4 (2014): e1003542. https://doi.org/10.1371/journal.pcbi.1003542.

Graul, R. J., and R. L. Stanley. "Specific Gravity Adjustment of Urine Analysis Results." *American Industrial Hygiene Association Journal* 43, no. 11 (1982): 863. doi:10.1080/15298668291410701.

Greely, Henry T. "Amgen Buys deCODE—Reflections Backwards, Forwards, and on DTC Genomics." *Stanford Law School: Law and Bioscience* (blog). December 13, 2012. https://law.stanford.edu/2012/12/13/lawandbiosciences-2012-12-13-amgen-buys-decode-reflections-backwards-forwards-and-on-dtc-genomics/.

Green, Douglas R. "Stress in Biomedical Research: Six Impossible Things." *Molecular Cell* 40, no. 2 (2010): 176–78. doi:10.1016/j.molcel.2010.10.007.

Greenberg, Jay R., and Stephen A. Mitchell. *Object Relations in Psychoanalytic Theory*. Cambridge, MA: Harvard University Press, 1983.

Greenhough, Beth, Cressida Jervis Read, Jamie Lorimer, Javier Lezaun, Carmen McLeod, Amber Benezra, Sally Bloomfield, et al. "Setting the Agenda for Social Science Research on the Human Microbiome." *Palgrave Communications* 6, no. 1 (2020): 18. https://doi.org/10.1057/s41599-020-0388-5.

Guarino, Ben. "The March for Science Began with This Person's 'Throwaway Line' on Reddit." *Washington Post*, April 21, 2017. https://www.washingtonpost.com/news/speaking-of-science/wp/2017/04/21/the-march-for-science-began-with-this-persons-throwaway-line-on-reddit/.

Gunter, Chris, and Daniel MacArthur. "Great Sequencing Power—Great Responsibility." *Spectrum | Autism Research News* (blog). June 6, 2014. https://www.spectrumnews.org/opinion/viewpoint/guest-blog-great-sequencing-power-great-responsibility/.

Hamilton, Jennifer A. "Revitalizing Difference in the HapMap: Race and Contemporary Human Genetic Variation Research." *Journal of Law, Medicine and Ethics* 36, no. 3 (2008): 471–77. https://doi.org/10.1111/j.1748-720X.2008.293.x.

Hammond, Samuel. "How Congress Ruined the Endless Frontier Act." Niskanen Center. May 20, 2021. https://www.niskanencenter.org/how-congress-ruined-the-endless-frontier-act/.

Hansson, Goran K., and Kristina Edfeldt. "Toll to Be Paid at the Gateway to the Vessel Wall." *Arteriosclerosis, Thrombosis, and Vascular Biology* 25, no. 6 (2005): 1085–87.

Haraway, Donna J. *Modest_Witness@Second_Millennium. FemaleMan_Meets_OncoMouse: Feminism and Technoscience*. London: Routledge, 1997.

Haraway, Donna J. *When Species Meet*. Minneapolis: University of Minnesota Press, 2008.

Hargreave, F. E., R. Leigh, and K. Parameswaran. "Asthma as a Disease Concept." *The Lancet* 368, no. 9545 (2006): 1415–16. https://doi.org/10.1016/S0140-6736(06)69595-0.

Hartigan, John. "Plant Publics: Multispecies Relating in Spanish Botanical Gardens." *Anthropological Quarterly* 88, no. 2 (2015): 481–507.
Hedgecoe, Adam. "Schizophrenia and the Narrative of Enlightened Geneticization." *Social Studies of Science* 31, no. 6 (2001): 875–911. https://doi.org/10.1177/030631201031006004.
Helmreich, Stefan. "Gravity's Reverb: Listening to Space-Time, or Articulating the Sounds of Gravitational-Wave Detection." *Cultural Anthropology* 31, no. 4 (2016): 464–92. https://doi.org/10.14506/ca31.4.02.
Helmreich, Stefan. *Sounding the Limits of Life: Essays in the Anthropology of Biology and Beyond*. Princeton, NJ: Princeton University Press, 2015. https://press.princeton.edu/books/hardcover/9780691164809/sounding-the-limits-of-life.
Hercher, Laura. Interview with Heidi Rehm. *The Beagle Has Landed* (blog). March 24, 2020. https://beaglelanded.com/podcasts/heidi-rehm/.
Hilgartner, Stephen. *Reordering Life: Knowledge and Control in the Genomics Revolution*. Cambridge, MA: MIT Press, 2017.
Hillyer, Richard. *Divided between Carelessness and Care: A Cultural History*. Gordonsville, VA: Palgrave Macmillan, 2013.
Hoffjan, Sabine, Dan Nicolae, and Carole Ober. "Association Studies for Asthma and Atopic Diseases: A Comprehensive Review of the Literature." *Respiratory Research* 4, no. 1 (2003): 14. https://doi.org/10.1186/1465-9921-4-14.
Hoffmeyer, Jesper. *Biosemiotics: An Examination into the Signs of Life and the Life of Signs*. Scranton, PA: University of Scranton Press, 2008.
Homer. *The Odyssey*. Translated by Emily Wilson. New York: W.W. Norton, 2018.
Hrdy, Sarah Blaffer. *Mother Nature: Maternal Instincts and How They Shape the Human Species*. New York: Ballantine Books, 2000.
Hrdy, Sarah Blaffer. *Mothers and Others: The Evolutionary Origins of Mutual Understanding*. Cambridge, MA: Harvard University Press, 2011.
Hrdy, Sarah Blaffer, and Judith M. Burkart. "The Emergence of Emotionally Modern Humans: Implications for Language and Learning." *Philosophical Transactions of the Royal Society B: Biological Sciences* 375, no. 1803 (2020): 20190499. https://doi.org/10.1098/rstb.2019.0499.
Huxley, T. H. *T. H. Huxley on Education*. Cambridge, UK: Cambridge University Press, 1971.
Hyde, Rob. "Erika von Mutius: Reshaping the Landscape of Asthma Research." *The Lancet* 372, no. 9643 (2008): 1029. doi:10.1016/S0140-6736(08)61428-2.
Ioannidis, John P. A. "Why Most Discovered True Associations Are Inflated." *Epidemiology* 19, no. 5 (2008): 640–48.
Ioannidis, John P. A. "Why Most Published Research Findings Are False." *PLOS Medicine* 2, no. 8 (2005): e124. https://doi.org/10.1371/journal.pmed.0020124.
Ioannidis, John P. A., Marta Gwinn, Julian Little, Julian P. T. Higgins, Jonine L. Bernstein, Paolo Boffetta, Melissa Bondy, et al. "A Road Map for Efficient and Reliable Human Genome Epidemiology." *Nature Genetics* 38, no. 1 (2006): 3–5.

Ioannidis, John P. A., Philip S. Rosenberg, James J. Goedert, and Thomas R. O'Brien. "Commentary: Meta-Analysis of Individual Participants' Data in Genetic Epidemiology." *American Journal of Epidemiology* 156, no. 3 (2002): 204–10. https://doi.org/10.1093/aje/kwf031.

Irigaray, Luce. *I Love to You: Sketch for a Felicity within History*. New York: Routledge, 1996.

Ishaq, Suzanne L., Francisco J. Parada, Patricia G. Wolf, Carla Y. Bonilla, Megan A. Carney, Amber Benezra, Emily Wissel, et al. "Introducing the Microbes and Social Equity Working Group: Considering the Microbial Components of Social, Environmental, and Health Justice." Edited by Jack A. Gilbert. *MSystems* 6, no. 4 (2021): e00471-21. https://doi.org/10.1128/mSystems.00471-21.

Jacob, François. *The Statue Within: An Autobiography*. Long Island, NY: Cold Spring Harbor Laboratory Press, 1995.

Johnson, Eddie Bernice. "A Bipartisan Vision for the Future of American Science." *Issues in Science and Technology* 37, no. 1 (2021). https://issues.org/bipartisan-vision-future-american-science-policy-eddie-bernice-johnson/.

Kahn, Jonathan. *Race in a Bottle: The Story of BiDil and Racialized Medicine in a Post-Genomic Age*. New York: Columbia University Press, 2012.

Kaiser, David, ed. *Pedagogy and the Practice of Science: Historical and Contemporary Perspectives*. Cambridge, MA: MIT Press, 2005.

Karikó, Katalin, Houping Ni, John Capodici, Marc Lamphier, and Drew Weissman. "MRNA Is an Endogenous Ligand for Toll-Like Receptor 3." *Journal of Biological Chemistry* 279, no. 13 (2004): 12542–50. https://doi.org/10.1074/jbc.M310175200.

Karikó, Katalin, Michael Buckstein, Houping Ni, and Drew Weissman. "Suppression of RNA Recognition by Toll-Like Receptors: The Impact of Nucleoside Modification and the Evolutionary Origin of RNA." *Immunity* 23, no. 2 (2005): 165–75. https://doi.org/10.1016/j.immuni.2005.06.008.

Karikó, Katalin, Hiromi Muramatsu, Frank A. Welsh, János Ludwig, Hiroki Kato, Shizuo Akira, and Drew Weissman. "Incorporation of Pseudouridine into MRNA Yields Superior Nonimmunogenic Vector with Increased Translational Capacity and Biological Stability." *Molecular Therapy: The Journal of the American Society of Gene Therapy* 16, no. 11 (2008): 1833–40. https://doi.org/10.1038/mt.2008.200.

Karkazis, Katrina, and Rebecca Jordan-Young. "Sensing Race as a Ghost Variable in Science, Technology, and Medicine." *Science, Technology, and Human Values* 45, no. 5 (2020): 763–78. https://doi.org/10.1177/0162243920939306.

Kay, Lily. *Who Wrote the Book of Life?: A History of the Genetic Code*. Stanford, CA: Stanford University Press, 2000.

Keating, Peter, and Alberto Cambrosio. "Too Many Numbers: Microarrays in Clinical Cancer Research." *Studies in History and Philosophy of Science Part C: Studies in History and Philosophy of Biological and Biomedical Sciences* 43, no. 1 (2012): 37–51. doi:10.1016/j.shpsc.2011.10.004.

Keller, Evelyn Fox. "Drosophila Embryos as Transitional Objects: The Work of Donald Poulson and Christiane Nüsslein-Volhard." *Historical Studies in the Physical and Biological Sciences* 26, no. 2 (1996): 313–46. https://doi.org/10.2307/27757764.
Keller, Evelyn Fox. *Making Sense of Life: Explaining Biological Development with Models, Metaphors, and Machines*. Cambridge, MA: Harvard University Press, 2009.
Keller, Evelyn Fox. "Pot-Holes Everywhere: How (Not) to Read My Biography of Barbara McClintock." *Writing about Lives in Science*, edited by Vita Fortunati and Elena Agazzi, 33–42. Göttingen: Vandenhoeck and Ruprecht Unipress, 2014. https://doi.org/10.14220/9783737002639.33.
Keller, Evelyn Fox. *Refiguring Life: Metaphors of Twentieth-Century Biology*. New York: Columbia University Press, 1995.
Keller, Evelyn Fox. *Reflections on Gender and Science*. New Haven, CT: Yale University Press, 1995.
Kenner, Alison. *Breathtaking: Asthma Care in a Time of Climate Change*. Minneapolis: University of Minnesota Press, 2018.
Khandekar, Aalok, Brandon Costelloe-Kuehn, Lindsay Poirier, Alli Morgan, Alison Kenner, Kim Fortun, Mike Fortun, and the PECE Design Team. "Moving Ethnography: Infrastructuring Doubletakes and Switchbacks in Experimental Collaborative Methods." *Science and Technology Studies* (Special Issue: Methodography of Ethnographic Collaboration) 34, no. 3 (2021): 78–102. https://sciencetechnologystudies.journal.fi/article/view/89782.
Khoury, Muin J. "The Case for a Global Human Genome Epidemiology Initiative." *Nature Genetics* 36, no. 10 (2004): 1027–28. https://doi.org/10.1038/ng1004-1027.
Khoury, Muin J. "Editorial: Emergence of Gene-Environment Interaction Analysis in Epidemiologic Research." *American Journal of Epidemiology* 186, no. 7 (2017): 751–52. https://doi.org/10.1093/aje/kwx226.
Kirby, Vicki. *Quantum Anthropologies: Life at Large*. Durham, NC: Duke University Press, 2011.
Kirby, Vicki. "Tracing Life: 'La Vie La Mort.'" *CR: The New Centennial Review* 9, no. 1 (2009): 107–26. https://doi.org/10.1353/ncr.0.0059.
Kittay, Eva Feder, and Ellen K. Feder. *The Subject of Care: Feminist Perspectives on Dependency*. Lanham, MD: Rowman and Littlefield, 2003.
Knorr Cetina, Karin. "The Care of the Self and Blind Variation: The Disunity of Two Leading Sciences." *The Disunity of Science: Boundaries, Contexts, and Power*, edited by Peter Galison and David J. Stump, 287–310. Palo Alto, CA: Stanford University Press, 1996.
Knorr Cetina, Karin. "Objectual Practice." In *The Practice Turn in Contemporary Theory*, edited by Theodore R. Schatzki, Karin Knorr Cetina, and Eike von Savigny, 175–88. New York: Routledge, 2001.
Koenig, Barbara A., Sandra Soo-Jin Lee, and Sarah S. Richardson. *Revisiting Race in a Genomic Age*. New Brunswick, NJ: Rutgers University Press, 2008.
Kollewe, Julia. "Covid Vaccine Technology Pioneer: 'I Never Doubted It Would Work.'" *The Guardian*, November 21, 2020. http://www.theguardian.com

/science/2020/nov/21/covid-vaccine-technology-pioneer-i-never-doubted-it-would-work.
Koopmans, Matthijs. "From Double Bind to N-Bind: Toward a New Theory of Schizophrenia and Family Interaction." *Nonlinear Dynamics, Psychology, and Life Sciences* 5, no. 4 (2001): 289–323. https://doi.org/10.1023/A:1009518729645.
Kuska, Bob. "Beer, Bethesda, and Biology: How 'Genomics' Came into Being." *JNCI: Journal of the National Cancer Institute* 90, no. 2 (1998): 93. https://doi.org/10.1093/jnci/90.2.93.
Lancet. "Asthma: Still More Questions than Answers." *The Lancet* 372, no. 9643 (2008): 1009. https://doi.org/10.1016/S0140-6736(08)61414-2.
Lander, Eric. Interview with the author. Whitehead Institute, Cambridge MA, 1992.
Lander, Eric, and Leonid Kruglyak. "Genetic Dissection of Complex Traits: Guidelines for Interpreting and Reporting Linkage Results." *Nature Genetics* 11, no. 3 (1995): 241–47. https://doi.org/10.1038/ng1195-241.
Lappé, Martine. "The Paradox of Care in Behavioral Epigenetics: Constructing Early-Life Adversity in the Lab." *BioSocieties* 13, no. 4:698–714. https://doi.org/10.1057/s41292-017-0090-z.
Lappé, Martine D. "Taking Care: Anticipation, Extraction and the Politics of Temporality in Autism Science." *BioSocieties* 9, no. 3 (2014): 304–28. http://dx.doi.org.libproxy.rpi.edu/10.1057/biosoc.2014.14.
Larkin, Brian. "The Politics and Poetics of Infrastructure." *Annual Review of Anthropology* 42, no. 1 (2013): 327–43. https://doi.org/10.1146/annurev-anthro-092412-155522.
Latour, Bruno, and Steve Woolgar. *Laboratory Life: The Construction of Scientific Facts*. Princeton, NJ: Princeton University Press, 1986.
Laurie, Timothy Nicholas. "Epistemology as Politics and the Double-Bind of Border Thinking: Lévi-Strauss, Deleuze and Guattari, Mignolo." *PORTAL Journal of Multidisciplinary International Studies* 9, no. 2 (2012). https://doi.org/10.5130/portal.v9i2.1826.
Lavau, Stephanie, and Nick Bingham. "Practices of Attention, Possibilities for Care: Making Situations Matter in Food Safety Inspection." *Sociological Review* 65 (2 suppl, 2017): 20–35. https://doi.org/10.1177/0081176917710526.
Lavin, Matthew. "Why Digital Humanists Should Emphasize Situated Data over Capta." *Digital Humanities Quarterly* 15, no. 2 (2021). http://www.digitalhumanities.org/dhq/vol/15/2/000556/000556.html.
Lavoie, Mireille, Thomas De Koninck, and Danielle Blondeau. "The Nature of Care in Light of Emmanuel Levinas." *Nursing Philosophy* 7, no. 4 (2006): 225–34. https://doi.org/10.1111/j.1466-769X.2006.00279.x.
Lemaitre, Bruno. *An Essay on Narcissism and Science: How Do High Ego Personalities Drive Research in the Life Sciences?* Rockville: MD: Federation of American Societies for Experimental Biology, 2015. https://www.epflpress.org/produit/974/9782839918411/an-essay-on-science-and-narcissism.
Lemaitre, Bruno. "Mixed Feelings about the Nobel Prize." 2011. http://www.behinddiscoveries.com/toll/mixedfeelings.

Lemaitre, Bruno. "The Road to Toll." *Nature Reviews Immunology* 4, no. 7 (2004): 521–27. https://doi.org/10.1038/nri1390.

Lemaitre, Bruno. "Science, Narcissism and the Quest for Visibility." *FEBS Journal* 284, no. 6 (2017): 875–82. https://doi.org/10.1111/febs.14032.

Lemaitre, Bruno, Emmanuelle Nicolas, Lydia Michaut, Jean-Marc Reichhart, and Jules A. Hoffmann. "The Dorsoventral Regulatory Gene Cassette Spätzle/Toll/Cactus Controls the Potent Antifungal Response in Drosophila Adults." *Cell* 86, no. 6 (1996): 973–83. https://doi.org/10.1016/S0092-8674(00)80172-5.

Leonelli, Sabina. "Documenting the Emergence of Bio-Ontologies: Or, Why Researching Bioinformatics Requires HPSSB." *History and Philosophy of the Life Sciences* 32, no. 1 (2010): 105–25.

Leonelli, Sabina, and Rachel A. Ankeny. "Re-thinking Organisms: The Impact of Databases on Model Organism Biology." *Studies in History and Philosophy of Science Part C: Studies in History and Philosophy of Biological and Biomedical Sciences* 43, no. 1 (2012): 29–36. doi:10.1016/j.shpsc.2011.10.003.

Lewontin, Richard. *The Triple Helix: Gene, Organism and Environment*. Cambridge, MA: Harvard University Press, 2000.

Liu, Yushi, Toru Nyunoya, Shuguang Leng, Steven A. Belinsky, Yohannes Tesfaigzi, and Shannon Bruse. "Softwares and Methods for Estimating Genetic Ancestry in Human Populations." *Human Genomics* 7, no. 1 (2013): 1. https://doi.org/10.1186/1479-7364-7-1.

Loogväli, Eva-Liis, Urmas Roostalu, Boris A. Malyarchuk, Miroslava V. Derenko, Toomas Kivisild, Ene Metspalu, Kristiina Tambets, et al. "Disuniting Uniformity: A Pied Cladistic Canvas of MtDNA Haplogroup H in Eurasia." *Molecular Biology and Evolution* 21, no. 11 (2004): 2012–21. https://doi.org/10.1093/molbev/msh209.

Lopez, German. "How Congress Wrecked Its Own Science Bill, Explained in 600 Words." *Vox*, June 4, 2021. https://www.vox.com/2021/6/4/22518923/endless-frontier-act-innovation-competition-act-china-congress.

MacArthur, D. G., T. A. Manolio, D. P. Dimmock, H. L. Rehm, J. Shendure, G. R. Abecasis, D. R. Adams, et al. "Guidelines for Investigating Causality of Sequence Variants in Human Disease." *Nature* 508, no. 7497 (2014): 469–76. https://doi.org/10.1038/nature13127.

Maddox, Brenda. *Rosalind Franklin: The Dark Lady of DNA*. New York: Harper Collins, 2013.

Manolio, Teri A., Francis S. Collins, Nancy J. Cox, David B. Goldstein, Lucia A. Hindorff, David J. Hunter, Mark I. McCarthy, et al. "Finding the Missing Heritability of Complex Diseases." *Nature* 461, no. 7265 (2009): 747–53. https://doi.org/10.1038/nature08494.

Marcus, George E. "Holism and the Expectations of Critique in Post-1980s Anthropology." In *Experiments in Holism*, edited by Ton Otto and Nils Bubandt, 28–46. Hoboken, NJ: John Wiley and Sons, 2010. https://doi.org/10.1002/9781444324426.ch3.

Marks, Jonathan. *Tales of the Ex-Apes: How We Think about Human Evolution*. Berkeley: University of California Press, 2015.
Marks, Jonathan. *What It Means to Be 98% Chimpanzee: Apes, People, and Their Genes*. Berkeley: University of California Press, 2003.
Martin, Aryn, Natasha Myers, and Ana Viseu. "The Politics of Care in Technoscience." *Social Studies of Science* 45, no. 5 (2015): 625–41. https://doi.org/10.1177/0306312715602073.
Martinez, Fernando D. "Gene-Environment Interactions in Asthma." *Proceedings of the American Thoracic Society* 4, no. 1 (2007): 26–31. https://doi.org/10.1513/pats.200607-144JG.
Mattern, Shannon. "Maintenance and Care." *Places Journal*, November 2018. https://placesjournal.org/article/maintenance-and-care/.
Mayer, Bruce J. "The Discovery of Modular Binding Domains: Building Blocks of Cell Signaling." *Nature Reviews Molecular Cell Biology* 16, no. 11 (2015): 691–98. https://doi.org/10.1038/nrm4068.
McAllister, Kimberly, Leah E. Mechanic, Christopher Amos, Hugues Aschard, Ian A. Blair, Nilanjan Chatterjee, David Conti, et al. "Current Challenges and New Opportunities for Gene-Environment Interaction Studies of Complex Diseases." *American Journal of Epidemiology* 186, no. 7 (2017): 753–61. https://doi.org/10.1093/aje/kwx227.
McElheny, Viktor K. *Watson and DNA: Making a Scientific Revolution*. New York: Basic Books, 2009.
McNamer, Sarah. "The Origins of the Meditationes vitae Christi." *Speculum* 84, no. 4 (2009): 905–55. http://www.jstor.org/stable/40593681.
M'charek, Amade, Victor Toom, and Lisette Jong. "The Trouble with Race in Forensic Identification." *Science, Technology, and Human Values* 45, no. 5 (2020): 804–28. https://doi.org/10.1177/0162243919899467.
Medzhitov, Ruslan. "Approaching the Asymptote: 20 Years Later." *Immunity* 30, no. 6 (2009): 766–75. https://doi.org/10.1016/j.immuni.2009.06.004.
Medzhitov, Ruslan, P. Preston-Hurlburt, and C. A. Janeway. "A Human Homologue of the Drosophila Toll Protein Signals Activation of Adaptive Immunity." *Nature* 388, no. 6640 (1997): 394–97. https://doi.org/10.1038/41131.
Mitchell, Stephen A. *Relationality: From Attachment to Intersubjectivity*. Hillsdale, NJ: Analytic Press, 2000.
Moffatt, Miriam F., Michael Kabesch, Liming Liang, Anna L. Dixon, David Strachan, Simon Heath, Martin Depner, et al. "Genetic Variants Regulating ORMDL3 Expression Contribute to the Risk of Childhood Asthma." *Nature* 448, no. 7152 (2007): 470–73. https://doi.org/10.1038/nature06014.
Moi, Toril. "Patriarchal Thought and the Drive for Knowledge." In *Between Feminism and Psychoanalysis*, edited by Teresa Brennan, 189–205. New York: Taylor and Francis, 1989.
Mol, Annemarie, Ingunn Moser, and Jeannette Pols, eds. *Care in Practice: On Tinkering in Clinics, Homes and Farms*. London: Transcript Verlag, 2015.

Mol, Annemarie, Ingunn Moser, and Jeannette Pols. "Care: Putting Practice into Theory." In *Care in Practice: On Tinkering in Clinics, Homes and Farms*, edited by Annemarie Mol, Ingunn Moser, and Jeannette Pols. London: Transcript Verlag, 2015.

Montoya, Michael J. "Bioethnic Conscription: Genes, Race, and Mexicana/o Ethnicity in Diabetes Research." *Cultural Anthropology* 22, no. 1 (2007): 94–128. https://doi.org/10.1525/can.2007.22.1.94.

Mooney, Chris. "Historians Say the March for Science Is 'Pretty Unprecedented.'" *Washington Post*, April 22, 2017. https://www.washingtonpost.com/news/energy-environment/wp/2017/04/22/historians-say-the-march-for-science-is-pretty-unprecedented/.

Murphy, Michelle. "Unsettling Care: Troubling Transnational Itineraries of Care in Feminist Health Practices." *Social Studies of Science* 45, no. 5 (2015): 717–37. https://doi.org/10.1177/0306312715589136.

Nadim, Tahani. "Data Labours: How the Sequence Databases GenBank and EMBL-Bank Make Data." *Science as Culture* 25, no. 4 (2016): 496–519. https://doi.org/10.1080/09505431.2016.1189894.

Nelson, Alondra. "Bio Science: Genetic Genealogy Testing and the Pursuit of African Ancestry." *Social Studies of Science* 38, no. 5 (2008): 759–83. https://doi.org/10.1177/0306312708091929.xyz

Nelson, Alondra. *The Social Life of DNA: Race, Reparations, and Reconciliation after the Genome*. Boston: Beacon Press, 2016.

Nelson, Karen E., and Bryan A. White. "Metagenomics and Its Application to the Study of the Human Microbiome." In *Metagenomics: Theory, Methods, and Applications*, edited by Diana Marco, 171–82. Norwich, UK: Horizon Scientific Press, 2010.

Nelson, Nicole C., Kelsey Ichikawa, Julie Chung, and Momin M. Malik. "Mapping the Discursive Dimensions of the Reproducibility Crisis: A Mixed Methods Analysis." *PLOS ONE* 16, no. 7 (2021): e0254090. https://doi.org/10.1371/journal.pone.0254090.

NIH (National Institutes of Health). "Analysis of Genome-Wide Gene-Environment (G x E) Interactions (R21)." Funding Opportunity Announcement PAR-13-382, October 30, 2013. http://grants.nih.gov/grants/guide/pa-files/PAR-13-382.html.

Nowak, Rachel. "Bacterial Genome Sequence Bagged." *Science* 269, no. 5223 (1995): 468–70. doi:10.2307/2887644.

Nüsslein-Volhard, Christiane, and Eric Wieschaus. "Mutations Affecting Segment Number and Polarity in Drosophila." *Nature* 287, no. 5785 (1980): 795–801. https://doi.org/10.1038/287795a0.

Ober, Carole, Nancy J. Cox, Mark Abney, Anna Di Rienzo, Eric S. Lander, Benjarat Changyaleket, Heidi Gidley, et al. "Genome-Wide Search for Asthma Susceptibility Loci in a Founder Population." *Human Molecular Genetics* 7, no. 9 (1998): 1393–98. https://doi.org/10.1093/hmg/7.9.1393.

Ober, Carole, and Donata Vercelli. "Gene-Environment Interactions in Human Disease: Nuisance or Opportunity?" *Trends in Genetics* 27, no. 3 (2011): 107–15. https://doi.org/10.1016/j.tig.2010.12.004.

Ogbunu, C. Brandon. "James Watson and the Insidiousness of Scientific Racism." *Wired*, January 29, 2019. https://www.wired.com/story/james-watson-and-scientific-racism/.

Olarte-Sierra, María Fernanda, and Tania Pérez-Bustos. "Careful Speculations: Toward a Caring Science of Forensic Genetics in Colombia." *Feminist Studies* 46, no. 1 (2020): 158–77. https://doi.org/10.15767/feministstudies.46.1.0158.

Patton, Paul. "Events, Becoming, and History." In *Deleuze and History*, edited by Jeffrey A. Bell and Clare Colebrook, 33–53. Edinburgh: University of Edinburgh Press, 2009.

Pennisi, Elizabeth. "A Low Number Wins the GeneSweep Pool." *Science* 300, no. 5625 (2003): 1484. doi:10.2307/3834429.

Penrose, L. S., and R. Penrose. "Impossible Objects: A Special Type of Visual Illusion." *British Journal of Psychology* 49, no. 1 (1958): 31–33. https://doi.org/10.1111/j.2044-8295.1958.tb00634.x.

Pickering, Andrew. *The Mangle of Practice: Time, Agency, and Science*. Chicago: University of Chicago Press, 1995.

Pielke, Roger. "A 'Sedative' for Science Policy." *Issues in Science and Technology* 37, no. 1 (2020): 41–47. https://issues.org/endless-frontier-sedative-for-science-policy-pielke/.

Pinel, Clémence, Barbara Prainsack, and Christopher McKevitt. "Caring for Data: Value Creation in a Data-Intensive Research Laboratory." *Social Studies of Science* 50, no. 2 (2020): 175–97. https://doi.org/10.1177/0306312720906567.

Plantin, Jean-Christophe, Carl Lagoze, Paul N. Edwards, and Christian Sandvig. "Infrastructure Studies Meet Platform Studies in the Age of Google and Facebook." *New Media and Society* 20, no. 1 (2018): 293–310. https://doi.org/10.1177/1461444816661553.

Polanyi, Michael. *Personal Knowledge: Towards a Post-Critical Philosophy*. Chicago: University of Chicago Press, 1962.

Pols, Jeannette. "Accounting and Washing Good Care in Long-Term Psychiatry." *Science, Technology and Human Values* 31, no. 4 (2006): 409–30. https://doi.org/10.1177/0162243906287544.

Pritchard, Jonathan K., Matthew Stephens, and Peter Donnelly. "Inference of Population Structure Using Multilocus Genotype Data." *Genetics* 155, no. 2 (2000): 945–59.

Pritchard, Jonathan K., Xiaoquan Wena, and Daniel Falush. Documentation for STRUCTURE Software: Version 2.3 (2010). http://pritch.bsd.uchicago.edu/structure.html.

Puig de la Bellacasa, Maria. "Making Time for Soil: Technoscientific Futurity and the Pace of Care." *Social Studies of Science* 45, no. 5 (2015): 691–716. https://doi.org/10.1177/0306312715599851.

Puig de la Bellacasa, Maria. "Matters of Care in Technoscience: Assembling Neglected Things." *Social Studies of Science* 41, no. 1 (2011): 85–106. doi:10.2307/40997116.

Puig de la Bellacasa, Maria. *Matters of Care: Speculative Ethics in More Than Human Worlds*. Minneapolis: University of Minnesota Press, 2017.

Puig de la Bellacasa, María. "'Nothing Comes without Its World': Thinking with Care." *Sociological Review* 60, no. 2 (2012): 197–216. https://doi.org/10.1111/j.1467-954X.2012.02070.x.

Reardon, Jenny. *Race to the Finish: Identity and Governance in an Age of Genomics*. Princeton, NJ: Princeton University Press, 2004.

Reddy, Deepa S. "Caught in Collaboration." *Collaborative Anthropologies* 1 (2008): 51–80.

Reddy, Deepa S. "Good Gifts for the Common Good: Blood and Bioethics in the Market of Genetic Research." *Cultural Anthropology* 22, no. 3 (2007): 429–72. https://doi.org/10.1525/can.2007.22.3.429.

Redish, A. David, Erich Kummerfeld, Rebecca Lea Morris, and Alan C. Love. "Opinion: Reproducibility Failures Are Essential to Scientific Inquiry." *Proceedings of the National Academy of Sciences* 115, no. 20 (2018): 5042–46. https://doi.org/10.1073/pnas.1806370115.

Reich, Warren T. "History of the Notion of 'Care.'" In *Encyclopedia of Bioethics*, 2nd ed., edited by Warren T. Reich, 319–31. New York: Simon and Schuster Macmillan, 1995. https://theology.georgetown.edu/research/historyofcare/classicarticle/.

Rheinberger, H.-J. "Experimental Systems—Graphematic Spaces." In *Inscribing Science: Scientific Texts and the Materiality of Communication*, edited by T. Lenoir and H. U. Gumbrecht, 285–303. Stanford, CA: Stanford University Press, 1998.

Rice, Clinton, Jugal K. Ghorai, Kathryn Zalewski, and Daniel N. Weber. "Developmental Lead Exposure Causes Startle Response Deficits in Zebrafish." *Aquatic Toxicology (Amsterdam, Netherlands)* 105, no. 3/4 (2011): 600–608. doi:10.1016/j.aquatox.2011.08.014.

Richardson, Sarah S., and Hallam Stevens. *Postgenomics: Perspectives on Biology after the Genome*. Durham, NC: Duke University Press, 2015.

Riedler, Josef, Charlotte Braun-Fahrländer, Waltraud Eder, Mynda Schreuer, Marco Waser, Soyoun Maisch, David Carr, Rudi Schierl, Dennis Nowak, and Erika von Mutius. "Exposure to Farming in Early Life and Development of Asthma and Allergy: A Cross-Sectional Survey." *The Lancet* 358, no. 9288 (2001): 1129–33. doi:10.1016/S0140-6736(01)06252-3.

Roach, Tom. *Friendship as a Way of Life: Foucault, AIDS, and the Politics of Shared Estrangement*. New York: SUNY Press, 2012.

Robbins, Paul, and Sarah A. Moore. "Ecological Anxiety Disorder: Diagnosing the Politics of the Anthropocene." *Cultural Geographies* 20, no. 1 (2013): 3–19, 10. https://doi.org/10.1177/1474474012469887.

Roberts, Leslie. "Plan for Genome Centers Sparks a Controversy." *Science* 246, no. 4927 (1989): 204–5. https://doi.org/10.1126/science.2799383.

Roberts, Roland G. "Deep Genealogy and the Dilution of Risk." *PLOS Biology* 11, no. 9 (2013): e1001660. doi:10.1371/journal.pbio.1001660.

Rock, Fernando L., Gary Hardiman, Jackie C. Timans, Robert A. Kastelein, and J. Fernando Bazan. "A Family of Human Receptors Structurally Related to Drosophila Toll." *Proceedings of the National Academy of Sciences* 95, no. 2 (1998): 588–93.

Rogers, Katie. "Biden's Top Science Adviser Resigns after Acknowledging Demeaning Behavior." *New York Times*, February 8, 2022. https://www.nytimes.com/2022/02/07/us/politics/eric-lander-resigns-white-house.html.

Roland Shearer, Rhonda. "Marcel Duchamp's Impossible Bed and Other 'Not' Readymade Objects: A Possible Route of Influence from Art to Science. Part 1." *Art and Academe* 10, no. 1 (1997): 26–62.

Ronell, Avital. *Crack Wars: Literature Addiction Mania*. Champaign: University of Illinois Press, 2004.

Ronell, Avital. *The Test Drive*. Champaign: University of Illinois Press, 2007.

Roosth, Sophia. "Turning to Stone: Fossil Hunting and Coeval Estrangement in Montana." *Res: Anthropology and Aesthetics* 69/70 (2018): 62–75. https://doi.org/10.1086/699900.

Rotman, Brian. *Ad Infinitum . . . The Ghost in Turing's Machine: Taking God Out of Mathematics and Putting the Body Back In. An Essay in Corporeal Semiotics*. Stanford, CA: Stanford University Press, 1993.

Rottenberg, Elizabeth. *For the Love of Psychoanalysis: The Play of Chance in Freud and Derrida*. New York: Fordham University Press, 2019.

Saint-Amour, Paul. "Weak Theory, Weak Modernism." *Modernism/Modernity Print Plus* 3, no. 3 (2018). https://modernismmodernity.org/articles/weak-theory-weak-modernism.

Salimi, Nima, and Randi Vita. "The Biocurator: Connecting and Enhancing Scientific Data." *PLOS Computational Biology* 2, no. 10 (2006): e125. doi:10.1371/journal.pcbi.0020125.

Schillmeier, Michael. "The Cosmopolitics of Situated Care." *Sociological Review* 65 (2 suppl 2017): 55–70. https://doi.org/10.1177/0081176917710426.

Schrader, Astrid. "Abyssal Intimacies and Temporalities of Care: How (Not) to Care about Deformed Leaf Bugs in the Aftermath of Chernobyl." *Social Studies of Science* 45, no. 5 (2015): 665–90. https://doi.org/10.1177/0306312715603249.

Seaver, Nick. "Care and Scale: Decorrelative Ethics in Algorithmic Recommendation." *Cultural Anthropology* 36, no. 3 (2021): 509–37. https://doi.org/10.14506/ca36.3.11.

Sedgwick, Eve Kosofsky. "Epidemics of the Will." In *Tendencies*, 130–42. Durham, NC: Duke University Press, 1993.

Sedgwick, Eve Kosofsky. *Touching Feeling: Affect, Pedagogy, Performativity*. Durham, NC: Duke University Press, 2003.

Sedgwick, Eve Kosofsky, and Adam Frank, eds. *Shame and Its Sisters: A Silvan Tomkins Reader*. Durham, NC: Duke University Press, 1996.

Shah, Esha. "Who Is the Scientist-Subject? A Critique of the Neo-Kantian Scientist-Subject in Lorraine Daston and Peter Galison's Objectivity." *Minerva* 55, no. 1 (2017): 117–38. https://doi.org/10.1007/s11024-017-9313-5.

Shapin, Steven. "Pump and Circumstance: Robert Boyle's Literary Technology." *Social Studies of Science* 14, no. 4 (1984): 481–520. https://doi.org/10.1177/030631284014004001.

Shepherdson, Charles. "Derrida and Lacan: An Impossible Friendship?" *Differences* 20, no. 1 (2009): 40–86. doi:10.1215/10407391-2008-016.

Shields, Alexandra E., Michael Fortun, Evelynn M. Hammonds, Patricia A. King, Caryn Lerman, Rayna Rapp, and Patrick F. Sullivan. "The Use of Race Variables in Genetic Studies of Complex Traits and the Goal of Reducing Health Disparities: A Transdisciplinary Perspective." *American Psychologist* 60, no. 1 (2005): 77–103. https://doi.org/10.1037/0003-066X.60.1.77.

Shostak, Sara. *Exposed Science: Genes, the Environment, and the Politics of Population Health*. Berkeley: University of California Press, 2013.

Shubin, Neil. *Your Inner Fish: A Journey into the 3.5-Billion-Year History of the Human Body*. New York: Vintage Books, 2009.

Silk, J. B. "Cooperation without Counting: The Puzzle of Friendship." In *The Genetic and Cultural Evolution of Cooperation*, edited by P. Hammerstein, 37–54. Dahlem Workshop Report 90. Cambridge, MA: MIT Press, 2003.

Silk, Joan. "Using the 'F'-Word in Primatology." *Behaviour* 139, no. 2 (2002): 421–46. https://doi.org/10.1163/156853902760102735.

Simone, AbdouMaliq. "Infrastructure: Introductory Commentary." Curated Collections, *Cultural Anthropology* website, November 26, 2012. https://culanth.org/curated_collections/11-infrastructure/discussions/12-infrastructure-introductory-commentary-by-abdoumaliq-simone.

Sipiora, Phillip, Janet Atwill, and Gayatri Chakravorty Spivak. "Rhetoric and Cultural Explanation: A Discussion with Gayatri Chakravorty Spivak." *Journal of Advanced Composition* 10, no. 2 (1990): 293–304.

Sirugo, Giorgio, Scott M. Williams, and Sarah A. Tishkoff. "The Missing Diversity in Human Genetic Studies." *Cell* 177, no. 1 (2019): 26–31. https://doi.org/10.1016/j.cell.2019.02.048.

Sleeboom, Margaret. "The Harvard Case of Xu Xiping: Exploitation of the People, Scientific Advance, or Genetic Theft?" *New Genetics and Society* 24, no. 1 (2005): 57–78. https://doi.org/10.1080/14636770500037776.

Smart, Andrew, Richard Tutton, Paul Martin, George T. H. Ellison, and Richard Ashcroft. "The Standardization of Race and Ethnicity in Biomedical Science Editorials and UK Biobanks." *Social Studies of Science* 38, no. 3 (2008): 407–23. https://doi.org/10.1177/0306312707083759.

Smith, David Roy. "Last-Gen Nostalgia: A Lighthearted Rant and Reflection on Genome Sequencing Culture." *Frontiers in Genetics* 5 (2014): 146. doi:10.3389/fgene.2014.00146.

Spivak, Gayatri Chakravorty. "Translation as Culture." *Parallax* 6, no. 1 (2000): 13–24. https://doi.org/10.1080/135346400249252.

Spivak, Gayatri Chakravorty. "Translator's Preface." In *Of Grammatology*, by Jacques Derrida. Baltimore, MD: Johns Hopkins University Press, 1976.

Spivak, Gayatri Chakravorty, and Sarah Harasym. *The Post-Colonial Critic: Interviews, Strategies, Dialogues*. London: Routledge, 2014.
Star, Susan Leigh. "The Ethnography of Infrastructure." *American Behavioral Scientist* 43, no. 3 (1999): 377–91. https://doi.org/10.1177/00027649921955326.
Stere, Paul. "The *Fushi Tarazu*." *Trends in Biochemical Sciences* 23 (1998): 500.
Sternberg, Richard V. "On the Roles of Repetitive DNA Elements in the Context of a Unified Genomic-Epigenetic System." *Annals of the New York Academy of Sciences* 981, no. 1 (2002): 154–88. https://doi.org/10.1111/j.1749-6632.2002.tb04917.x.
Stevens, Hallam. *Life Out of Sequence: A Data-Driven History of Bioinformatics*. Chicago: University of Chicago Press, 2013.
Stiegler, Bernard. "Care." In *Telemorphosis: Theory in the Era of Climate Change, Vol. 1*, edited by Tom Cohen. London: Open Humanities Press, 2012. http://quod.lib.umich.edu/o/ohp/10539563.0001.001/1:6/—telemorphosis-theory-in-the-era-of-climate-change-vol-1?rgn=div1;view=fulltext.
Stiegler, Bernard. "Relational Ecology and the Digital Pharmakon." *Culture Machine* 13 (2012). https://culturemachine.net/wp-content/uploads/2019/01/464-1026-1-PB.pdf.
Stiegler, Bernard. *Taking Care of Youth and the Generations*. Stanford, CA: Stanford University Press, 2010.
Strachan, D. P. "Family Size, Infection and Atopy: The First Decade of the 'Hygiene Hypothesis.'" *Thorax* 55, no. 1 (suppl, August 2000): S2–10.
Strauss, Anselm. "Work and the Division of Labor." *Sociological Quarterly* 26, no. 1 (1985): 1–19. https://doi.org/10.1111/j.1533-8525.1985.tb00212.x.
Subramaniam, Banu. "Cartographies for Feminist STS." *Engaging Science, Technology, and Society* 7, no. 2 (2021): 65–69. https://doi.org/10.17351/ests2021.817.
Subramaniam, Banu, and Angela Wiley. "Introduction to Science out of Feminist Theory Part One: Feminism's Sciences." *Catalyst: Feminism, Theory, Technoscience* 3, no. 1 (2017): 1–23.
Sullivan, Patrick F., L. J. Eaves, K. S. Kendler, and M. C. Neale. "Genetic Case-Control Association Studies in Neuropsychiatry." *Archives of General Psychiatry* 58, no. 11 (2001): 1015–24. doi:10.1001/archpsyc.58.11.1015.
Sulston, John. With Georgina Ferry. *The Common Thread*. London: Transworld, 2010.
Sunder Rajan, Kaushik, and Sabina Leonelli. "Introduction: Biomedical Trans-Actions, Postgenomics, and Knowledge/Value." *Public Culture* 25, no. 3 (71) (2013): 463–75. https://doi.org/10.1215/08992363-2144607.
Terrell, John Edward. *A Talent for Friendship: Rediscovery of a Remarkable Trait*. Oxford: Oxford University Press, 2014.
Thompson, Vanessa E. "Repairing Worlds: On Radical Openness beyond Fugitivity and the Politics of Care: Comments on David Goldberg's Conversation with Achille Mbembe." *Theory, Culture and Society* 35, no. 7/8 (2018): 243–50. https://doi.org/10.1177/0263276418808880.
Tilghman, Shirley. "Keynote Address." National Human Genome Research Institute Symposium. Interviewed by NHGRI Director Eric Green, November 25, 2019, Bethesda, MD. https://www.youtube.com/watch?v=7C4HZsWkhs4.

Tomkins, Silvan S. *Affect Imagery Consciousness: The Complete Edition*. New York: Springer, 2008 [1962].
Tomšič, Samo. “Better Failures: Science and Psychoanalysis.” In *Lacan Contra Foucault: Subjectivity, Sex, and Politics*, edited by Nadia Bou Ali and Rohit Goel, 81–108. London: Bloomsbury Academic, 2018.
Tomšič, Samo. “Mathematical Realism and the Impossible Structure of the Real.” *Psychoanalytische Perspectieven* 35, no. 1 (2017): 9–34.
Traweek, Sharon. “Generating High-Energy Physics in Japan: Moral Imperatives of a Future Pluperfect.” In *Pedagogy and the Practice of Science*, edited by David Kaiser, 357–90. Cambridge, MA: MIT Press, 2005.
Tronto, Joan C. “Beyond Gender Difference to a Theory of Care.” *Signs* 12, no. 4 (1987): 644–63.
Tronto, Joan. *Moral Boundaries: A Political Argument for an Ethic of Care*. New York: Routledge, 1993.
Trundle, Catherine. “Tinkering Care, State Responsibility, and Abandonment: Nuclear Test Veterans and the Mismatched Temporalities of Justice in Claims for Health Care.” *Anthropology and Humanism* 45, no. 2 (2020): 202–11.
Tsilidis, Konstantinos K., Orestis A. Panagiotou, Emily S. Sena, Eleni Aretouli, Evangelos Evangelou, David W. Howells, Rustam Al-Shahi Salman, Malcolm R. Macleod, and John P. A. Ioannidis. “Evaluation of Excess Significance Bias in Animal Studies of Neurological Diseases.” *PLOS Biology* 11, no. 7 (2013): e1001609. doi:10.1371/journal.pbio.1001609.
Tyler, Stephen A. “Post-Modern Ethnography: From Document of the Occult to Occult Document.” In *Writing Culture: The Poetics and Politics of Ethnography*, edited by James Clifford and George E. Marcus, 122–40. Berkeley: University of California Press, 1986.
Ureta, Sebastian. “Ruination Science: Producing Knowledge from a Toxic World.” *Science, Technology, and Human Values* 46, no. 1 (2021): 29–52. https://doi.org/10.1177/0162243919900957.
US Department of Energy. “Report of the Invitational DOE Workshop on Genome Informatics, 26–27 April 1993.” http://web.ornl.gov/sci/techresources/Human_Genome/publicat/miscpubs/bioinfo.shtml.
US Senate. *The Human Genome Project*. Subcommittee on Energy Research and Development, Committee on Energy and Natural Resources, 1st session, July 11, 1990.
Venter, J. Craig. *A Life Decoded: My Genome, My Life*. London: Penguin, 2007.
Vercelli, Donata. “Discovering Susceptibility Genes for Asthma and Allergy.” *Nature Reviews Immunology* 8, no. 3 (2008): 169–82. https://doi.org/10.1038/nri2257.
Vermeulen, Niki, John N. Parker, and Bart Penders. “Understanding Life Together: A Brief History of Collaboration in Biology.” *Endeavour* 37, no. 3 (2013): 162–71. doi:10.1016/j.endeavour.2013.03.001.
Vinsel, Lee, and Andrew L. Russell. *The Innovation Delusion: How Our Obsession with the New Has Disrupted the Work That Matters Most*. New York: Penguin Random House, 2020.

Viseu, Ana. "Caring for Nanotechnology? Being an Integrated Social Scientist." *Social Studies of Science* 45, no. 5 (2015): 642–64. https://doi.org/10.1177/0306312715598666.
Visscher, Peter M., Matthew A. Brown, Mark I. McCarthy, and Jian Yang. "Five Years of GWAS Discovery." *American Journal of Human Genetics* 90, no. 1 (2012): 7–24. https://doi.org/10.1016/j.ajhg.2011.11.029.
Visscher, Peter M., Naomi R. Wray, Qian Zhang, Pamela Sklar, Mark I. McCarthy, Matthew A. Brown, and Jian Yang. "10 Years of GWAS Discovery: Biology, Function, and Translation." *American Journal of Human Genetics* 101, no. 1 (2017): 5–22. https://doi.org/10.1016/j.ajhg.2017.06.005.
Visser, Max. "Gregory Bateson on Deutero-Learning and Double Bind: A Brief Conceptual History." *Journal of the History of the Behavioral Sciences* 39, no. 3 (2003): 269–78. https://doi.org/10.1002/jhbs.10112.
Vogel, Gretchen. "A Scientific Result without the Science." *Science* 276, no. 5317 (1997): 1327. https://doi.org/10.1126/science.276.5317.1327.
von Mutius, E., C. Fritzsch, S. K. Weiland, G. Röll, and H. Magnussen. "Prevalence of Asthma and Allergic Disorders among Children in United Germany: A Descriptive Comparison." *BMJ: British Medical Journal* 305, no. 6866 (1992): 1395–99.
Vukmirovic, Ognjenka Goga, and Shirley M. Tilghman. "Exploring Genome Space." *Nature* 405, no. 6788 (2000): 820–22. https://doi.org/10.1038/35015690.
Wade, Nicholas. "A Decade Later, Genetic Map Yields Few New Cures." *New York Times*, June 12, 2010. https://www.nytimes.com/2010/06/13/health/research/13genome.html.
Walford, Antonia. "Data Moves: Taking Amazonian Climate Science Seriously." *Cambridge Journal of Anthropology* 30, no. 2 (2012): 101–17.
Walker, Alexis. "Diversity, Profit, Control: An Empirical Study of Industry Employees' Views on Ethics in Private Sector Genomics." *AJOB Empirical Bioethics*, April 18, 2022, 1–13. https://doi.org/10.1080/23294515.2022.2063993.
Watson, James D. *Avoid Boring People: Lessons from a Life in Science*. Oxford: Oxford University Press, 2007.
Watson, James D. "Remarks at the Academy of Arts and Sciences, Cambridge, MA, February 14, 1990." Author's recording and transcript.
Watson, Matthew C. "Derrida, Stengers, Latour, and Subalternist Cosmopolitics." *Theory, Culture and Society* 31, no. 1 (2004): 75–98.
Watson, Matthew C. "Listening in the Pakal Controversy: A Matter of Care in Ancient Maya Studies." *Social Studies of Science* 44, no. 6 (2014): 930–54. https://doi.org/10.1177/0306312714543964.
Weissmann, Gerald. "Pattern Recognition and Gestalt Psychology: The Day Nüsslein-Volhard Shouted 'Toll!'" *FASEB Journal* 24, no. 7 (2010): 2137–41. https://doi.org/10.1096/fj.10-0701ufm.
Williams, Hywel, Colin Robertson, Alistair Stewart, Nadia Aït-Khaled, Gabriel Anabwani, Ross Anderson, Innes Asher, et al. "Worldwide Variations in the Prevalence of Symptoms of Atopic Eczema in the International Study of

Asthma and Allergies in Childhood." *Journal of Allergy and Clinical Immunology* 103, no. 1 (1999): 125–38. https://doi.org/10.1016/S0091-6749(99)70536-1.

Wills-Karp, Marsha, and Susan L. Ewart. "Time to Draw Breath: Asthma-Susceptibility Genes Are Identified." *Nature Reviews Genetics* 5, no. 5 (2004): 376–87. https://doi.org/10.1038/nrg1326.

Wilson, Elizabeth A. "Another Neurological Scene." *History of the Present* 1, no. 2 (2011): 149–69. https://doi.org/10.5406/historypresent.1.2.0149.

Wilson, Elizabeth A. "Gut Feminism." *Differences: A Journal of Feminist Cultural Studies* 15, no. 3 (2004): 66–94.

Wilson, Elizabeth A., and Adam Frank. *A Silvan Tomkins Handbook*. Minneapolis: University of Minnesota Press, 2020. https://manifold.umn.edu/projects/a-silvan-tomkins-handbook.

Wilson, James F., et al. "Population Genetic Structure of Variable Drug Response." *Nature Genetics* 29 (2001): 265–69.

Wolfe, Audra J. *Freedom's Laboratory: The Cold War Struggle for the Soul of Science*. Baltimore, MD: Johns Hopkins University Press, 2018.

Xu, Xin, Zhian Fang, Binyan Wang, Changzhong Chen, Wenwei Guang, Yongtang Jin, Jianghua Yang, et al. "A Genomewide Search for Quantitative-Trait Loci Underlying Asthma." *American Journal of Human Genetics* 69, no. 6 (2001): 1271–77.

Yanai, Itai, and Martin Lercher. "Improvisational Science." *Genome Biology* 23, no. 1 (2022): 4. https://doi.org/10.1186/s13059-021-02575-w.

Yanai, Itai, and Martin Lercher. "Night Science." *Genome Biology* 20, no. 1 (2019): 179. https://doi.org/10.1186/s13059-019-1800-6.

Yanai, Itai, and Martin Lercher. "Novel Predictions Arise from Contradictions." *Genome Biology* 22, no. 1 (2021): 153. https://doi.org/10.1186/s13059-021-02371-6.

Yanai, Itai, and Martin Lercher. "The Two Languages of Science." *Genome Biology* 21 (1): 147. https://doi.org/10.1186/s13059-020-02057-5.

Zamel, N., P. A. McClean, P. R. Sandell, K. A. Siminovitch, and A. S. Slutsky. "Asthma on Tristan Da Cunha: Looking for the Genetic Link. The University of Toronto Genetics of Asthma Research Group." *American Journal of Respiratory and Critical Care Medicine* 153, no. 6 (pt. 1, 1996): 1902–6. https://doi.org/10.1164/ajrccm.153.6.8665053.

Zegura, Ellen, Carl DiSalvo, and Amanda Meng. "Care and the Practice of Data Science for Social Good." In *Proceedings of the 1st ACM SIGCAS Conference on Computing and Sustainable Societies*, 1–9. Menlo Park, CA: ACM, 2018. https://doi.org/10.1145/3209811.3209877.

Zimmer, Carl. "Yet-Another-Genome Syndrome." *The Loom* (blog). *Discover Magazine*, April 2, 2010. https://www.discovermagazine.com/planet-earth/yet-another-genome-syndrome.

Zorich, Diane M. "Data Management: Managing Electronic Information: Data Curation in Museums." *Museum Management and Curatorship* 14, no. 4 (1995): 430–32. doi:10.1016/0260-4779(96)84690-5.

INDEX

Printed and bound by CPI Group (UK) Ltd, Croydon, CR0 4YY

09/06/2025

14685750-0001